Scott King | *Following the Earth Around*

BY THE SAME AUTHOR

NATURAL HISTORY
Rice County Odonata Journal: Volume One (2008)
Rice County Odonata Journal: Volume Two (2009)
Rice County Odonata Journal: Volume Three (2010)
Rice County Odonata Journal: Volume Four (2011)
A Photographic Guide to Some Common Wasps and Bees of Minnesota
Dragonflies and Damselflies of Minnesota: Atlas and Annotated Checklist

POETRY
Leftover Ordinary
Where the Water Falls
Lida Songs
All Graced In Green
Nine
Dragonfly Haiku (with Ken Tennessen)
Brevities

TRANSLATIONS OF YANNIS RITSOS
Stones
The Wavering Scales (with Martin McKinsey)
The Shadows of Birds
Petrified Time (with Martin McKinsey)

TRANSLATIONS OF FERYEDOUN FARYAD
Heaven Without a Passport

EDITOR
Perfect Dragonfly: A Commonplace Book of Poems Celebrating a Decade & a Half of Printing & Publishing at Red Dragonfly Press

Scott King

Following the Earth Around

❖

Journal of a Naturalist's Year

Thistlewords Press

Text copyright © 2019 by Scott King
All rights reserved

ISBN 978-1-545508-40-4

Many entries were first posted as journal entries at iNaturalist.org, sometimes in different versions.

Special thanks to Lisa and Lida, loving followers.

Cover and section photographs by Scott King
Cover photo legend (clockwise from upper left corner)
1. December 19: Snow Midge at Cowling Arboretum
2. January 11: Sumac at St Olaf Natural Lands
3. February 5: Red Oak as St Olaf Natural Lands
4. March 27: Lichen Moth caterpillar at McKnight Prairie
5. April 27: Grote's Sallow at moth light
6. May 4: Spring Beauty Andrena Bee at Nerstrand State Park
7. June 18: Banded Pennant in Florida
8. July 14: Desert Scrub Oak at Zion National Park
9. August 7: Square-headed Wasp at Cowling Arboretum
10. September 13: New England Aster at St Olaf Natural Lands
11. October 11: Whorled Milkweed at Cowling Arboretum
12. November 23: Golden-eye Lichen at Maplewood State Park

Designed and typeset at Red Dragonfly Press
using Warnock Pro digital type

Published by Thistlewords Press
307 Oxford Street
Northfield, MN 55057

Contents

Preface 7

JANUARY 9
FEBRUARY 59
MARCH 93
APRIL 137
MAY 185
JUNE 239
JULY 289
AUGUST 337
SEPTEMBER 379
OCTOBER 415
NOVEMBER 455
DECEMBER 493

General Index 533
Species Index 533

PREFACE: BEGINNING MY STUDIES

> "Beginning my studies, the first step pleas'd me so much,
> The mere fact consciousness, these forms, the power of motion,
> The least insect or animal, the senses, eyesight, love,
> The first step I say awed me and pleas'd me so much,
> I have hardly gone and hardly wish'd to go any farther,
> But stop and loiter all the time to sing it in extatic songs."
> – Walt Whitman, from *Inscriptions*

I was inspired, to begin with, by Sam Kieschnick, an urban wildlife biologist for Texas Parks and Wildlife, who posted at least one observation on iNaturalist.org every day in 2016 (and every day since!). I felt disposed to attempt the same thing, with the difference of adding a journal entry for each day. In short, I followed the earth for an entire year, season by season, day by day, camera in hand, pen in pocket. The plan was simple; the result complex (and extensive!), a melange of serendipitous encounters and phenological fluctuations in flora and fauna across the ever-changing days and seasons.

When the year was over, I'd submitted 1,905 observations to iNaturalist for 991 separate species, the majority being insects. In addition to our city lot and backyard where many of the observations were made, I visited several local nature reserves repeatedly: St Olaf Natural Lands (133 visits, 546 observations), Cowling Arboretum (55 visits, 278 observations), and McKnight Prairie (6 visits, 60 observations). The year included short trips to Florida, Utah, northern Minnesota, and Wisconsin. Throughout the year, I'd managed to post less than half of the daily journal entries, leaving a thick pile of notebook pages to be typed and edited, hence the full-year delay in gathering them all together in this book.

Often, these daily journal entries anchor at a single observation, relating what I knew or didn't know about a particular insect or plant, a place I chose to "stop and loiter." At times, the text begins off topic and stays off topic, but usually, it retains (or suggests) some tangential relation to the day's observations, weather conditions, or history. Other times, the journal entry hardly goes farther than a simple enumeration of what was observed. Obviously, much much more could have been written; a lot of additional information could have been included; many interesting plants and animals, themes and ideas, places and people were overlooked—resulting in an almost endless number of missed opportunities. Despite these omissions, I think the book reflects abundance more than paucity and gives an honest account of what caught my naturalist's eye across the span of a single year.

Several literary precedents exist. The foremost influence on this book is Henry David Thoreau's *Journals*. I read, day-to-day as I made my own observations, his entries for the year 1854. This particular year was chosen for two reasons: the near-daily content (545 pages in the edition I have) and the alignment of the calendar days for 1854 and 2017. A second influence is the writing of Edwin Way Teale. Two of his books, *Circle of the Seasons* and *A Walk Through the Year*, provided the model for the structuring of this book. While neither of those books follows the events of a single, specific year, both include entries for every day of the year. Teale's entries are less hurried than my own, more discursive, probably more readable.

The following authors and books influenced this book in various ways, sparking my curiosity and heightening my poetic sensibilities: *Italian Journey* by Goethe, *Journey to Portugal* and *The Notebook* by José Saramago, *No Time To Spare* by Ursula Le Guin, the object poetry of Francis Ponge, and *Swampwalker's Journal* by David M. Carroll.

I would have preferred to include all the photos. There were too many, so a simple list of observations takes their place. The daily temperatures are from a weather station in Minneapolis about 30 miles north. The winter lows are often a few degrees cooler than in the city.

❖ ❖ ❖

One of our devastating flaws is to be so inconsiderate of Nature while, at the same time, exerting our dominion over its resources. This book has been my small attempt at being more mindful and less destructive, a record of one cycle through the seasons, a strange loop through the year. My hope is that this book, by providing a measure of the natural history of a single year, will be of use to others in the future, either as a model or (more hopefully!) a reassurance, and not become a record of what has been lost.

January

Spring Creek – Cowling Arboretum – January 16

January 1   32°

BLACK-CAPPED CHICKADEE   21°

The congenial chickadee, certainly one of Minnesota's mainstay birds during the winter and a splendid way to start off the new year on iNaturalist. This bird has been a personal favorite since childhood, probably one of the first birds I knew by name. As evidence of this long-standing admiration, around thirty years ago, as a college student, I made a bookmark, which I still have and use, from a piece of birchbark, adding a simple ink drawing of a chickadee and this tiny, two-line poem by Henry David Thoreau (with his spelling).

> The chicadee
> Hops near to me.

Poecile atricapillus – Black-capped Chickadee

33°
—
28°

January 2

I WENT FOR A WALK IN THE RIVER

With the temperature hovering right around freezing, it made sense to take advantage of the warm weather and do something ambitious. I decided to go for a walk in the river. In Northfield, I see that the Cannon River is entirely frozen, but at successive crossings, the river begins to open, black slices of rushing water divide iced-over expanses from the snow-covered banks. And when I park at the canoe landing some miles further upstream the river is nearly free of ice.

Clear, cold, flowing swiftly due to last weeks unusual winter rains, I hesitate to enter into the water. Once in, the force of moving water grabs my legs (and my attention!) and I quickly determine the boundaries beyond which I should not step.

I kick-net for half an hour. Using a net is one way to explore an otherwise unreachable reality. The surprise and immediacy of netting some unfamiliar animal is exhilarating and almost as fantastical as being able to reach through the night sky and pull down a star or two for close observation. I peek into the net and examine each haul, then empty the contents onto the shore ice, picking through the catch. Though I'd come looking for dragonflies, I find instead a variety of other wonders: Blackside Darters, Midwest Salmonfly larvae, Watersnipe Fly maggots, crayfish, caddisflies, and scuds.

Here are the opening lines of a favorite long poem that sprang to mind today when I was "summoned" to the river:

I went for a walk by the river,
summoned by a drift or turn in the air,
unsure I didn't hear a flock
of ghostly birds disturb the trees.
Or perhaps it was the hollow whirls of water,
vortexes turning on emptiness,
that called me.

– Dale Jacobson, from *A Walk By The River*

Agnetina flavescens – Midwestern Stone
Pteronarcys pictetii – Midwestern Salmonfly
Gammarus pseudolimnaeus – Amphipod
Atherix sp. – Fly larva
Coenagrionidae – Narrow-winged Damselflies
Orconectes virilis – Virile Crayfish
Acroneuria sp. – Stoneflies
Percina maculata – Blackside darter
Etheostoma nigrum – Johnny Darter

32°
—
2°

January 3

ABOVE THE SNOW

Tuesday, January 3, 1854—one hundred and sixty-three years ago to the day—Henry David Thoreau had this to say about the winter in Concord, Massachusetts: "It is now fairly winter. We have passed the line, have put the autumn behind us, have forgotten what these withered herbs that rise above the snow here and there are, what flowers they ever bore." He overstates a little; just four days later he will describe in knowing detail the spidery seeds of the Gray Goldenrod as they spread across the snow, but I certainly understand how the memory of color begins to slip away at this time of year. It's mainly a black and white season, with the light-absorbing shadows and long dark nights predominating.

Tuesday, January 3, 2017—12 degrees F with winds over 20 mph, putting the windchill well below zero. So only a short hike in the St Olaf Natural Lands today, maybe twenty minutes out and back. The wind turbine overlooking the big pond emits a rather disagreeable noise, slicing the dense winter air, an intermittent whoosh as though a giant was jumping rope slowly. At the far end of the pond, a college student (I assume!) has cleared a speed skating track and is skating laps in his lycra suit. He has to be cold! I follow the shore a short distance then cut into the woods to escape the wind. Ice from yesterday's freezing rain glistens on branches. Overnight, the rain changed to wet snow which had frozen solid by morning. Now the "withered herbs that rise above the snow" hold a burden of frost and spattered flakes; a beautiful rime decorates the remains of last years flowers. The red bark of the dogwood catches my eye as I pass. Then, just as I turn toward the parking lot, I notice the bare thorny branches of a hawthorn, hidden some distance off the trail. I look forward to revisiting this tree when it blooms this coming spring.

Cornus sericea – Red Osier Dogwood
Crataegus sp. – Hawthorns

January 4

METAPHORS

Edwin Way Teale, increasingly forgotten and unread nowadays, wrote many wonderful books. In 1966, *Wandering Through Winter*, the fourth in a series of seasonal travelogues, won the Pulitzer Prize. Both naturalist and photographer, he was in many ways a brilliant predecessor to those of us involved with iNaturalist. For instance, his *Circle of the Seasons: The Journal of a Naturalist's Year* (1953) collects his written nature observations for each day of one entire year.

Interestingly, his entry for this day, January 4th, refers to the common image of the new year being a blank book yet to be written. He continues, "Nature's year is also a book to be written." From the perspective of midwinter, the year's entire story is yet to come. Extending that metaphor, the previous year has been printed and bound, and these sparse winter days stand equivalent to the blank pages between the end of the text and the back cover, a little quiet space for contemplation, a pause.

I like the idea that the end is past and that the beginning yet to come. Standing near the center of a frozen pond at the still point of the turning year, the temperature at midday exactly zero degrees, I pause momentarily. Like a hand clutching many different sized rings together, this particular point in time arrives at a unique convergence of many natural cycles—diurnal cycles, seasonal cycles, climactic cycles caused by solar orbit wobbles, and so forth. I might even be standing at the cross that marks the very center of some great unregistered infinity sign. I move on. My attention was drawn back from the music of the spheres to the surface of the pond, where I follow a line of fox prints toward the shore. Whether it was a Red Fox or a Grey Fox can no longer be determined; its presence both recorded and erased.

A number of years later, Teale would improve upon the idea of the year as a circle, a simple lap to be run, when he published *A Walk through the Year* (1978). The metaphor becomes a little less abstract, richer in details and surprises whether we envision a simple daily walk or an extended journey. Of course, there are other metaphors we might consider. The year might be considered to be a labyrinth, where we enter get lost and, with any luck, find a way back to the start. Or the combination of circles and time that creates spiral- or helix-shaped years. Or following the motion of waves—crest to trough to crest—we might drift through the years. Take your pick.

Canidae – fox tracks

January 5

EMBRACING THE COLD

According to E. C. Pielou, in *After the Ice Age: The Return of Life to Glaciated North America*, the timing is such that we are nearing the end of the current interglacial period, that the earth should be due to enter into its next period of glaciation, that the continental ice sheets are scheduled to return. However, the change that humankind is affecting and will inflict upon the climate may break those predictions, dramatically, even tragically. No longer must we consider time periods tens of thousands of years long, rapid climate change is occurring on a scale of centuries and decades. Based upon ice core data, "The baseline CO_2 value for interglacials is approximately 290 parts per million. On 9th May 2013 the concentration of atmospheric CO_2 exceeded 400 parts per million for the first time since the balmy conditions of the Pliocene when the sea level was more than 20 m higher than today." (quote from *The Ice Age: A Very Short Introduction* by Jamie Woodward)

Strangely serendipitous reading this morning led me to the following passage in *The First Wash of Spring* by Scottish writer George Mackay Brown, a comment upon on a series of mild winters in the Orkney Islands in January of 1993. "Let's hope it's only laziness or indifference, up there with the snow giants in North Greenland or Spitzbergen, and not something more sinister, like global warming or the greenhouse effect, that's pared their teeth or their claws, or curbed their ferocious gleefulness. For, however we dislike snow and blizzards, to be walking mouth deep in a tepid ever-rising sea is too hideous to think about."

This morning the temperature was -7 degrees F, nine degrees below average for this date. Granted the temperature on any given day doesn't mean much in the greater scheme of things, but it does cross my mind, during truly cold weather, that these cold days may become rarer, that I should embrace the cold and enjoy it while it lasts. An odd sentiment, to view these below zero days as a special occasion or gift, but I give it a go and bundle up—long underwear, snow pants, wool hat, layers, and heavy winter boots—for a midwinter hike at the local Arboretum. It's colder than yesterday's hike. Mallards huddle in what little open water remains. Wisps of frozen white vapor drift around them and above them. I follow a faint set of mink tracks to the entrance of a hole among the roots of a streamside tree. I study the same large bird tracks I studied yesterday, either Bald Eagle or a Heron that didn't make it south. The sun drops below the horizon and snow turns blue before I leave. More than anything, I'm happy I didn't stay inside.

Sorghastrum nutans – Indiangrass
Anas platyrhynchos – Mallard

January 6

A FEW RECALCITRANT ROBINS

The cold weather continues. For the third day in a row, I visit the Cowling Arboretum for a short afternoon hike. It's sunny and a few degrees above freezing at 3 pm. This morning, however, the temperature dipped to -10 degrees F and as a result, there's even less open water than yesterday. The number of Mallards congregating in the flowing water of the creek has increased or it's the same number more concentrated. Jagged ice flows, jammed and jumbled at the confluence of two swift channels in the Cannon River, give the scene a truly arctic cast. And at one point amid slabs of frozen water tilted vertically forming white and blue fins a black branch juts up as though some explorer had planted it there. I take a second look half expecting to see a flag attached at the top.

Across the main channel, arrayed along the lip of ice edging the moving water, a flock of American Robins perches and drinks. I'm always a little surprised to see these birds midwinter. But then again I do see them most winters here in Northfied. According to *The Birds of Minnesota* (1932) by T. S. Roberts, Robins do winter in Minnesota "in limited numbers, chiefly in sheltered places, in the southern part of the state." So while the vast majority have moved on to warmer climes, a recalcitrant few tough it out here in the north.

Here are a few apt lines by Welsh poet Gillian Clarke from her winterish collection, *Ice* (2012), and the poem 'Winter':

The river froze, and broke, and froze,
its heart slowed in its cage,
the moon a stone
in its throat.

Arctium lappa – Greater Burdock
Turdus migratorius – American Robin

9° / -2°

January 7

THE PRAIRIE LOOP

Lisa joined me on today's outing. After nearly nine hours of volleyball spectatoring and family, we took the dog and headed for a short hike around the big pond at the St Olaf Natural Lands. We've hiked this loop many times, out along the wooded south side of the pond then back through the wide open restored prairie. Today may well have been one of the colder visits, but the prairie colors in winter were rich and vibrant and worth the wind-bitten nose and cheeks.

As we walked out onto the prairie our dog ran some distance ahead then stopped. It appeared, at first, the two other big dogs were joining him. Only they weren't dogs. A pair of White-tailed Deer walked directly toward the dog and us, stopping at a distance of about thirty yards. The deer stretched their necks, raising their heads and focusing their ears until they suddenly realized what company they'd fallen in with. One bounded away and then the other, white-tailing it through the tallgrass prairie.

Driving home we crossed campus. And just as we were commenting that we'd seen not a single bird on this hike, Lisa spotted a Red-tailed Hawk. The hawk was perched low in a tree directly over a sidewalk. I took a few hurried photos out the car window, with the view cluttered by branches. Despite the obstructed, I was pleased and considered it a triumph since it was the first hawk I'd ever managed to photograph.

Buteo jamaicensis – Red-tailed Hawk
Lespedeza capitata – Round-headed Bush Clover
Odocoileus virginianus – White-tailed Deer

January 8
"WHAT A LICHENIST HE MUST BE!"

18°
—
-5°

Thoreau never had to journey via freeway, zipping place to place at seventy miles per hour. This thought crosses my mind today, while on our way to visit friends in a western suburb of Minneapolis, an eighty-five mile round trip from our home in Northfield. As a result of traveling and visiting, outdoor time was limited to an afternoon dog walk, joined by our friends and their dogs.

Unfortunately, it was one of those bare midwinter days when the world seems worn through and bereft of anything vibrant, reduced almost entirely to inanimate elements. And in the suburbs where we walked today those elements consisted of asphalt streets white and dusty from road salt, spills and spatters of coarse-grained sidewalk salt, silent slips of ice, snow, inanimate houses, dormant trees, and cars, lots of cars. Luckily, the company was good and the dogs were happy.

During our walk, we get a glimpse of a woodpecker high in a tree. Then for a while a few of us did our best to name the winter trees along the way. At one point we stopped to examine several dead branches arching above the sidewalk, the undersides of the branches encrusted with Polypore fungi. Near the end of our walk, a noisy throng of House Sparrows craftily flew and hopped inside the thicket of a yardside hedge.

That distant glimpse of a woodpecker would be as close as I'd come to matching Thoreau's observation for this same date:

"Stood within a rod of a downy woodpecker on an apple tree. How curious and exciting the blood-red spot on its hindhead!

I ask why it is there, but no answer is rendered by these snow-clad fields. It is so close to the bark I do not see its feet. It looks behind as if it had on a black cassock open behind and showing a white undergarment between the shoulders and down the back. It is briskly and incessantly tapping all round the dead limbs, but rarely twice in a place, as if to sound the tree and so see if it has any worm in it, or perchance to start them. How much he deals with the bark of trees, all his life long tapping and inspecting it! He it is that scatters those fragments of bark and lichens about on the snow at the base of trees. What a lichenist he must be! Or rather, perhaps it is fungi makes his favorite study, for he deals most with dead limbs. How briskly he glides up or drops himself down a limb, creeping round and round, and hopping from limb to limb, and now flitting with a rippling sound of his wings to another tree!"
– Henry David Thoreau, *Journals* (Sunday, January 8, 1854)

Irpex lacteus – Milk-white Toothed Polypore
Populus deltoides – Eastern Cottonwood
Passer domesticus – House Sparrow

January 9   27°

THE SHADOW OF A SMELL   17°

Today I walked for close to an hour at the St Olaf Natural Lands, enjoying the somewhat milder weather—the temperature near twenty above, the wind stiff but not piercing, the skies overcast and only just beginning to let loose tiny snowflakes (there'd be a lot more in the afternoon). I thought for sure I'd see a few birds today but I didn't, not a single bird. It might be the time of year. It might be that they're all hanging out near bird feeders in town. Or, suddenly aware of the sound of the wind turbine overhead, it might be that the noise and motion of the wind turbine blades make the land in its immediate vicinity disagreeable. Looking up at the turbine I notice (with some amount of concern) a giant black mark that snakes from the hub of the turbine half-way down the length of one blade. How many gallons of oil does it take to make a mark that large!?

Most of my walk is through and around two cattail wetlands. The first wetland is choked with cattails. The second is more diverse with areas of other vegetation and more water and currently home to a thriving population of muskrats. There are at least a half-dozen push-ups of various size scattered around the wetland. Seeing the muskrat popups, I flashed back to the lethal young naturalist I was as a child. For several years, as a teenager, I trapped muskrats, quite successfully.

I remember the red plastic snow sled I pulled for miles along various trap lines. In the sled, a hardwood pack basket held traps and stakes for making the sets. I also brought along a hatchet (which I still have), an ice chisel, and large black rubber gloves for reaching inside the muskrat houses and working in the water. Not a glamorous picture, probably. But I guarantee you I thought it was as exciting. And I logged an extraordinary number of hours outside.

I could go on. For this memory leads me to wonder what I did with all the bodies. I know that I did the work of skinning and stretching the trapped muskrats myself, but I can't remember doing it. Though as I try to remember more, the shadow of a smell stirs and rolls over in its slumber, somewhere down deep in a back corner of my mind. So I think I'll stop here before it wakes fully.

Polyporales – Shelf Fungi
Ondatra zibethicus – Muskrat
Typha – Cattails

January 10   28°

SNOWED IN   4°

Today it snowed, from the morning to early afternoon. And the time I'd allowed for a walk was needed for shoveling. So no outdoors photo for today. To make up for it, I decided to open up a goldenrod gall gathered from the prairie during a walk a few days ago. Inside I found a flaccid, unfreezable little blob. Not much to look at, and yet this nondescript larva is part of a remarkable life cycle. It weathers winter inside its vegetable capsule, not unlike us humans inside our wooden houses, except the gall fly larva doesn't have central heating. It survives by removing water from its body and by producing the antifreeze glycerol.

At some point, late winter or early spring, the larva will pupate and the adult fly will emerge a few weeks after, just as the goldenrod begins to grow in late spring.

Eurosta solidaginis – Goldenrod Gall Fly

10°
—
4°

January 11

THE GLIDING SNAIL OF PERCEPTIVENESS

Virginia Woolf in her essay 'Street Haunting: A London Adventure', describes the transformative moment of leaving the house as the shucking of an oyster. First, she sets the stage for this dramatic image by describing the common and familiar surroundings of our home where "we sit surrounded by objects which perpetually express the oddity of our own temperaments and enforce the memories of our own experience." Then, at the beginning of the next paragraph when her walk gets underway, she hits us with this: "The shell-like covering which our souls have excreted to house themselves, to make for themselves a shape distinct from others, is broken, and there is left of all these wrinkles and roughnesses a central oyster of perceptiveness, an enormous eye."

Being an inveterate walker (I almost wrote invertebrate!), I understand what she's trying to convey. We are changed when we leave our houses and enter the larger world, whether the destination is a neighborhood, a city, or a wilderness. I would have preferred the use of a less violent image (though I trust she had her reasons). Perhaps a different class of Molluscs, Gastropoda instead of Bivalvia, would have worked just as well. Wouldn't the image of a snail do the trick? That moment it extends out of the safety of its shell, stretching out its tentacles, eyes at their very tips. The gliding snail of perceptiveness.

Reading this morning from *Slugs and Snails* by Robert Cameron, a recent title in the New Naturalist Series, I felt my curiosity about real snails intensify. So, while out walking today I thought to look for snail shells along the creek bank in Cowling

Arboretum. Even with the temperature dropping back toward zero, with three days worth of new powdery snow, I succeeded in finding two snail shells without much effort. One shell that of a tiny Ramshorn Snail (3mm). The other a dextrally spiraled Pleurocerid Snail (27mm). Both freshwater snails.

[I'd like to note here, that I really know next to nothing about snails. Fortunately for me and others posting observations of snails, there are several vigilant malacologists on iNaturalist that never seem to tire of lending their aid. Thanks in particular to Susan Hewitt, your help has been much appreciated.]

Pleuroceridae – Pleurocerid Snails
Planorbidae – Ramshorn Snails

January 12

A SPLASH OF COLOR

Being a naturalist in winter can present some challenges. Some days, like today, it's all about perseverance and fighting those voices that suggest staying inside by the fire with a good book. For the first time this week it didn't snow. But with the temperature at 10 degrees and the wind whipping out of the west at 30 plus mph, the fallen snow took on new life, drifting and traveling about.

Out in the open, it was less than pleasurable. I strapped on my snowshoes and quickly headed for the shelter of the woods, entering a bright landscape, all solitude, and sunshine. Black and white rule this landscape, but there are surprise splashes of color here and there. Among the flashiest are the Sunburst Lichens adorning the bare trunks and branches. It's hard not to admire lichens, their hardiness, their indifference to weather, their pigments catching the eye like paint samples.

Xanthoria – Sunburst Lichens

January 13

LONG AND SHORT PEDICELS

9°
—
-7°

> The pieces of unprofitable land
> are what I like, best seen in winter...
> their failure's proof of reclamation,
> their vigour justifies all wastes and weeds.
> – Molly Holden, from 'Pieces of Unprofitable Land'

Sometimes I wonder how it is that I learn anything. Other times I worry I may never learn a thing. Today, though, I felt it happen. It began with a mistake, an oversight, well actually, an undersight since I'd failed to actually examine what I was looking at and made a false assumption. Luckily, someone on iNaturalist with better vision in this regard noticed my mistake.

On January 6th, I'd posted a photo of what I assumed to be Lesser Burdock. The one reference I'd checked listed only this single species for Minnesota and it's the one and only species I'm familiar with. If I'd been less confident about my knowledge and consulted *Minnesota Flora* by Steve Chadde, I'd have quickly seen that there are three known species for the state—Lesser, Greater, and Downy—with Lesser Burdock being the most common. All three species are Eurasian in origin and all three are familiar champions of ruined and unprofitable lands: field edges, road ditches, developments and any disturbed soil in general.

This morning iNaturalist member Erika Mitchell pointed out the January 6th observation was Greater Burdock, not Lesser Burdock, noting the difference in pedicel length. Greater Burdock has single flowers on long pedicels. Lesser Burdock has clumps of flowers on short pedicels. A difference easy to see once it's been pointed out. And one accentuated in winter.

By happenstance, leaving the St Olaf library this afternoon, I noticed some burdock plants just outside the building. I wouldn't have given the plants a second thought if it wasn't for my recent misidentification. Now, however, the recognition of it being burdock triggered the question: "What kind was it?" A second look showed the burs clumped about the stem—short pedicels—therefore Lesser Burdock. I stepped off the sidewalk, partway into the window well to take a photo with my phone. The temperature being in the single digits my phone died after a single, not-so-perfect photo, even still it's evidence enough that I actually did learn something.

Arctium minus – Lesser Burdock

January 14

SNOW MIDGES

24°
—
4°

A truly wonderful winter day. While most everyone else was out cross-country skiing, I went for a winter walk. After a week of below average temperatures, a day with little wind and temperatures in the low twenties felt almost summery in the afternoon sun. Apparently, the Snow Midges thought so as well! I've known that there are insects that emerge in winter—wingless gall wasps, winter stoneflies, winter crane flies and midges—but I'd never encountered them before today.

I saw a total of four midges. One of them flew a few inches when disturbed. Otherwise, they just crawled along on top of the snow. The small creek, from where they emerged, flowed and chimed through a short stretch of riffles directly below. It really is mind-bending to see insects out in January. Of course, winter trout fishermen know all about these winter hatches, with a number of midge flies that imitate both pupae and adults.

Diamesa – Snow Midge
Arctium lappa – Greater Burdock
Arctium minus – Lesser Burdock

January 15

"EARTH IN ITS WINGED SEEDS"

Found this tiny, single-winged spruce seed on its epic voyage crossing the snow of our backyard. Just another of the little, magnificent occurrences that go on around us all the time.

"And the earth in its winged seeds, like a poet in his words, travels...." – St.-John Perse, from *Anabasis* (Part V)

From the traveling seed to the art of noticing, I cannot help but think of the poet W. S. Merwin and the palm forest he has nurtured into existence, seed by seed. He gives this good advice in his poem 'Exercise':

> First forget what time it is
> for an hour
> do it regularly every day

Picea – Spruces

January 16 32°

QUADRATURE OF THE LUNE 14°

When they snagged the tips of my snowshoes a few days ago, I cursed them. Today, in a more expansive mood, I stopped to admire their faded purple color and curvilinear form. Black Raspberries (*Rubus occidentalis*). The fruit a favorite treat in summer. The canes form an impenetrable thicket, arching and looping every which way. And in winter, without leaves, the canes create against a backdrop of snow a kind of geometer's junkyard filled with snippets of circles and straight lines. This reminds me of the rather arcane essay about Hippocrates' quadrature of the lune I'd read earlier in the day.

A lune is a crescent formed by the arcs of two different-sized circles, the small one peeking out from behind the big one. Hippocrates of Chios (470 - 410 BCE) was able to square this lune, that is he was able to construct a square with the same area using only a straight edge and compass. This was the first curved area to have its area calculated using geometry. One hopes Hippocrates might have been rewarded with Mt Ida Raspberries (*Rubus idaeus*).

"On the typographic bushes of the poem down a road leading neither out of things nor to the mind, certain fruits are composed of an agglomeration of spheres plumped with a drop of ink."   – Francis Ponge, from 'The Blackberries'

Solidago – Goldenrods
Rubus occidentalis – Black Raspberry
Faxonius virilis – Virile Crayfish

35°
—
29°

January 17

JANUARY RAINS

When it rains in January in Minnesota there can be problems. Like yesterday when it arrived in the late afternoon as freezing drizzle that quickly glazed roads and sidewalks and encased stems and branches. When I went outside to assess the situation, I slid down the incline of our driveway like a curling stone. It was dangerous to walk. And even more dangerous to drive.

That freezing rain turned to wet snow overnight and we woke to immaculate winter scenery and canceled schools (due to the ice underneath the snow). After shoveling snow/slush/ice, I decided to take a late afternoon hike in the rain.

Happily today's rain was nothing more than a fine mist. With the temperature above freezing, the mist caused no trouble whatsoever. The woods, nearly soundproof already due to several inches of soft, melting snow, were muffled even further by the mist dampening and saturating the calm air.

White-tailed Deer moved quietly about the woods, unconcerned at my presence. The moss on rocks and tree trunks seemed especially vibrant, flush with new moisture and warmer temps; beneath diffuse and darkening skies the mosses almost glowed.

Setaria – Foxtails And Bristlegrasses
Solidago flexicaulis – Broad-leaved Goldenrod
Bryopsida – True Mosses
Odocoileus virginianus – White-tailed Deer

January 18

JANUARY THAW

41°
—
25°

After the irresolute weather of the last two days—rain with temperatures below freezing, snow with temperatures above freezing—a true January thaw has settled in, sunshine and temperatures well above freezing. As the day warmed up the half-inch sheath of ice began to lose its grip on the cement of our driveway and I set about loosening it further. While plying the heavy steel ice spud to the driveway ice, a loud bird call overhead caught my attention. Looking up, I saw a Red-bellied Woodpecker on a branch high in our Red Elm. I set aside the ice spud and fetched my camera from inside, returning just in time for a single photograph before the bird flew.

The Red-bellied Woodpecker is not an uncommon visitor to our city lot. However, prior to identifying one for the first time several years ago, I was not aware of the existence of this bird. Having grown up near the Red River Valley in northwestern Minnesota, this species would have been far less common than here in southeastern Minnesota. I was familiar with Flickers which are about the same size and shape. Because of this, I believe my unsophisticated mind skimmed over them and mistook them for Flickers….for an embarrassingly long time. I didn't possess name or knowledge and therefore the bird simply was subtracted from my perception of the world. A blindspot.

"It is not, on the whole, that natural phenomena and entities themselves are disappearing; rather that there are fewer people able to name them, and that once they go unnamed they go to some degree unseen. Language deficit leads to attention deficit." – Robert Macfarlane, from *Landmarks* (2015)

According to *The Birds of Minnesota* by Thomas S. Roberts (1932), the Red-bellied Woodpecker extended its range northward into Minnesota a little over a century ago, the first documented sighting came in 1893. According to the iNaturalist map, the range of the Red-bellied Woodpeckers now extends throughout the state, even into Manitoba.

Melanerpes carolinus – Red-bellied Woodpecker

January 19   37°

"SPRINGTIME IN WINTER"   29°

Thomas McGrath—Rhodes Scholar, communist, North Dakotan—was one of the great poets of MacCarthy era darkness. So, on the eve of the inauguration, it's no surprise that lines and poems of his visited my thoughts.

MY POEMS

Totally out of date
According to the bourgeois seasons.
Springtime in winter—
And winter most of the year.
    – Thomas McGrath

A second day of thawing, of "springtime in winter" as McGrath puts it (though he was referring allegorically to cultural tastes and politics). I find it somewhat fitting to have found a Darkling Beetle on this day. The family name, Tenebrionidae, means roughly "inhabitant of dark places." The dark place I peeked into to find this beetle was the space under loose bark on a dead tree. It's quite astonishing to find an insect alive and kicking mid-January, especially considering that just a few days ago the temperature was close to sixty degrees colder.

There are occasions when I question the efficacy of observations such as this. How can a photo of a beetle from a rotting log help preserve the world's diversity against the threats of global commerce and the resource depletion and habitat destruction that follows in its wake? Other times I question (as others have) the unusual life I lead—a stay-at-home father, hit-and-miss writer, and zealous naturalist. In most instances, however, I rally quickly from such doubts, convinced of the value of these pursuits, angered into action by the ongoing injustices and acts of stupidity at large in the world, determined to do my part in opposition, no matter how trivial and isolate in the face of the grand threats.

"Politics is the continuation of war by other means."
– Thomas McGrath, from the poem 'A Momentary Loss of Belief in the Wisdom of the Common People and a Curse on the Bastards Who Own and Operate Them'

Alobates pensylvanica – False Mealworm Beetle
Quercus bicolor – Swamp White Oak

January 20

WINTER WILDFLOWERS

37°
—
34°

Seed capsules, tattered leaves, broken stems. The beauty of winter wildflowers is the beauty of ruins. Without leaves or petals obscuring the underlying form, structure takes precedence. We see the lines of force, the arch or flying buttress mandated by live weight and wind. We see the strength that makes possible the eye-catching delicacy of summer blooms.

Here's a small poem by the Greek poet Yannis Ritsos (1909 - 1990), a political poet who would have despised the President being sworn into office today:

> Τα χερια του αγγειοπλαστη
> τα δικα σου χερια
> πηλινα πραγματα
> ευθαυστα
> τι ωραια που καθρεφτιζουν
> το αθραυστο
>
> The hands of the potter
> your own hands
> earthen things
> the breakable
> so beautifully mirrors
> the unbreakable

from *Χειροποιητος / Handmade* (1977)

Oenothera biennis – Common Evening-primrose
Larix laricina – Tamarack
Verbena hastata – Blue Vervain

38°
—
35°

January 21

WATER BEARS IN THE TREES

Some years ago, while reading *Wasp Farm* by Howard Ensign Evans, I remember being stung a little bit by his self-satisfaction at having seen a Tardigrade—I too wanted to see a Tardigrade. So this morning when iNaturalist member Lee Elliott posted a photo of a Tardigrade I decided to go looking for one.

How does one find a Tardigrade? Well, it's easier than you might think. Lee's observation included a useful bit of information: his Tardigrade had been found on lichen. It was a simple task to retrieve a piece of lichen-encrusted bark from under our backyard White Ash. This was placed in a petri dish with some water. After letting it soak for a couple hours, the lichen was scraped off the bark into the water and the water examined under the microscope. The first moving things I spotted were nematodes or some other kind of tiny, clear worms. Then, after a little more searching, a Tardigrade.

Water Bear is a good name. However, after watching this one move in water for a while, I'm inclined to describe it more as a clumsy, eight-legged water hamster. They've also been called Moss Pigs. No matter what the name, Tardigrades are legendary for the extreme conditions they are known to survive. For starters, they can withstand temperatures as low as -200 °C (-328 °F) and as high as 150 °C (300 °F). Apparently, they can also survive the vacuum of outer space. And they are radiation proof. And they survive near absolute desiccation. Pretty much indestructible.

After today's successful search, I now know that there are Water Bears in the trees (and pretty much everywhere else). And if someone asks me if I've ever seen a Tardigrade, I can now answer, Yes!

"Tardigrades: there is a frontier for you. Have you ever seen one? I may not know where to find the distributor in my car; I may stumble over the laws of thermodynamics; but I have seen a tardigrade!"
– Howard Ensign Evans, from *Wasp Farm*

Tardigrada – Tardigrades

40°
—
34°

January 22

URBAN COYOTE

Some days you just get lucky and nature finds you. For instance, today, enjoying brunch at the relatives', sitting bloody mary in hand, my back to the windows, I hear my niece ask her mom, "Is that a dog? or a wolf? or a coyote?" Interestingly enough, it was the latter, a coyote making its way upstream on Minnehaha Creek on what was left of the ice after the recent spell of warm weather.

We all watched it walk upstream then double back into some brush at the side of the yard. The coyote stood motionless for a while and then suddenly jumped and pounced on something in the snow, but unsuccessfully. Then it crossed the yard, dug for a little bit in one place, and continued on its way, returning to the ice on the creek. I didn't have my camera at hand, but my daughter was kind enough to part with her iPhone for a few minutes so that I could take a couple pictures.

My father-in-law questioned how a coyote could survive in the suburbs. I thought it more surprising that all these people manage to survive here without growing any food or raising any livestock. Like deer and crows and roaches, the coyotes are one of the few species that prosper in the wake of human progress. If they eat a few feral cats...so be it.

Canis latrans – Coyote

January 23   35°

"A BIRD OF BAD MORAL CHARACTER"   33°

Benjamin Franklin famously criticized the Bald Eagle in a letter to his daughter, Sarah:

"For my own part I wish the Bald Eagle had not been chosen the Representative of our Country. He is a Bird of bad moral Character. He does not get his Living honestly. You may have seen him perched on some dead Tree near the River, where, too lazy to fish for himself, he watches the Labour of the Fishing Hawk; and when that diligent Bird has at length taken a Fish, and is bearing it to his Nest for the Support of his Mate and young Ones, the Bald Eagle pursues him and takes it from him."
– Benjamin Franklin, January 26, 1784

If not pleased by this bird's behavior, what should we think of people who behave in a like manner or worse? Those at the top of the income food chain likely didn't attain that wealth by simple, hands-on labor and minimum wage. Isn't our economy based upon the theft of labor, the exploitation of workers and natural resources? Maybe the Bald Eagle isn't such a bad choice after all. For my own part, I believe no bird deserves to be saddled with the responsibility of being an emblem for a vast and varied nation.

And speaking of moral murkiness, not so long ago the Bald Eagle was on the endangered species list, nearly vanishing from the continental United States. Persecuted by farmers and fisherman, shot indiscriminately for government bounties, and inadvertently poisoned by the misuse/overuse of the pesticide

DDT, a persistent organic pollutant. (By the way, DDT is still used in many parts of the world for indoor residual spraying to combat malaria and for agricultural pests in India.) Happily, the eagle population has made a strong come back in recent years following federal protection and the banning of DDT and they are once again a common sight here in Minnesota.

Haliaeetus leucocephalus – Bald Eagle

January 24

AWAITING THE SNOW

34°
—
32°

On a gray afternoon
awaiting the snow
I fall asleep—
quiet books, a still pencil,
some empty pages
fill the dream space around me—
but then wake
to big flakes falling sideways
filling the air with brightness,
the way words can
overwhelm the mind
with their light.

A short simple walk in the late afternoon. A thickly overcast day, the trees, the animals, even the landscape itself quietly awaiting the evening snow.

Prunus americana – American Plum
Quercus macrocarpa – Bur Oak

32°
—
29°

January 25

BURNS NICHT

In honor of the Scottish poet Robert Burns (1759 - 1796) and the annual celebration of his birthday, I read some poems in the morning and did some thistling in the afternoon. Despite a number of wet, heavy snowfalls and branch-breaking ice storms, a few thistles still stand or at least lean slightly above the snow. I found one large Bull Thistle and a number of native Field Thistles during a short walk at the St Olaf Natural Lands. I gathered two flower heads (one from each species) and brought them home to photograph the seeds and the spines. While the seeds were somewhat similar, the spines on the bracts are quite distinct between the two species. There's an elegance, and a valor, to these winter wildflowers; as fond as I am of the purple blooms in summer, I'm just as pleased with the sight of thistles in winter.

Up in the Morning Early

Cauld blaws the wind frae east to west,
   The drift is driving sairly;
Sae loud and shrill's I hear the blast,
   I'm sure it's winter fairly.

Up in the morning's no for me,
   Up in the morning early;
When a' the hills are cover'd wi' snaw,
   I'm sure its winter fairly.

The birds sit chittering in the thorn,
   A' day they fare but sparely;

And lang's the night frae e'en to morn,
   I'm sure it's winter fairly.

Up in the morning's no for me,
   Up in the morning early;
When a' the hills are cover'd wi' snaw,
   I'm sure its winter fairly.

– Robert Burns

Cirsium discolor – Field Thistle
Cirsium vulgare – Bull Thistle

30°
23°

January 26

PIGEON FANCIER

The pigeon far more than the finch takes a leading role in Charles Darwin's *On the Origin of Species*. He explains the important role of domestic pigeons in the first chapter, 'Variation Under Domestication. "Believing that it is always best to study some special group, I have, after deliberation, taken up domestic pigeons." He kept an aviary and studied pigeons and the breeding of pigeons in great depth. Darwin also seems to have enjoyed this work far more than he must have expected to at the outset and became very passionate about these birds.

"Darwin loved his pigeons . . . he spent hours reading self-help manuals and books by breeders to make sure he was doing the right thing and visiting shows and exhibitions to see what was available. He found it very entertaining hobnobbing with breeding experts and trying to exude an air of practical knowledge as he leaned over cages of absurdly ruffled feathers. The esoteric world of pigeon fanciers seemed to him delightfully fresh and curious."
—Janet Browne, from *Charles Darwin: Voyaging*

Even though the days of rooftop aviaries have passed, feral pigeons are a common sight in cities and small towns the world around. The once fancy domestic birds, the pouters and tumblers, have survived by rewilding the rooftops, instinctually loving tall building ledges the way their ancestors once loved cliffs.

"Great as the differences are between the breeds of pigeons, I am fully convinced that the common opinion of naturalists is correct, namely, that all have descended from the rock-pigeon (Columba livia)..."
–Charles Darwin, from *On the Origin of Species*

Columba livia domestica – Feral Pigeon

31°
—
17°

January 27

BINDWEED STEM GALL MIDGE

Finding a plant gall I haven't seen before is always a pleasure. Today, while looking at the remains of last summer's wildflowers along the edge of a small catchment pond at St Olaf, I noticed the nifty spiral of a Field Bindweed stem wrapped around the thin trunk of a Sumac. And toward the bottom of the bindweed stem were several elongated nodules. Plant galls.

Clearly a must-have volume for every curious naturalist is the one-of-a-kind guidebook by Charley Eiseman and Noah Charney, *Tracks & Sign of Insects and Other Invertebrates* (2010). Included among the diverse chapters on droppings, leaf mines, sign on twigs and rocks and shells is a chapter on plant galls. Eiseman and Charney give quick mention of the Bindweed Stem Gall, but also give the general sage encouragement to collect and rear the galls to discover the identity of the gall makers: "If a gall is inhabited, it is often possible to collect it and see what emerges. Bear in mind, however, that what emerges may not be the insect responsible for the gall. Numerous parasitoids and inquilines are often 'guests' in the galls of other insects." For me, this potential for the parasitoid wasps is all the more reason to gather galls and shelve them in vials.

The other indispensable reference is Margaret Redfern's New Naturalist volume, *Plant Galls* (2011). Redfern has studied plant galls for much of her life and this book presents a dazzling breadth and depth of knowledge on the subject, including discussion of the complex and not always well-understood biochemistry involved in the formation of galls as well as including information on many of the interesting and bizarre life cycles of the gall makers.

"Understanding how galls develop is a major problem that has exercised cecidologists for years. The gall causer redirects normal growth and development of the plant, the details probably varying in different gall-causing groups. In bacterial galls the processes are fairly well understood but not so in animal galls; in these, several 'morphogens' have been suggested but none have been shown conclusively to cause galls. The gall causer manipulates the plant to ensure an adequate, reliable food supply and also to protect itself from adverse environmental conditions including unfavorable climate and predators. How this happens is the subject of active research today."

Plant galls are fascinating. For an overwintering larva, a gall functions as a space capsule, traversing the inhospitable seasons. For a growing larva, the gall provides protection and food. It's as if instead of building a house we could inject some instructive chemical into the trunk of a living tree and it would proceed to grow us a room complete with a well-stocked pantry.

Neolasioptera convolvuli – Bindweed Stem Gall Midge
Convolvulus arvensis – Field Bindweed

28° | 23°

January 28

THE CEDAR HEDGE

Along the north edge of our backyard, a row of four ornamental cedar trees, decades overgrown, provide a kind of living fence. The cedar hedge also provides cover for many birds, both for residents and for those passing through. And through the winter months, the hedge provides some welcome color. Especially welcome is the bright red Cardinal perched in its dusky green branches.

Thuja sp. – Whitecedars / Arborvitae

January 29   29° / 18°

DRAGONFLY WITHDRAWL

When one works with dragonflies spring summer and fall, winter can try your soul. This lapse in sensory stimulation can be depressing and can result in some rather fringe behaviors.

Currently, I have a strong attachment for a *Leucorrhinia* nymph overwintering in our refrigerator. I take it out for periods of sunlight and to feed it. Nothing like a dormant dragonfly nymph to provide some mid-winter cheer, to supply some needed connection to the dragonfly realm, the plastic container a surrogate for a summer pond.

Other remedies for dragonfly withdrawal include trips to the tropics, armchair traveling, dreaming, painting, processing data, writing grants for the upcoming summer, organization and identification of specimens.

Passer domesticus – House Sparrow

42°
—
15°

January 30

NEIGHBORHOOD CROWS

Light snow in the morning. Later, after noon, the sun came out and the temperature rose and the newly fallen snow melted—first from the asphalt street, then the cement sidewalks, and then from other lighter colored surfaces. Because the street was the first to melt it had the first puddles. Our three neighborhood crows dropped down for a drink and to look for edible bits.

Corvus brachyrhynchos – American Crow

January 31

"A-BUDDING"

38°
—
30°

A very warm winter day. Unfortunately I was confined to being indoors for the day and limited to just a few minutes in the backyard looking at a several kinds of buds, just as Thoreau did on this date, the last day of January 1854:

"In the winter, when there are no flowers and leaves are rare, even large buds are interesting and somewhat exciting. I go a-budding like a partridge. I am always attracted at this season by the buds of the swamp-pink, the poplars, and the sweet-gale."
– Henry David Thoreau, from *The Journal of Henry D. Thoreau*

Vinca minor – Lesser Periwinkle
Rhododendron – Rhododendrons And Azaleas
Forsythia – Forsythia
Cornus sericea – Red Osier Dogwood

February

Jack Pine – St Olaf Natural Lands – February 16

February 1　　30°/12°

OUR STREET ELM

For the longest time, I assumed this was an American Elm, one of the lucky ones to have survived Dutch Elm Disease. But a few years back two people suggested it was a Red Elm. One of the tells between these two species is how near the ground the trunk splits into branches: American Elms, the once stately avenue trees, branched high up, the trunks forming simple and elegant columns; Red Elms, like the tree in our front yard, branch earlier.

Shortly after moving to Northfield, I bought a pamphlet at a local garage sale entitled *The Trees of Northfield* self-published by Harvey E. Stork of Carleton College in 1948. This pamphlet contains a survey of the trees found along Northfield streets. Since this was thirteen years before the first reported case of Dutch Elm Disease in Minnesota, American Elms were the most numerous tree.

Ulmus rubra – Red Elm
Fungi – Fungi Including Lichens

17°
—
8°

February 2

DER WASSERPAPILLON

Groundhog Day. At the risk of starting a new tradition, I removed an overwintering dragonfly nymph from the refrigerator, set it on the counter under the window, let it see its shadow in the bright sunshine of this clear winter day.

By odd happenstance, I picked up an old book, *A Study of Goethe* by Brian Fairly, and in the first few pages came across this dragonfly poem which contains the surprising metaphor of a dragonfly being a water-butterfly. So I looked up the entire poems and translated it.

> Die Freuden
>
> Da flattert um die Quelle
> Die wechselnde Libelle,
> Der Wasserpapillon,
> Bald dunkel und bald helle,
> Wie ein Chamäleon;
> Bald rot und blau, bald blau und grün.
> O daß ich in der Nähe
> Doch seine Farben sähe!
>
> Da fliegt der Kleine vor mir hin
> Und setzt sich auf die stillen Weiden.
> Da hab' ich ihn, da hab' ich ihn!
> Und nun betracht' ich ihn genau,
> Und seh' ein traurig dunkles Blau.
>
> So geht es dir, Zergliedrer deiner Freuden!
> (written 1767-69)

Enjoyment

Fluttering over the pool
the ever-changeful dragonfly,
a water-butterfly,
sometimes dark, sometimes bright,
like a chameleon.
If only I might have a closer look
and view its true colors.

The little thing flies past
and lands on the quiet grass.
There, I have him in my net. I've got him!
But examining the dragonfly closely I find
its color to be a sorrowful dark blue.

So it is, for all you dissectors of joy!

Goethe was, without doubt, one of the great observers of the natural world. An early poem by Goethe and an adaption of a French poem about the damage done to the beauty of butterflies when handled, the choice of changing the subject from a butterfly to a dragonfly suggests he'd observed dragonflies closely, even perhaps having held them in hand. The problem with this switch is that a dragonfly's appearance isn't harmed by being handled, it doesn't have powdery scales on its wings that can be rubbed off. Instead, Goethe isolates the difference in observing a living dragonfly in flight against the static up-close observation of a captive insect. There's some truth to his perspective, though almost always I'm astonished by the close-up colors, the patterns, the detail in the eyes, abdomen, and wing venation that are only visible when the dragonfly is caught in a photograph or held in hand.

Leucorrhinia intacta – Dot-tailed Whiteface

24° February 3

7° ## GREEN ASH

The centerpiece of our backyard, this large Green Ash has likely stood here as long as the house which is closing in on seventy years. One of the last trees to leaf out in the spring and one of the first to let go of its yellow leaves in the fall, it still hosts a diversity of animal and insect life in its canopy, including the caterpillars of the large and showy swallowtails.

In the aftermath of Dutch Elm Disease, Green Ash became the most commonly planted tree, along with maples and hackberry, along the city streets of Northfield. Unfortunately, with the threat of the Emerald Ash Borer, these trees may become scarce as well.

The bark on the main trunk is uniform and shows the diamond pattern characteristic of this species. In contrast to the clean, columnar trunk, the upper branches and twigs are a mess, scraggly and swollen at the joints like arthritic fingers.

❖ ❖ ❖

Yesterday, Donald Trump authorized the Army Corps of Engineers to proceed with the Dakota Access Pipeline, bypassing permitting and environmental impact assessments.

Fraxinus pennsylvanica – Green Ash

February 4

FUYUGOMORI

35°
13°

A winter illness—whether the flu, the common cold, or some other ailment—compounds the season's difficulties. Short as the daylight hours are, sleeping off a fever shortens them further. Scratchy eyes, runny nose, and hacking cough disturb the day's routine, interrupt any long night's sleep.

As we rest and recover, an old instinct grips us; we long like hedgehog or chipmunk to curl up and wait out the remainder of winter snug in a safe den. The Japanese have the word fuyugomori which translates roughly as the seclusion of winter or winter isolation, but I suppose the terms hibernation, winter blues, and snowbound would come close as well. No matter how we translate the word, it needs to include the sense of being cut-off from neighbors and friends and from the natural world.

> "The dog pressed to the door
> rattles it as he turns in sleep—
> the seclusion of winter."
> – Yosa Buson, from *Haiku Master Buson*

Luckily, the recent go-round with the flu is over. And today, for the first time in nearly a week, I broke free of the seclusion of winter for a short while, doing a couple of errands and going for a short but satisfying walk at a nearby nature area. It felt great to be let outdoors. Un-snowbound.

Verbascum thapsus – Great Mullein
Planorbidae – Ramshorn Snails
Sphaeriidae – Pea Clams

32°

18°

February 5

FEBRUARY'S DECEPTIVE SMILE

A cold sunny day, like a smile without warmth.

Acer negundo – Boxelder Maple
Sparganium eurycarpum – Big Bur-reed
Alisma triviale – Northern Water Plantain
Quercus rubra – Northern Red Oak
Phalaris arundinacea – Reed Canary Grass

February 6   33°/18°

TURNING STONES

For as long as I can remember, I've been turning over stones to see what lived underneath them. As a child, hundreds of summer hours were spent snorkeling in the shallows of Lake Lida where my family owned some lakeshore. I gave most of my attention to the bigger creatures, crayfish and mudpuppies, but I was equally captivated by darter minnows and large leeches that moved like magic carpets through the water. That strangely quiet and often bizarre world took hold of my imagination and has been a source of enjoyment and mystery ever since.

Even today, decades later, curiosity still beckons and I find myself at the edge of the small creek reaching down to flip over a flat rock. A damselfly nymph clings to the underside of the rock. From the stout body shape and wide caudal fins, I'm fairly certain it is a dancer nymph (genus *Argia*). In the summer, at this location, Blue-fronted Dancers (*Argia apicalis*) outnumber the Blue-tipped and Powdered Dancers at least one hundred to one. So judging solely on odds, this nymph is likely to be a Blue-fronted Dancer as well. It would need to be reared to know for sure.

Argia is a species-rich genus, especially in the tropics where well over a hundred species are known, with many more yet to be described and discovered. In Minnesota things are a little simpler; only five species have been recorded here: *Argia apicalis, Argia tibialis, Argia moesta, Argia fumipennis,* and *Argia plana*, the latter known from a single site in the southwest corner of the state.

Argia sp. – Dancers

32°

12°

February 7

BENTHIC INVERTEBRATE PHOTO BOOTH

Months ago I purchased a box of microscope slides with the idea of using them to construct an inexpensive, micro-aquarium for use as a benthic invertebrate photo booth. And last night I finally got around to assembling a prototype. Using silicone sealant a stack of three slides were stuck one by one on top of each other to a plastic base to make the bottom. Then another slide was added, on edge instead of flat, to create the sides. From above the micro-aquarium looked like a fat capital "I" with really long serifs. Silicone sealant was added to all the corners and around the base and then allowed to set overnight.

So today I returned to the creek to find new subjects. Unfortunately, today was not nearly as pleasant a day as yesterday; a freezing drizzle and a raw wind made it very cold to be playing in the creek without gloves. I stayed just long enough to find a couple creatures to photograph.

Physa acuta – Acute Bladder Snail
Simuliidae – Black Flies

February 8

INDOOR WILDLIFE

14°
—
1°

It's not surprising that other animals take advantage of our habitations, especially here in the temperate northlands where long winters require the fauna to have good strategies for surviving the cold. One strategy, common among a number of non-native species, is to take up residence inside our heated houses. On frigid days like today with the high temperature in the single digits, it's not too likely to encounter an insect or spider outside, so it makes sense to go looking for indoor wildlife—just one of the many benefits of having spiders and other insects inside the house. Besides, there's more than enough heat to go around, and I certainly can't begrudge them space.

Of all the arthropod residents in our house, Cellar Spiders must be the most numerous. Since they prefer out of the way ceiling corners, they go about their business unnoticed and undisturbed for the most part. The exception is when they venture into my daughter's room; she has a zero-tolerance policy against spiders (despite all my best efforts to sway her opinion). Today's photo is of a spider found hanging out above our kitchen table.

These spiders, if fallen or knocked down from their webs, look quite ungainly when forced to use their filamentous legs to walk upon the ground. Most of the time, however, they dangle upside down from their loosely woven webs. If disturbed while hanging on their webs they shake back-and-forth so frenetically that they appear to blur or seem to spin, a wild and unforgettable behavior once observed.

Pholcus phalangioides – Long-bodied Cellar Spider

February 9

ENTOMOLOGICAL HOUSEKEEPING

Winter affords time for straightening the office, cleaning off the desk, and tidying up some stray specimens. Today's observation is a photograph of a Keyhole Wasp (genus *Trypoxylon*) that was reared from a cocoon found in a trap nest. Unfortunately I didn't notice the wasp when it emerged and was unable to get live photos and release the adult. So it will be pinned and labeled and added to the entomology collection.

Keyhole Wasps provision their nests with spiders, lots and lots of spiderlings are stuffed into each nest cell. While most male Hymenoptera take no part in provisioning, constructing, or tending the nests, male Keyhole Wasps help out the female, mostly by guarding the entrance to the nest while the female is away hunting. This protects the nest from parasitic wasps, such as cuckoo wasps, that would otherwise sneak into an unguarded nest.

The wasp in the photo emerged from the cocoon in the innermost nest cell. The wasps in the other two cocoons never emerged. The spiders photographed are from the third cell from the entrance. The cells with the spiders are failed cells; either an egg wasn't laid in these cells or the egg or the larva didn't survive to eat the spiders.

Trypoxylon collinum – Keyhole Wasp

February 10   46°/20°

ALGAE MONSTER, SCALE INSECT, AND SNOW MIDGES

An afternoon walk at the Cowling Arboretum, the upper part along Spring Creek. Warm and Sunny. At the newly renovated bridge over the creek, many Snow Midges were basking on the orange stucco of the railings. And more were seen all along the creek. It was a pleasure to see flying insects again.

At one of the riffles in the creek, I turned over several stones. Under the edge of one of the stones, I found a little heap, like a living tangle of roots, of Ebony Jewelwing nymphs piled on top of each other. The Calopterigidae, the broadwing damselflies, are the twiggiest nymphs. If you're going to live among the roots and stems in the water beneath the undercut banks of a stream it's no doubt useful to blend into the surroundings and become a stick, especially if there happen to be trout in the stream as well.

Adult Ebony Jewelwings are abundant along this stretch of the creek in the summer. River Jewelwings and American Rubyspots can also be found here, but usually in much lower numbers. The nymphs of all three species look rather similar. The labrum, head shape, and first antennae segment length separate *Calopteryx* from *Hetaerina*. The nymph photographed, having stayed relatively motionless for the winter months algae has grown rather thick all over the nymph's exoskeleton, obscuring the details of the head and prothorax, making it a kind of algae monster.

On my way out of the Arboretum, I stopped to photograph the buds of a Hackberry. I noticed along one of the branches what appeared to be dried flowers or stamens at a number of leaf scars or buds. Closer inspection showed these to be tiny fungal spore-bearing bodies. Looking at the photographs afterward, a scale insect was noticed on the branch, victim of the entomophagous fungi, *Cordyceps clavulata*.

Coccidae – Tortoise Scales
Calopterygidae – Broad-winged Damselflies
Celtis occidentalis – Common Hackberry
Diamesa – Snow Midges
Cordyceps clavulata – Entomopathogenic Fungi

February 11 — 43°/36°

A MAGGOTY AFFAIR

Another day of melting, the light dreary, the temperature warm. In the afternoon, I took a short walk to the woods at Lashbrook Park. As I approached the woods, I could hear a male Cardinal singing somewhere high in the treetops, celebrating the spring-like weather.

I peeked beneath the loose bark on several downed trees, hoping to find a bright red bark beetle. Instead, I uncovered a variety of fly larvae and pupae, a real maggoty affair. The biggest larvae (20mm and 30mm) were Crane Flies. Another was *Phaonia*, a fly larvae that is a predator of other insect larvae. The others, some as small as 2mm, I couldn't immediately identify. Several have been kept for rearing, so with any luck, they might be identified when they emerge as adults.

Tipulomorpha – Crane Flies
Tipulidae – Large Crane Flies
Phaonia sp. – Predatory Muscid Flies
Sambucus racemosa – Red-berried Elder
Coleoptera – Beetles

40°
—
29°

February 12

SUNSHINE ON THE BLUFFS

A family outing to the Cannon River Wilderness Area (west unit), what my daughter and wife refer to as the "up-and-down park" because of the upper hiking trail along bluffs cut by a series of ravines. Windy, temperature in the mid 30s, the roads muddy, the trails icy, and the off-trail snow nothing but slush.

Thoreau, in his journal entry for this day in 1854, says that "gray rocks or cliffs with a southwest exposure attract us now, where there is warmth and dryness." With this in mind, we climbed up one of the bluffs at the park to bask in the day's warmth.

Out of the wind and catching the late afternoon sun, the bluffs were warm indeed. At the base of one of the limestone and sandstone outcroppings, I spotted the year's first spider and first fly. Liverwort covered one area of the bluff face.

Continuing my mid-winter antics of bark lifting, I explored a rotten log and uncovered a firefly larva. "Good things come to those who peek under bark," I quipped to my wife as I showed her what I'd found. These larvae are very strange creatures, outrageously armored as if the insect were modeled upon just the tail of an alligator with its embedded bony plates, a kind of crazy block-fault mountain range in miniature, or a series of living wedges. Oh yeah, and they're pink and black and possibly glow in the dark.

Lampyridae – Fireflies
Schizophora – Flies
Araneoidea – Sheet Weaver Spider
Marchantiopsida – Complex Thalloid Liverworts
Tilia americana – Basswood

February 13   48°

A TALE OF TUBE-TAILED THRIPS   28°

Ever since my early acquaintance with French poet Francis Ponge's object poems, initially through the work of Robert Bly who translated and interpreted some of these poems, subsequently through the work of poets like Thomas R. Smith who took up the form in English and through the work of other translators (especially those with less poetic agenda than Bly who tends to bend the poems and interpretations—somewhat—to his own purpose, stressing a Jungian connection between poet and object).

The following passage from the poem 'Banks of the Loire,' the opening poem of Ponge's *Mute Objects of Expression* (Archipelago Press, 2012). As the leadoff poem in this collection, it takes on the role of a poetic manifesto. A manifesto not only for this kind of Pongian text but for the objects of natural history.

"The object is always more important, more interesting, more capable (full of rights): it has no duty what so ever toward me, it is I who am obliged to it."
– Francis Ponge (translated by Lee Fahnestock)

The Scottish philosopher David Hume, writing upon a different topic, that of pride and humility, discusses objects from a different, yet I think relevant, angle, that of joy and pleasure. But also the fickleness of our valuation of objects, and the often mistaken evaluation of an object's true merit.

" 'Tis a quality observable in human nature, and which we shall endeavour to explain afterwards, that every thing, which is of-

ten presented, and to which we have been long accustom'd, loses its value in our eyes, and is in a little time despis'd and neglected. We likewise judge of objects more from comparison than from their real and intrinsic merit; and where we cannot by some contrast enhance their value, we are apt to overlook even what is essentially good in them. These qualities of the mind have an effect upon joy as well as pride; and 'tis remarkable, that goods, which are common to all mankind, and have become familiar to us by custom, give us little satisfaction; tho' perhaps of a more excellent kind, than those on which, for their singularity, we set a much higher value. But tho' this circumstance operates on both these passions, it has a much greater influence on vanity. We are rejoic'd for many goods, which, on account of their frequency, give us no pride. Health, when it returns after a long absence, affords us a very sensible satisfaction; but is seldom regarded as a subject of vanity, because 'tis shar'd with such vast numbers."
– David Hume, from *A Treatise of Human Nature*

I prefer to consider natural history observations and the objects of those observations, the plants and animals, whether written out or recorded by camera, to agree in broad ways with these two viewpoints of David Hume and Francis Ponge. While I try, like Ponge, to put the object first, to subtract the self from the equation and avoid issues of pride, I also go all in for the sensory experience, the joy of close observation and the satisfaction of learning.

Even though it is little valued, possibly even "despis'd and neglected," Common Mullein has to be one of our fanciest weeds, and certainly our softest. Some of its many common names are tiny object poems, the likes of which Francis Ponge might approve: Woolly Mullein, Poor Man's Blanket, Feltwort, Flannel Plant, and Fleece Weed (my own contribution). The basal rosettes of the Common Mullein I found this afternoon were warm to the touch, sun-drenched. I fingered the fuzzy petals, pulling back the messy outer ones to reveal the protected, lotus-like inner petals. Tiny, black specks moved slowly across the soft, fibrous leaves—Tube-tailed Thrips, dozens of them. The largest of these minuscule insects had to stretch to measure two millimeters. We don't "overlook even what is essentially good in them" but mostly overlook them entirely due to their small size.

Populus deltoides – Eastern Cottonwood
Phlaeothripidae – Tube-tailed Thrips
Verbascum thapsus – Great Mullein

43°
—
26°

February 14

MUD BETWEEN THEIR TOES

Spring-like things are beginning to happen after several weeks of above average temperatures. The leaves of the first wildflowers are green and growing. The birds are more active and more vociferous. The raccoons have come out and felt the mud between their toes.

"There were two nuthatches at least, talking to each other. One hung with his head down on a large pitch pine, pecking the bark for a long time,—leaden blue above, with a black cap and white breast. It uttered almost constantly a faint but sharp quivet or creak, difficult to trace home, which appeared to be answered by a baser and louder gnah gnah from the other."
– Henry David Thoreau, from *The Journal of Henry D. Thoreau* (February 14, 1854)

Faxonius virilis – Virile Crayfish
Enemion biternatum – False Rue Anemone
Sitta carolinensis – White-breasted Nuthatch
Procyon lotor – Common Raccoon
Trombidiidae – True Velvet Mites

February 15 36°

19 OF 42 19°

See the open mouth of my suitcase
Sayin' leave this place
– Soul Asylum, from 'Leave Without A Trace'

Minnesota Department of Natural Resources information on the mussels of the Minnesota River states that 19 of 42 native species are extirpated from the watershed. This shocking statistic is a direct result of land use in the watershed. Today's observation, a Giant Floater (*Pyganodon grandis*), comes from an adjacent watershed, that of the Cannon River. A recent survey of the Cannon River lists fifteen species as still being present; it's uncertain what species have been lost. Just as in the song lyric quoted above, a dead mussel, the two parts of the shell still attached at the hinge but open, resembling an empty suitcase, presents a strong image of departure.

This particular species, however, is not yet threatened. Giant Floaters are still very common because they are generalists and can be found in small streams, impoundments, and lakes and, in addition, they are relatively tolerant of pollutants. They are one of the few mussels that can thrive in habitats with little or no current and with muddy or silty bottoms. The glochidia of the Giant Floater, the parasitic larval stage of the mussel, is not limited to a single host fish but hitches a ride in the gills of most of the fish species that live in the river, which further explains this mussel's hardiness and the wide distribution.

Pyganodon grandis – Giant Floater Mussel

42°
—
27°

February 16

THE HOUSE VACATED

Observed, beneath the eaves of a small tool shed, the vacated nest of a paper wasp. It brought to mind these lines:

> Who knows
> how long this house will remain vacant
> or whether it will crumble on its own
>   without another tenant.
>
> – Yannis Ritsos, from the poem 'The House Vacated'

Polistes – Umbrella Paper Wasps

February 17   63°

PRAIRIE VIEWING   30°

The highway sign read "PRAIRIE VIEWING" and the short loop off the interstate, like the road to a rest area only without the buildings, swung off and transected a small grassy area. A surprising number of other cars had pulled off here as well: a trucker sleeping, two cars stopped to walk their dogs, and some others, perhaps like us, who wanted a look and a short walk. The sign for the prairie remnant trail professed that it was maintained by some Story County group but what we discovered was a path that resembled a rarely used game trail through thick trees, mostly Eastern Red Cedar, that ended at a fence at the edge of a cornfield. Back at the parking lot, a closer look at the open area (not much different at first glance than the road ditches along the interstate) in front of the car revealed a few prairie plants: Rattlesnake Master, Indiangrass, Little Bluestem, Mountain Mint and a few others I didn't recognize.

Such a tiny island of native plants in an ocean of Ag-industry farmland. A monument to what has been lost. A sobering reminder of how little is left.

"Disturbed areas represent opportunity. Too much order represents the opposite."
- John Janovy, Jr., from the essay 'Disturbed Areas' in *Back in Keith County*

Passer domesticus – House Sparrow
Eryngium yuccifolium – Rattlesnake Master
Sorghastrum nutans – Indiangrass
Schizachyrium scoparium – Little Bluestem
Pycnanthemum – Mountain Mint

68°
—
32°

February 18

A HOT DAY IN OMAHA

After a successful morning of volleyball spectating (Lida's team won their pool), we had the afternoon free. The absolutely stunning weather—clear blue skies, sunshine, and summery temperatures—allowed us to hit the trails at Elwood Park in short-sleeved shirts. The trail we took followed the eastern edge of the University of Nebraska Omaha campus. Certainly, the highlight of our outing was seeing a Question Mark butterfly—I'm pretty sure this is the first time I've ever seen a butterfly in February. A couple of plants caught my attention as well, including the lush-green, parsley-like basal rosettes of Poison Hemlock. Quite a lot of birds were out as well: Robins, Chickadees, Cedar Waxwings, and Cardinals. Walking a side street back to the car we stopped to admire a giant Sycamore towering above the neighborhood houses.

Polygonia interrogationis – Question Mark

February 19

SOUNDS GOOD

72°
—
39°

I've tried several times to describe the call of male Red-winged Blackbirds—never with complete success. A trill, a squirt gun that shoots music, or maybe the acoustic equivalent of a glance at a bright, possibly even triumphant, flag, rippling in the wind. No matter what it sounds good.

Agelaius phoeniceus – Red-winged Blackbird
Columba livia domestica – Feral Pigeon

64° / 41°

February 20

RAINY DAY TRAVEL

After the third day of volleyball, which ended mid-afternoon, we headed home, driving from Omaha to Northfield in the rain. I watched the world scroll by outside the car windows. A few hawks along the interstate, here and there some remnant oaks, and as far as the eye can see stubble fields blurring under slate gray clouds.

Passer domesticus – House Sparrow
Zea mays – Field Corn

February 21

SANDHILL CRANES

62°
—
32°

In both 2016 and in 2015, I saw Sandhill Cranes on March 19th. Thomas S. Roberts, in *The Birds of Minnesota* (1932), gives the earliest date as March 22, 1889, in Wright County and the average fourteen migration arrival dates as March 30th. So, despite the record warm weather, I was very surprised to hear and then see several migrating Sandhill Cranes today, February 21st.

> Cranes fly overhead, calling
> Each other; so here, I stop my
> Worrying, just letting
> My mind roam back over thoughts
> Of happier days.
> – Tu Fu, from 'A Summer Night'

Antigone canadensis – Sandhill Crane
Sepsis – Black Scavenger Fly
Sciomyzidae – Marsh Flies
Pardosa – Thin-legged Wolf Spiders
Delphacidae – Delphacid Planthoppers
Nabis – Damsel Bug
Cecidomyiinae – Gall Midge
Alticini – Flea Beetles
Bembidion – Ground Beetles
Epinotia vertumnana – Olethreutine Moth

59°
—
37°

February 22

JUST THE FUSELAGE

After nearly a week of record-breaking temperatures, four of the last seven days setting new daily high temperatures, a fair number of insects are making an appearance even though it is still February. Over the last two days, I've observed several beetles, a moth, an ant, three species of fly, a planthopper, and a damsel bug. Last year a similar warm up the second week of March brought out some of these same insects. This year they are two weeks ahead of that. With one to two feet of snow predicted over the next two days, this parade of spring weather and spring creatures will pause for a while.

Looking around the front yard and the gardens, I came across the remains of a dragonfly, a darner of some kind. Lodged among the snipped stems of last year's Coneflowers, the dragonfly resembled the wreckage of a minuscule plane, just the fuselage and a few twisted remnants of a wing.

Aeshna – Mosaic Darners
Lasius – Citronella Ants
Pollenia – Cluster Flies
Phormia regina – Black Blow Fly

February 23

NOCTURNAL ESCAPADES

39°
—
29°

A Yellow Sac Spider, a species introduced from Eurasia, was found trapped in our sink this morning. I captured it, but before letting it go outside, I cooled it in the refrigerator and then took its photo.

This is a common spider in our house, and one often found wandering the walls and fingering the ceilings, especially after dark. Occasionally, they are found ensconced in their silken, namesake sacs, hung in the high corners like mountain climber tents.

These pale-legged spiders prowl nightly for prey. According to *Common Spiders of North America* by Richard Bradley, spiders of this genus roam over vegetation at night, overtaking their prey. Curiously, these nightly escapades are subsidized by a diet of plant nectar. More curiously, then, one wonders what kind of sports drink these spiders substitute for plant nectar in a house without plants?

❖ ❖ ❖

National Guard evicted remaining protesters at Standing Rock in North Dakota.

Cheiracanthium – Longlegged Sac Spiders

29°
—
20°

February 24

SNOW DAY

Snow days are still exhilarating. Despite the improvements to weather forecasting that take some of the surprise from the event—schools closed twelve hours in advance, salt mixtures applied to the streets in anticipation, computer-generated snowfall maps available at the touch of a finger—there's no escaping the dazzlement of waking to a half foot of new snow.

Philodromidae – Running Crab Spiders

February 25 34°

A TALE OF TWO CITIES

14°

With yesterday's snowfall, the springing of spring was unsprung, the phenological clock reset to its proper position of winter in February, the green grass, the early insects all hidden under six inches of new snow. And with the disappearance of the insects and early greenery, my naturalist activities were relegated to a few photos of the buds on front yard bushes, the Honeysuckle and the Hydrangea.

The curious thing about this particular snowfall is how abruptly the accumulation tapered off from south to north. Ten miles south of town a foot of snow fell; ten miles north of town no snow fell. Driving from Northfield to Burnsville was like driving from one season to the next, winter to spring, from snow everywhere you looked to a landscape of dormant grasses turning green. A tale of two cities, on two sides of a snowstorm. Not nearly as abrupt as those summer downpours where it is raining in the backyard and dry and sunny in the front yard, but almost.

> "It was the season of Light, it was the season of Darkness, it was the spring of hope, it was the winter of despair."
> – Charles Dickens, from *A Tale of Two Cities*

Lonicera sp. – Honeysuckles
Hydrangea – Hydrangeas

February 26

YOUR NAME

A bright, wintry day. Lisa and I walked to St Olaf in the late afternoon to return a library book. Along the way, we noticed a small woodpecker high in a Maple Tree. A look at the photos later showed the shortish bill that identifies it as a Downy Woodpecker, separating it from the lookalike, but larger, species the Hairy Woodpecker.

Earlier this day, iNaturalist member Sam Kieschnick posted the question, *Why do you like to name things?* I've wondered about this as well. And I've attempted to answer it a number of times. The most concise answer I've come across is this statement by odonatologist K. D. Dijkstra: "Names introduce species to humanity... All awareness, conservation and research starts with the question: What species is that?"

Learning someone's name, psychologically and socially, is a meaningful act. Conversely, not remembering someone's name or not bothering to learn someone's name is an act of dismissal, they don't matter to us. Learning the common names and scientific names of other species is an ethical act.

There is also something intimate about a name. Which is why so many love songs contain names. To the point of not wanting to share a name with others, of keeping it to our selves.

> "And I won't tell 'em your name."
> – Goo Goo Dolls, from 'Name'

Picoides pubescens – Downy Woodpecker

February 27

WARM ENOUGH

44°
—
22°

Between dropping off boxes at the Post Office and driving players to practice in Burnsville, I snuck in a ten-minute walk at the upper Arboretum. At the footbridge over Spring Creek, it was warm enough that the snow midges were active, landing here and there on the railing and on sunlit patches of snow. Looking closely at the rail, I noticed a dark lump near the edge. Not certain what it was I nudged it with my finger and it readjusted its legs. Aha, a spider.

According to *Common Spiders of North America* by Bradley, spiders of the genus Eustala "rest curled up on branches, looking like a piece of bark. If disturbed, the spider may run a short distance then resume this cryptic pose." Instead of a portable folding chair, a rather bizarre portable folding spider.

Optimistically I'd brought along my zoom lens, hoping for a bird photo, leaving my macro lens at home, not optimistic enough to foresee insects or spiders. Not having a vial along I had to improvise, fashioning a triangle sampling envelope from a five dollar bill in my wallet. After scooching the spider onto the bill, I folded it up and placed it in my coat pocket so that I could take pictures of it later.

Eustala sp. – Spider
Orthocladiinae – Snow Midge

47°
—
32°

February 28

DREAM OBSERVATIONS

Today's photo observation, the bud of the invasive Common Buckthorn, comes from the banks of the Cannon River, but even before waking this morning I was busy with dream observations.

I don't know where the dream began but there was a game resembling badminton or chess. The rules weren't quite right and definitely not in my favor. I wasn't winning so I slipped across the street to take some photos of nature.

The place was a kind of bus stop, possibly. People waited around. And there were big blocks of stone. Looking up I photographed a stocky palm tree. Then a few curious weeds that resembled pressed plants. A legless lizard caught my attention as it slowly progressed toward the undulating green and white leaves of a plant that seemed to crawl off and cover the lizard. Tree roots metamorphosed into a snake with a head shaped like the head of a Spiny Softshell Turtle. I looked it in the eye and turned to go (thinking to myself that I'd remember their names when I woke up...that was a mistake!). As I moved away I felt something strike me on the back, not a push, or a bite, but more like a firm hand placed on the shoulder. And I opened my eyes.

> O sir, you are old,
> Nature in you stands on the very verge
> Of his confine."
> – Shakespeare, from *King Lear*

Rhamnus cathartica – Common Buckthorn

March

Lyman Lake – Cowling Arboretum – March 14

March 1   37°

SNOW-BOUND   23°

The first day of March is more wintry than the last day of February due to an overnight snowfall. Snowbound, but only for a short while.

> "And, when the second morning shone,
> We looked upon a world unknown,
> On nothing we could call our own.
> Around the glistening wonder bent
> The blue walls of the firmament,
> No cloud above, no earth below,—
> A universe of sky and snow!
> The old familiar sights of ours
> Took marvellous shapes; strange domes and towers
> Rose up where sty or corn-crib stood,
> Or garden-wall, or belt of wood"

– John Greenleaf Whittier, from 'Snow-Bound: A Winter Idyl'

Pterostichini – Woodland Ground Beetles

31°
—
17°

March 2

MARCH COLD

A clear brisk early March day. The temperature, in the upper 20s, just a little under the average for the day. The wind solidly out of the west. As I entered the woods I could hear crows, a lot of crows. At one spot at least twenty were perched in the treetops overhead. There was little doubt they were up to something.

Underfoot, hard surfaces of snow and ice alternated with stretches of mud and softening dirt. The sun was beginning to melt the frost in the earth. March cold isn't like the cold that stings our soul in January. It's no longer sustainable. It's temporary. The warming that leads to spring has begun.

Flowers on the maple trees have begun to open despite the freezing temperatures. Sap has begun to rise and run. Tiny spears of green have begun to ascend through the dark.

Acer – Maples
Ulmus pumila – Siberian Elm
Prunus serotina – Black Cherry
Carex – True Sedges

March 3

SAPCICLES

31° / 13°

Another cold March day with the high temperature not making it to 30 degrees. The warm spell prior to this cold spell started the sap flowing in the Sugar Maples. And because of the low temperatures, any leaks in the trees due to damaged bark or snapped branches froze into icicles of sap, sapcicles. I had the pleasure of sampling a few of these sweet, sticky icicles today.

Acer saccharum – Sugar Maple
Salix interior – Interior Sandbar Willow
Salix discolor – Pussy Willow

43°
—
26°

March 4

A LATE WINTER WALK

Today I hiked to the oak savanna and fen at the Cannon River Wilderness Area. Curiously I've never visited this site in the winter when snow-covered. Now that it is March, the opportunity for winter walks will diminish quickly, especially with warmer weather in the immediate forecast.

While more of a swampwalker, in the spirit of David M. Carroll, I also have a long history of woods wandering, in the spirit of Walt McLaughlin. Best is when both occur on a single hike. Because of the remnant big woods, the floodplain forest, the small acreage of oak savanna and because of the spring-fed seeps and fen, this hike has both and is one of my favorites.

The woods, as I started out, were snowy and quiet. The trail, neither slippery nor too deep to follow, was easy walking the whole way. It didn't take long to reach the fen at the end of the trail. The water between tussocks of sedge was still frozen, making it easy to move around and explore. I walked out to the middle and stood among the red osier and willow shrub, bog birch and hazel thickets. The winter remains of last summer's Turtlehead blossoms caught my attention. A close look showed that the delicate white blooms had altered, leaving sepals and seed pods that resembled the open mouths of young robins held wide for food.

On the open slopes of the oak savanna above the fen, I noticed large circles clear of snow. Each circle was positioned on the sunny side of standing Red Oaks. The leaf litter, on the sunny side of the trees—a deep, leather-brown color—must aid in melting the snow by capturing the radiant heat of the sun. Plus it's this area that isn't shadowed by the leaves remaining on the tree. It was comfortably warm inside these oak circles, while cool and crunchy snow remained outside them. Exploring one such circle, I found a small wolf spider on top of the leaves. Turning over a piece of wood I found a much larger spider, a Funnel Web Spider, surrounded by a loose web. This spider had long orange-and-black legs pulled in tightly to its body, the knee joints coming together above the cephalothorax in a ring, making it difficult to see the eyes of the spider. After being in the sun for a few minutes it loosened and scrambled off, disappearing into the surrounding leaf litter.

These seemingly spot-lit circles were charming and transported visitors slightly ahead in time, providing a premonition of spring. Two steps and I was back to winter and the long, enjoyable walk back to my car.

Hypogastrura – Springtails
Cladonia – Pixie Cup Lichens
Polytrichum juniperinum – Juniper Polytrichum Moss
Lycosidae – Wolf Spiders
Coras – Funnel Web Spiders
Corinnidae – Antmimic And Ground Sac Spiders
Cladonia – Pixie Cup Lichens
Quercus rubra – Northern Red Oak
Cordyceps clavulata – Entomopathogenic Fungi
Chortophaga viridifasciata – Green-striped Grasshopper
Cornus sericea – Red Osier Dogwood
Chelone glabra – White Turtlehead
Betula papyrifera – Paper Birch
Onoclea sensibilis – Sensitive Fern
Oedipodinae – Band-winged Grasshoppers
Salix discolor – Pussy Willow
Urocyon cinereoargenteus – Gray Fox

59°
—
35°

March 5

ASPENS

Imagine Thoreau doing a stint as the greeter at Wallmart and you'll have a sound idea of my day. Not that I'm anywhere near the naturalist that Thoreau was nor my job as unfriendly as smiling for a big corporation but a six-hour shift as door security at a sporting facility is a little out of character. Luckily my shift started early enough that I had a chance to take a short hike at a nearby park, Murphy-Hanrehan Park Reserve, after I was done.

Driving past Hanrehan Lake I could see wide traces of open water, and walking the trails of the park most of the smaller wetlands were partly open as well. If the prediction for rain and possibly thunderstorms holds the remaining ice will go quickly. The narrow gaps of open water between shore and the main body of lake ice reminded me of a few spring ice-fishing outings when I was young when makeshift bridges were thrown out across this gap while the main lake ice was still sound enough to fish. I'm not certain I would try that anymore, especially on these shallow southern Minnesota lakes.

A few people walked their dogs, a few people hiked, but mostly the park trails were empty and quiet and I enjoyed the comfortable weather and the moments of quiet. Before leaving I took photos of two kinds of phytopathogenic fungi—Aspen Bracket and Black Knot. Aspen Bracket is the fruiting body of *Phellinus tremulae* which causes Aspen trunk rot. The brackets have always reminded me of steps or the artificial climbing holds used on climbing walls and no doubt could be used to climb the trees if they were abundant enough or spread out along the trunk. Black Knot is a fungus that affects trees of the genus *Prunus* like Chokecherry and Wild Plum. According to one website, Black Knot resembles burnt marshmallows on a branch, which is a pretty good description.

Apiosporina morbosa – Black Knot
Phellinus tremulae – Aspen Bracket

65°
—
39°

March 6

INK WITH ICE CRYSTALS

This morning I opened an old translation of *The Story of the Ere-Dwellers* (also know as *Eyrbyggja Saga*). I can think of no prose more elemental than that of the Icelandic Sagas; it's as if they were penned in black ink with ice crystals. This language of barren fells, blood feuds, Viking ships and the lingering presence of the Norse gods drift into the sentences of today's observation.

There was a bird, a Brown Creeper, that visited the great Ash this morning. Now the bird fell to climbing the trunk and searched out food from the crevices found therein the bark. This bird betakes itself to climb head upwards always ascending, never down. So the Brown Creeper moves from tree to tree.

"But Thorod made a bargain that winter with Thorgrima Witch-face that she should bring a storm on Bjorn as he went over the heath; and on a day Bjorn fared to Frodis-water, and in the evening when he was ready to go home the weather waxed thick, and somewhat it rained, and he withal was rather late ready; but when he came upon the heath cold grew the weather, and the snow drave down, and so dark it was that he might not see the road before him. Then came on a storm, with such hail that he might scarce keep his feet, and his clothes, which before had got wet through, took to freezing on him, and he was so wildered withal that he knew not which way he turned." — from *The Story of the Ere-Dwellers*, done into English from the Icelandic by William Morris and Eirikr Magnusson (1892)

Certhia americana – Brown Creeper
Anatis labiculata – Fifteen-spotted Lady Beetle
Oxytelinae – Rove Beetle

March 7   49°

EARLY MOTHS AND TORNADOES   29°

Yesterday evening, following a day of warm temperatures, the warmest day so far in 2017, a line of thunderstorms swept across the state. In some places, the storms also brought the first hail and the first tornadoes. The tornado that touched down near Anna Lake, west of Zimmerman and near several natural areas I visit most summers, was, according to the news, the earliest tornado on record for Minnesota by about two weeks. I later learned that this same tornado, touching down a few miles further north, vacuumed the ice off a section of Little Elk Lake.

After the thunderstorms pushed through Northfield (thankfully no tornadoes here), I collected the first Spring Cankerworm Moth which had landed by the front door light. Always one of the first moths of the year, this is the earliest date I've recorded. Last year the first was observed one day later, on March 8th. In 2015, the first observation was on March 16th. And the year before that, when I was obviously less vigilant, the date I first saw the moth was April 12th (probably not an accurate early date). Only the males of this species have wings. The females are flightless and kind of look like midget walruses.

In the afternoon, I attended a volunteer appreciation meeting for the Wasp Watchers Program, a biosurveillance program for the detection of the invasive Emerald Ash Borer. While not an official volunteer to the program, I had submitted an observation after finding a *Cerceris fummipennis* nest in Northfield which was enough to make the invitation list since it's the wasp being watched. The meeting was held at Hodson Hall on the University of Minnesota St Paul Campus. To start things off, three entomology graduate students shared their 'outreach arthropods' with us—Madagascar Hissing Cockroaches, Death-feigning Beetles, Darkling Beetles and larvae, giant millipedes, and a Rose Tarantula, providing lively accompaniment to cookies and coffee. Jennifer Schultz, director of the Wasp Watchers program, led the meeting, summarizing the previous year's results.

After Jennifer's presentation, we visited the U of MN insect collection and met the curator, Robin Thomson. This is a large collection, rows and rows of metal cabinets, nearly floor to ceiling. Each metal cabinet contained many wooden drawers filled with pinned insects. How satisfying to finally visit this collection. Many friends and odonatologists had spent time working here and I had missed several chances to help them. The wonder of looking at the insects in the collection intertwined with nostalgia for those long-ago student days when I spent many many hours in the nearby Entomology Library, around thirty years ago now.

From the U of MN, I drove south to give a slide presentation on wasps and bees and dragonflies to the Red Wing Master Gardeners. Curiously, many of the people were up-in-arms, out-of-sorts, and bent-out-of-shape about the recent discovery of the Emerald Ash Borer in a tree on Barn Bluff. Several people complained about the stupidity and irresponsibility of local residents who must have broken the firewood quarantine and introduced the beetle locally. Of course, this was a possibility, but they seemed to forget that these beetles can fly and might have arrived there on their own.

Paleacrita vernata – Spring Cankerworm Moth

March 8

OSSI DI SEPPIA

36°

24°

The ice is nearly out at the big pond at the St Olaf Natural Lands when I visit in the afternoon. I look closely for ducks in the dark patches of open water amidst the remnant ice but find none. I walk to the woods at the west end of the pond. The Quaking Aspen do more than tremble and quake in the cold, gale-force winds. Some of the top branches have actually snapped off and lay scattered on the trail.

Over twenty years ago I discovered the work of the Italian poet Eugenio Montale, a poet whose rugged elemental landscapes mixed complexly with human emotions. I especially remember pouring over the volumes translated and annotated by William Arrowsmith, for many many months. Looking at poems from the early collection *Ossi di Seppia* today, after many years, I recognize more easily the force that so overpowered me when I first encountered them.

Something about today's howling wind and the bare branches of the treetops whipping together until they sounded like the antlers of battling deer and the cold waves cutting a wedge into the last of the ice on the pond nudges me in the direction of Montale's poems this afternoon. One of the first poems I look at includes a few lines about poplar trees. A few poems later I encounter the Italian transmontane wind, a cold forceful wind gusting out of the north, which is even more apropos of today's weather. The poet reminds us that there are days when it's a blessing to be rooted in place:

"What looks like leaves or birds fly through
the sky dome and are lost.
And you, all shaken by the whips
of the unleashing wingd,
you who clutch to yourself
arms that are swelling with flowers yet unborn,
how hostile are the spirits
that overflow the convulsed earth in swarms,
my tender life, and how beloved are
your roots today."

– Eugenio Montale, from 'North Wind' (trans. by Ned Condini)

Populus tremuloides – Trembling Aspen

March 9

"BRIGHTSUN LOVE"

31°
—
18°

I drove to the local pond where the ice was going off yesterday only to find that it had frozen over anew. So for the second day in a row, there were no migrant ducks or shorebirds to see. One of these days. Before leaving I took a photograph of the dried fruit of the Wild Cucumber, resembling to some degree a small pufferfish. One end has blown open, it's four large seeds shot and fallen to the earth.

A very pleasant gift arrived in today's mail, a book by Floyce Alexander entitled *Sundown*, over two hundred pages of selected and new poems. Browsing it I encounter many old favorites and sample several new poems. There can be few poets more human, more personable, more erotic, or more honest than this poet. Especially now, after the cold celibacy of winter, after months of snowfall and thickening ice, these poems are welcome as would be a little of that "brightsun love" he mentions in the poem 'Blood Rivers.'

> "Lakes of blood rising high as watertable,
> overflowing in early spring, becoming rivers.
> Your fingers, your twining legs, your lips and their salt
> against the eyes of my closed lids, your brightsun love."
> — Floyce Alexander, from the poem 'Blood Rivers'

Echinocystis lobata – Wild Cucumber

19°
—
5°

March 10

SIXTY DEGREES COOLER

The temperature this morning was 7 degrees F, sixty degrees cooler than the high temperature on Monday, five days ago. Quite a reversal in weather, but certainly not unusual for springtime in Minnesota.

When I see Ostrich Ferns, especially the dried fertile fronds in winter, I think of fiddleheads. And when I think of fiddleheads I recall a wonderful lunch at the home of John and Edith Rylander near Gray Eagle, Minnesota. I was traveling with the Persian poet Fereydoun Faryad, on our way to a reading in Fargo, North Dakota, mid-May, 2011. Luckily we had enough time for this side trip to visit with John who spoke Farsi and Edith another poet. That splendid lunch included fiddlehead ferns, tuna-pasta salad spiced with wild onion, and homemade bread. Then fresh rhubarb sauce over vanilla ice-cream for dessert.

Matteuccia struthiopteris – Ostrich Fern

March 11

ANNIVERSARIES

21°
—
8°

Only yesterday did I learn that this year, the year 2017, marks the bicentennial of Henry David Thoreau's birth. What do such anniversaries mean? What do the passing years measure? In our capitalist world, its consumerism, a big date is useful for promoting new products and for winning grants for related projects. I see that at least two new biographies of Thoreau are slated for the upcoming months; there will surely be more.

Just last year it was the 100th Anniversary of Yellowstone Park. The 100th anniversary of the birth of the North Dakota poet Thomas McGrath. Eight years ago it was the Sesquicentennial of the publication of Charles Darwin's *On the Origin of Species*. The list multiplies easily. Tacitus supposedly finished *The Annals* in 117, one thousand nine hundred years ago, but I don't believe we have a name for that great span of years.

So what anniversaries should be celebrated. I, for one, am interested in following the earth around, keeping track of the yearly arrivals of flowers and insects, each a one year anniversary of uncertain date. A lifetime's labor, to fill in some of the blanks on the phenological calendar. This seems important to me. Where I grew up, ice-out was more talked about than Easter.

Trochosa sp. – Wolf Spider

23°
—
13°

March 12

CEILING RUNNER

After a slow trip home from a day of volleyball, twenty-five miles and three inches of new snow, we arrived home after dark. Turning the lights on I catch sight of a small black spider with a white stripe on the living room ceiling. Then I do what all good housekeepers do, I captured it alive, took a few photos, and released it back to the wilds of our living room.

A Parson Spider, a common and recognizable inhabitant in our home, though I have on several occasions encountered it outdoors also. Both its common name and scientific name derive from the shape of the light-colored stripe on its abdomen resembles the fancy necktie, a cravat, worn by eighteenth-century clergy.

In its natural habitat, this species hides beneath stones and logs. When encountered indoors this ground spider transforms into a notable ceiling runner, a contrarian of sorts. But then, who's to say they don't race along upside down on those low ceilings, the underbellies of stones and logs.

Herpyllus ecclesiasticus – Eastern Parson Spider

March 13

ANOTHER PUZZLE PIECE

27°
—
14°

To begin I assumed our big streetside elm was an American Elm, not really giving it much thought. Then over the years, two people have suggested it was a Red Elm and I began to suspect they were correct. I've never taken a close look at the buds or the flowers or the fruit to confirm which species, until today. After yesterday's snowfall and the windstorm earlier in the week a couple branches snapped off in the high canopy giving me an opportunity to take some macro photos of the buds. Another puzzle piece helping to complete the picture. Comparing photos from a few websites, I'm fairly certain this is an American Elm. The final puzzle pieces will be filled in later this spring: the flowers of the American Elm are stalked, those of the Red Elm are not.

Ulmus americana – American Elm

29°
—
9°

March 14

A SOUR-SWEET DAY

Minus 2 degrees this morning, then clear sky and melting snow in the afternoon. A day of good and bad. A sudden fever and sore throat, yet a few calm hours with a new book. Later in the afternoon, Lyman Lake is completely iced-over when I visit hoping for an easy photograph of a goose, not feeling well and not having the energy to take a hike. The lake had been open a few weeks back. I thought I saw open water and a gathering of ducks and geese up the trail, so I set off only to find I was mistaken. But along the way, I spotted the Bittersweet vine wound into the branches of a small tree along the shore. I always enjoy seeing this plant in winter. And it always reminds me of the fine poem of the same name. A fitting find for a sour-sweet day.

    Bitter-Sweet

    Ah, my dear angry Lord,
    Since thou dost love, yet strike;
    Cast down, yet help afford;
    Sure I will do the like.

    I will complain, yet praise;
    I will bewail, approve;
    And all my sour-sweet days
    I will lament and love.

    — George Herbert

Celastrus scandens – American Bittersweet

March 15

TOUCHES FLAME

32°
—
12°

This morning a pair of Northern Cardinals landed in the branches right outside our front window. While these bright red birds add a touch of flame to almost any branch, the branch being lit at the moment was a Winged Burning Bush. The two birds ate several of the orange berries, then left.

Winged Burning Bush (*Eunoymus alatus*), a popular landscaping tree, is currently banned in over twenty states, including adjacent Wisconsin. The reason for these bans has to do with this plant's potential for becoming a problem invasive. Like Buckthorn, the seeds of Winged Burning Bush are efficiently dispersed by the many birds that eat the berries. In addition to the morning's Cardinals, I've watched many other birds dine upon these berries: Robins, Hermit Thrushes, and Cedar Waxwings to name but a few. Interestingly, Winged Burning Bush belongs to the Bittersweet Family, Celastraceae. Having just observed the family's namesake plant the day before, it was easy to recognize the similarity in berries. So similar in color and the design, I was surprised I hadn't made the connection before.

On late autumn days, especially at dusk, the leaves of this bush almost glow, like smoky sunsets or well-banked coals. Even so, the responsible thing to do would be to depose of this and other non-natives, replacing them with native species. Smooth Sumac, with its crimson fall foliage, might be a good alternative.

Euonymus alatus – Winged Euonymus

40° / 20°

March 16

TWO SERVINGS

While it is difficult to do much of anything when one has the flu, I find writing to be particularly difficult when one has no energy. Add to that a fever and medicines and one is lucky to get more than a blurry word or two. A visit to the doctor, today, revealed that I had two servings of sickness: strep throat and the flu. And the thought of missing out on a spring day doubled that double misery, even if the weather was rather runny and cold.

Pinus strobus – Eastern White Pine
Taxus canadensis – Canadian Yew

March 17
"ONE OF THE TEETH OF TIME"

43°
—
29°

I found this little creature on the tile floor of the basement today. Being we have more books in our house than a small monastery, the Silverfish is not really a welcome guest. But, in reality, we aren't infested with them, and they don't eat much a letter here a letter there. Seeing it I'm reminded of a wonderful quote by the early champion of the microscope, Robert Hooke, and an Old English riddle that I've always enjoyed.

"It is a small white Silver-shining Worm or Moth, which I found much conversant among Books and Papers, and is suppos'd to be that which corrodes and eats holes through the leaves and covers; it appears to the naked eye, a small glittering Pearl-colour'd Moth, which upon the removing of Books and Papers in the Summer, is often observ'd very nimbly to scud, and pack away to some lurking cranney, where it may the better protect itself from any appearing dangers....This Animal probably feeds upon the Paper and covers of Books, and perforates in them several small round holes, finding, perhaps, a convenient nourishment in those hulks of Hemp and Flax, which have pass'd through so many scourings, washings, dressings and dryings, as the parts of old Paper must necessarily have suffer'd; the digestive faculty, it seems, of these little creatures being able yet further to work upon those stubborn parts, and reduce them into another form. And indeed, when I consider what a heap of Saw-dust or chips this little creature (which is one of the teeth of Time) conveys into its intrals, I cannot chuse but remember and admire the excellent contrivance of Nature."
– Robert Hooke, from *Micrographia* (Observation LII), 1665

And this Old English riddle from The Exeter Book:

Riddle 42

Moððe word fræt me þæt þuhte
wrætlicu wyrd þa ic þæt wundor gefrægn
þæt se wyrm forswealg wera gied sumes
þeof In þystro þrymfæstne cwide
⊠ þæs strangan staþol stælgiest ne wæs
wihte þy gleawra þe he þam wordū swealg

A moth ate words. To me it seemed
a remarkable fate, when I learned of the marvel,
that the worm had swallowed the speech of a man,
a thief in the night, a renowned saying
and its place itself. Though he swallowed the word
the thieving stranger was no whit the wiser.
(Alfred John Wyatt, translation)

Lepisma saccharina – Common Silverfish

March 18

A MURMURATION OF STARLINGS

43°
30°

Not quite a murmuration of European Starlings, those immense mesmerizing flights of thousands of birds, but dozens of these iridescent birds visited our yard this afternoon. The snow having just melted from the yard, the Starlings and a number of other birds took a sudden interest in the newly unveiled leaf litter.

Turdus migratorius – American Robin
Sturnus vulgaris – European Starling
Picoides pubescens – Downy Woodpecker
Junco hyemalis – Dark-eyed Junco
Corylus americana – American Hazelnut

55°
—
28°

March 19

OPEN WATER

Still convalescing from the flu, strep, and a bad reaction to the medicine prescribed for the strep, I hadn't strayed far from the couch for four days, letting the illnesses run their course. As the symptoms passed, what seemed to be side-effects of the antibiotics increased—sleeplessness, nausea, and (the worst by far!) a non-stop, unrelenting horrid taste in the mouth. Every morsel of food or drink tasted bitter, like an alkaloid poison excreted from the knees of lady beetles or the aftertaste of sipping a shot of turpentine. I slept most of the morning, tried to get up, slept some more, and only truly got up around one in the afternoon.

A few hours later, Lisa drove me to the large pond at the St Olaf Natural Lands. The pond was about ninety percent open, with just a small amount of ice hugging the curve of the southwest corner of the pond. Several dozen Mallard pairs. One Ring-necked Duck pair. We heard Sandhill Cranes passing overhead. A hatch of black midges flew at the shore where I also encountered several leaf-roller moths.

Driving home, we spotted a pair of Hooded Mergansers in the small catchment below the St Olaf science building. Lisa stopped the car and I was able to get out and manage a couple of photos of the male.

Lophodytes cucullatus – Hooded Merganser
Olethreutinae – Olethreutine Leafroller Moths
Aythya collaris – Ring-necked Duck
Anas platyrhynchos – Mallard
Trochosa terricola – Ground Wolf Spider

March 20

LADYBUGSTER

56°
—
34°

Wandering around in the blessed sunshine on this first day of spring, I came across a small congregation of Spotted Lady Beetles in the leaf litter and rocks in our backyard. Near a dozen of these fine looking native lady beetles roamed about near the base of a dead Weigela bush, no doubt not far from where they gathered to overwinter. So for a few minutes this afternoon I was pleased to observe their peaceful commotions and take on the role of "Ladybugster," a name coined by one of my favorite entomologists, the well-known Canadian nature nut, John Acorn. Acorn wrote the very first field guide to North American Lady Beetles in 2007—*Ladybugs of Alberta: Finding the Spots and Connecting the Dots*.

While the Spotted Lady Beetle is not an Albertan species and therefore not treated in the Acorn book, I still spent some time re-reading parts of the book. One of the important topics of this guide is Acorn's discussion and consideration of the invasive species question. Certainly, my own righteous zeal against invasive and non-native species has been duly tempered by knowing Acorn's thoughts on the subject. Though I admit to keeping a slight favoritism for encountering native species such as the Spotted Lady Beetle above the non-native species.

"Conservation, to me, is about sustaining the living world around us—people included—not about returning to a mythical golden era when we, or our ancestors, or someone else's ancestors, were 'in harmony with nature.' The biggest threat to conservation I see is the loss of credibility that comes from

'crying wolf,' and I think ladybugs provide a case in point. The second biggest threat to conservation is loss of habitat, not introduced species."
– John Acorn

Coleomegilla maculata – Spotted Lady Beetle

March 21

COLD SUNSHINE

40° — 25°

A cold walk. Cold sunshine, wind out of the north, a raw early February feel, hardly fit for a spring day, the temperature only a few degrees above freezing.

The small pond in the woods, now nearly clear of ice, has a few migrant ducks sailing the open water—the flashy white of male Buffleheads, the artsy decoy-like shape on Northern Shovelers. It's my first sighting of both of these species this year. They are joined by Canada Geese and Mallards and a single Ring-necked Duck.

In the adjacent woods, a male Cardinal hunches against the cold on a low branch and belts out its song as though it could raise the temperature a few degrees just by singing.

Plantago lanceolata – Ribwort Plantain
Cardinalis cardinalis – Northern Cardinal
Spatula clypeata – Northern Shoveler
Branta canadensis – Canada Goose
Bucephala albeola – Bufflehead

40°
—
21°

March 22

CRAB SPIDERS

Today, while my daughter had a training session, I had an hour to walk at Murphy-Hanrehan Park Reserve in Savage. Because of the weather—cold, overcast, windy—very few people were out and nearly all that were were running dogs in the dog park. I had the trails to myself. A few geese on the lake, a few chickadees in the trees, other than that not much going on. I had to resort to lifting bark.

Examining a small dead tree, still standing but with most of its bark fallen away, I found a Running Crab Spider stuck flat against the inner side of one of the remaining sections of bark, flat as a wall-hanging. Under the bark of a fallen tree, I found a bark crab spider. This spider was so cryptic and so flat I wasn't certain it was alive. Small, dark, and stout it resembled, almost, a dog tick.

Running Crab Spiders, family Philodromidae, are recognized by the longest legs being the second pair of legs. The Crab Spiders, family Thomisidae, the first two pairs of legs are of nearly equal length. Spiders from both families have been given the name of crab spiders because their flat bodies and widespread appendages resemble the sidling posture and form of ocean crabs.

Bassaniana – Crab Spider
Philodromus – Running Crab Spider

March 23

THE FIRST OF THE WIND BIRDS

40°
—
36°

I visited the pond in the afternoon, with the hope of spotting more migrant ducks. Each day this week as I've felt better and better after a recent illness, the weather has gotten worse and worse. Today, a nagging wind out of the southeast but as cold as any wind from the opposite point of the compass, low thick clouds, with rain beginning before I'd finished hiking the loop around the pond. And the temperature stuck in the mid 30s.

All the ducks on the pond are Mallards. A single pair of geese fly in and land, honking and barking as if they thought they might suddenly take possession of the place. As their raucous entry subsides another sound is heard, winging back and forth above the pond a Killdeer, the first of the wind birds encountered this year. Then on the far side of the pond, the second wind bird, a Wilson's Snipe nearly unnoticeable at the edge of the shore. Amazingly, this small lump of a bird (a near match for the clod of mud it is stationed next to) in flight reaches speeds of 60mph.

Here's a passage from Peter Matthiessen's *The Wind Birds: Shorebirds of North America*, a beautiful introduction to these interesting birds: "The restlessness of shorebirds, their kinship with the distance and swift seasons, the wistful signal of their voices down the long coastlines of the world make them, for me, the most affecting of wild creatures. I think of them as birds of wind, as 'wind birds.' "

Gallinago delicata – Wilson's Snipe
Charadrius vociferus – Killdeer

45°
—
38°

March 24

THORNY CHALICES

Motherwort, a member of the mint family Lamiaceae, is a common plant of shady waste areas, trail edges, farmyards, and other disturbed woody locations. This plant originated in Eurasia and probably was introduced to North America through cultivation because of its reputation as a simple. It is now naturalized here, and often considered an invasive.

The smell of Motherwort, distinctively more astringent than the more pleasant mints, brings to mind vivid images of the lane and pasture and feedlot around our family's small dairy barn. Despite its prevalence then, I did not know the name of this plant until recent years. Where I live now, it is noteworthy as being among the first green leaves one sees every spring.

Today I took a close look at the dried, overwintered remains of the flowering plant. Each calyx, after the flower is gone, resembles a thorny chalice (the five spines a sharp abstraction of the five-petalled flowers). Inside each calyx are four nutlets containing each a single seed. The nutlets are obpyramidal in shape, two millimeters in length, with a pubescent top. No doubt these spiny constructions are meant to catch in animal fur and disperse their seeds when and where they drop off their carrier. Equally obvious is the success of said design given the worldwide distribution of this plant.

Leonurus cardiaca – Common Motherwort
Agelaius phoeniceus – Red-winged Blackbird
Antigone canadensis – Sandhill Crane
Cathartes aura – Turkey Vulture

March 25

SAND LEAVES AND PLANT FEATHERS

43°
—
37°

Thoreau, seeing sand formations resembling leaves, became interested in the possibility of a transcendent connection between inorganic and organic nature. A living theory of forms. I felt much the same way today admiring beds of moss, vivid green in the low light and thin rain, some of the outgrowths looking for all the world like plant feathers, or looking closer, like tiny knitted ferns.

"That sand foliage! It convinces me that Nature is still in her youth, —that florid fact about which mythology merely mutters,—that the very soil can fabulate as well as you or I. It stretches forth its baby fingers on every side. Fresh curls spring forth from its bald brow. There is nothing inorganic. This earth is not, then, a mere fragment of dead history, strata upon strata, like the leaves of a book, an object for a museum and an antiquarian, but living poetry, like the leaves of a tree,—not a fossil earth. but a living specimen."
– Henry David Thoreau, from *Journals*, February 5, 1854

Ptilium crista-castrensis – Ostrich-plume Moss
Bryopsida – True Mosses
Silphium laciniatum – Compass Plant
Gleditsia triacanthos – Honey Locust
Rhus glabra – Smooth Sumac

March 26

HORSETAIL FORESTS

Walked to St Olaf from 3 to 5 pm. Temperature in the 40s.

I saw several bright feathers scattered across the dirt around home base on the practice field. My first thought was that they belonged to a Meadowlark, but these were feathers of a Northern Flicker. A piece of wing and a few individual feathers didn't give much of an indication as to how or when the bird died.

Two sections of hollow log, placed upright at the edge of a trail, contained in their hollow interiors magnificent growths of Cladonia lichen. Odd, but I don't recall ever seeing these logs before, even though I must have walked by them dozens of times during the last decade.

Equisetum. Puzzlegrass. Horsetail. Scouring Rush. I've admired this plant since childhood when I was most fond of the way the thallus segments popped when disunited. Occasionally, while camping I've resorted to using it for cleaning pots and pans (although this is not recommended for anything other than cast iron because the silica encrusted in the rush will scratch most surfaces). Now I like them because they grow where there is groundwater near the surface, revealing a secret of the hidden geology below. And I like them because this genus with relatively few species is a remnant of a once immense family of plants that dominated the landscape during the Carboniferous Era.

Equisetum hyemale – Rough Horsetail
Colaptes auratus – Northern Flicker
Cladonia – Pixie Cup Lichens

March 27   61°

ALMOST NOT COLD   37°

All day the promised sunshine appeared tantalizingly close, but the clouds held fast, at least in Northfield. Lisa and Lida, returning from a trip to Burnsville only twenty miles away, said it was sunny there the whole time. I'd worked all morning and whiled away the day, hoping the clouds would move off. Eventually, later in the afternoon, I went for a hike at McKnight Prairie sans sunshine.

Walking the ridge trail I saw no signs of the Pasque Flowers or other early plants. The clouds, thin as threadbare fabric, let in a soft light and slight warmth so that it felt almost not cold. At the top of the prairie bluff directly above the large blowout, I turned over a few of the small limestone shards. Surprisingly, I found an abundance of caterpillars. Quite a number were found on the underside of the stones, but as soon as I had noticed these I also began to see many others on top of the stones. Watching them, they appeared to be grazing the moss and lichen that covered the stones. I didn't make the connection at the time, but examining the photos later they were easily identified as the caterpillars of lichen moths. Of course.

Having seen photos and read descriptions of tiny terrestrial snails, I've been keen to find them in the field, realizing I'd simply overlooked them due to their minuscule size. So today, turning over a small piece of limestone not much bigger than a piece of a broken dinner plate, I was delighted to spot a small white spiral shell. Photographing it in the field proved impossible so I brought it home where I could use a better macro lens and a flash. With the help of iNaturalist snail expert Susan Hewitt, the snail was determined to be of the genus *Gastrocopta*. Because of the arrangement of teeth inside the aperture these snails have been given the common name of Snaggletooth Snails.

The final curiosity of the visit was a Thread-legged Bug which ambled out of a crevice in a small piece of cedar found on the ground. These insects look to be part walking stick part mantis, though much smaller, perhaps the size of long mosquito if there were such things as long mosquitoes. The Thread-legged Bugs move very slowly and meticulously, using only their rear four legs for walking. According to one account I happened across, these insects use their front legs to rob food from spider webs.

Gastrocopta – Land Snail
Emesinae – Thread-legged Bugs
Lycosidae – Wolf Spiders
Heteroptera – True Bugs
Hypoprepia – Lichen Moth caterpillar
Araneae – Spiders
Reptilia – Reptile egg

March 28

SPRING UNBOUND: BUTTERFLY SHADOWS

64°
—
35°

Spring was launched today. An entire day of sunshine with temperatures reaching 60 degrees lit the green fuse and launched a multitude of insects. The flood of events—emerging plants and blooming flowers—is now unstoppable and just ahead. Like Prometheus, Spring is now unbound. Or as Milton put it in *Paradise Lost* "At once came forth whatever creeps the ground."

After seeing what was likely a Honey Bee leave our backyard garden, I couldn't shake the thought of possibly finding the first native bee of the year. The best chances would be at the sandy slope of the oak savanna in the Cannon River Wilderness Area, so I went there to look. It's a lengthy hike, and challenging. But almost always worth the effort.

While I didn't succeed in finding that first native bee, I did see plenty of other things: a spectacular Rove Beetle that flew by and landed on a patch of moss, several kinds of terrestrial snail shells, a tangerine-striped leafhopper, Ichneumon Wasps under slabs of limestone, Virginia Wood Cockroaches, Winter Fireflies, a Wolf Spider, a Dwarf Spider, and two butterflies.

Edwin Teale writes in *Circle of the Seasons* on March 25 about bird shadows: "Without looking up, without hearing the bird call, I could have identified the shadow. How many birds can we recognize by their shadows alone?" I had the same experience today while walking in the woods, but with a butterfly shadow. Seeing a large shadow moving erratically across the leaves of the forest floor I knew without looking up that a Mourning Cloak was flying overhead.

Nitidotachinus – Rover Beetle
Anemone acutiloba – Sharp-lobed Hepatica
Arion – Slug
Anguispira alternata – Flamed Tigersnail
Ichneumon annulatorius – Ichneumon Wasp
Goodyera pubescens – Downy Rattlesnake Plantain
Parcoblatta virginica – Virginia Wood Cockroach
Ellychnia corrusca – Winter Firefly
Lycosidae – Wolf Spiders
Arboridia – Leafhopper
Webbhelix multilineata – Striped Whitelip
Dytiscidae – Predaceous Diving Beetles
Deroceras reticulatum – Milky Slug
Araneae – Spiders
Erigoninae – Money Spiders
Nymphalis antiopa – Mourning Cloak

March 29

51°

STOWAWAY

42°

I had forgotten about a snail shell in my pocket. Snapped into a plastic vial yesterday and brought home to be photographed—let this laggard, my forgetfulness, fill in for today's observation, I thought, especially as the weather had turned cold and rainy. As I tipped the shell from the vial a tiny black speck moved on the outside of the shell. It was so small I couldn't tell what it was. Appearing to fly from place to place, my first guess was a small fly or wasp. But then I noticed the splayed legs and recognized it as a spider, a minum member of the Salticidae, a jumper.

First, though, I set about documenting the snail shell. The spiral is not maintained for long, just one or two twists and the shell widens majestically, the tight beginnings a small knot opposite the large fragile globe. There are a number of land snails that look like this spread across several genera but all given the common name of Ambersnail because of their translucent, amber-colored shells. The shell is very thin, fragile as baked phyllo dough, and already some of the rim of the aperture has crumbled away.

After documenting the shell, attention turned to the stowaway. Because of its small size, I added an extension tube to the macro lens on the camera, gaining some magnification, giving away some depth-of-field. Now the spider could be viewed quite clearly, even the eyes, and jumping spiders have astonishing eyes. I always get the impression that they must be inquisitive creatures, especially when they look in your direction, huge eyes wide open. Their look is one of perpetual astonishment, philosophers or explorers, the poets of the spider world. (And, somewhat fittingly, this tiny jumping spider escaped in the vicinity of my typewriter.)

Jumpers often retreat at night and in cool weather to silk shelters in hidden places, in this instance, the spider ensconced itself inside an empty shell. Spiders will use other hollow objects as well, empty seed pods, folded and curled leaves, even the cast-off skins left behind by emerging dragonflies get co-opted as temporary dwellings.

Since we're on the subject of interesting spider feats, I recently learned of a species of spider in Madagascar that commandeers empty snail shells and, instead of simply occupying them, the spider hoists them off the ground to an overhanging branch by using its silk as a kind of block and tackle, creating a suspended abode that looks something like an ornament hanging from a Christmas tree branch.

Neon sp. – Jumping Spider
Succineidae – Amber Snails

March 30

"GO! . . . BALL! . . ."

50°
—
39°

Today I set off for the wooded pond on a utilitarian endeavor, that is to procure some benthic invertebrates as food for a dragonfly nymph I'm rearing. On my way, I walked by the St Olaf ball fields where the college men's team was practicing. One group repeated an infielder's drill where a coach shouts "Go!" and two players sprint away, then "Ball!" and the players look back for the thrown baseball and one calls it and makes the catch without breaking stride. For most of the time I'm out I hear these calls repeated. "Go! ... Ball! .... Go! ... Ball!"

At the pond, the first scoop through the shallow water yielded dozens of Phantom Midge larvae or *Chaoborus* (also widely known as Glassworms). A few more scoops yielded more of the same, but eventually, a few other creatures were observed as well: several leeches, a diving beetle, snails, and a damselfly nymph. I collected about a dozen *Chaoborus* and turned for home.

My first encounter with *Chaoborus* occurred while in college during a visit to the Trout Lake Research Station in northern Wisconsin. One night, I volunteered to help some biologists sample larval perch. This involved going out well after dark and using a push-net in front of a boat. In addition to the nocturnally active larval perch the nets captured masses of wriggling, clear jelly, a concentration of thousands of *Chaoborus*. I'd sampled this same lake during the day on numerous occasions and had never seen this insect. I was amazed.

One of the world's most common insects and yet one of the least observed due to its diurnal migration from sediments during daylight to the surface at night. Right away I admired the scientific name, how the word "chaos" was bundled into it. And I was mesmerized many times watching the springy way the larvae moved, how they would be absolutely still then suddenly coil into something like the letter "C" then just as suddenly straighten back into the long lowercase letter "l" and glide away.

Seven years ago, in 2010, I made the following observation at the same pond on nearly the same date: "I stopped at the edge of the pond to have a look at the welcome sight of open water. After walking and covering a lot of ground, it takes a few minutes for the body and the eyes to adjust to being still. Eventually, I begin to see movement in the shallow water, snails, zooplankton, a few small water beetles. And then a procession of phantom midge larvae appear, of course they were there all along, their slender transparent bodies are nearly invisible until betrayed by the pairs of air sacs that look like black segments of an orange located at the front and back of their bodies. They move in a most surprising manner, gliding like tiny pike then spastically they double themselves up and unfold, launching themselves like an arrow on another glide."
— *Rice County Odonata Journal: Volume Three* (Thistlewords Press, 2013)

Dytiscidae – Predaceous Diving Beetles
Coenagrionidae – Narrow-winged Damselflies
Chaoborus – Phantom Midge

March 31

FIRST FROG CALLS

57° / 31°

he final day of March began cool and overcast, but by late afternoon the clouds had dissipated. When I set out for a walk at the St Olaf Natural Lands, the sky was clear, the sun bright, and the temperature slightly optimistic, crossing the 50 degree mark.

At the catchment pond at the corner of campus, a single Chorus Frog called from the cattail debris. The first frog call of the year. Of course, it's impossible to spot the amphibian singer. And, of course, it stops calling the moment I try. Last spring I had studied the Minnesota frog and toad calls with the hope of applying for a survey route around Northfield, only to find that all the routes were filled. Nevertheless, I benefited from the effort. The call of each species is distinct and easily differentiated with a little practice, making it possible to identify which frogs and toads are present in a wetland just by listening. And the best time to listen is at night. Even though some of the singing carries over to the daylight hours, the real performance, the grand amphibian opera, takes place beneath the stars.

Pseudacris maculata – Boreal Chorus Frog

April

Armored Mayfly – April 25

April 1

BYRD MASSES

66°
—
35°

I'm not always optimistic, but today I definitely felt it. The warmest day so far this spring, the first of April, clear skies, and sunshine—it seemed possible that the first migrant dragonflies could show up on a day like this. So I set out to check the local ponds for Common Green Darners.

Sitting on the bench watching the pond in hopes that the first dragonfly of the year might fly by, I begin listening to the voices. A Blue Jay. Boreal Chorus Frogs. And new today, Wood Frogs. This early in the spring there is still a simplicity to the composition, a pared-down part song. The William Byrd Masses for 3, 4, and 5 Voices come to mind.

Nothing like the gravity of old church music to add ballast to an unchecked ebullience. I didn't see any dragonflies. They're not here yet. But, in addition to the Wood Frogs, I observed a few other firsts: the first Eastern Comma, the first Golden Dung Fly, the first Painted Turtle, and the first Northern Paper Wasp. And, later in the evening, I turned on the mothing light for the first time as well.

Lithobates sylvaticus – Wood Frog
Paleacrita vernata – Spring Cankerworm Moth
Phigalia strigataria – Small Phigalia Moth
Sylvicola sp. – Wood Gnat
Orthosia hibisci – Speckled Green Fruitworm Moth
Polistes fuscatus – Dark Paper Wasp
Scathophaga stercoraria – Yellow Dung Fly
Polygonia comma – Eastern Comma
Viburnum lentago – Nannyberry

55°
—
42°

April 2

"EQUILIBRIST"

From 7 A.M. to 12:30 P.M., I'm stationed at the door to the Midwest Volleyball Warehouse, a part-time job to help pay for my daughter's club season. Not long after I take my post and the doors are opened and players and spectators start entering the building, I noticed a moth on the outside of the entranceway glass. Eventually, during a lull, I had a chance to step out there and take a photo with my phone. A White-spotted Cankerworm Moth.

> I see an adder and, a yard away,
> a butterfly being gorgeous. I switch the radio
> from tortures in foreign prisons
> to a sonata of Schubert....
>
> Noticing you can do nothing about.
> It's the balancing that shakes my mind.
> What my friends don't notice
> is the weight of joy in my right hand
> and the weight of sadness in my left.
>
> – Norman MacCaig, from the poem 'Equilibrist'

Paleacrita merriccata – White-spotted Cankerworm Moth

April 3

TWICE STABBED

53°

47°

Damp and foggy throughout the morning. The early mists thickening into true April showers by afternoon.

Today's observation is a Twice-stabbed Lady Beetle. This small beetle, shiny like a drop of black ink, has a blood-red spot on each wing cover. It's native to eastern North America. I've found it most often at the base of Maple trees, which is where this one was found as well. (Though my very first encounter with this species was not so typical, when I noticed it roaming the brim of my hat.) They are thought to prey upon scale insects.

According to John Acorn, in his guide *Ladybugs of Alberta*, lady beetles of the genus *Chilochorus* may be the worst tasting of the lady beetles. I, for one, am willing to take his word upon this matter and am not tempted in the least to repeat his experiment.

Chilocorus stigma – Twice-stabbed Lady Beetle

63°
—
44°

April 4

THE DICTION OF SPRING

The diction of spring changes daily, sometimes from hour to hour as clouds build or sun shines. The words that tumble into the mind, sometimes habitual, sometimes surprising, arrive like waves of migrant birds. Snowmelt. Mud puddles. Leaf buds. Flowers. Bees. Butterflies. Yard work. Gardens. Sunlight. Open windows. Screen doors. And the first, first, first of everything.

Walking the dog, the first mining bee of the year takes flight from a patch of Siberian Squill. From the look of the black and shiny abdomen, I'm tempted to guess it's *Colletes*, a Cellophane Bee, but I couldn't be sure from the one quick glimpse. The bee flies off, making a partial loop as it leaves. A sure sign of spring: blue squill flowers and mining bees.

Still no dragonflies. I looked for them at four ponds at the St Olaf Natural Lands. There were, however, a number of new arrivals among the birds: Song Sparrows, Tree Swallows, Bluebirds, and Eastern Phoebes. At the wooded pond, the calls of the Chorus Frogs and Wood Frogs reverberated at about ten times the decibels of the previous visit three days ago. Obviously, the membership of this particular choir has increased dramatically. I had the good fortune to spot one of the singing Wood Frogs, even capturing a short video of it as it called. This was the first time I've actually watched one call and I was surprised that it didn't have a vocal sac beneath its chin, rather a pair of lateral vocal sacs that inflated on both sides toward the rear of its head.

Veronica arvensis – Corn Speedwell
Lithobates sylvaticus – Wood Frog
Chrysemys picta – Painted Turtle
Zanthoxylum americanum – Prickly Ash
Oligochaeta – Earthworms

Melospiza melodia – Song Sparrow
Tachycineta bicolor – Tree Swallow
Othocallis siberica – Siberian Squill
Sialia sialis – Eastern Bluebird
Sayornis phoebe – Eastern Phoebe

April 5   56°
JUMPING BEETLES AND BOG TALK   38°

Today I met with artist Meg Ojala, a St Olaf professor who has embarked on a study of bogs. Twice last year we visited a small local bog together. Now I was getting a first look at images of the many other bogs she'd visited, the bogs of northern Minnesota and the bogs of Finland. Photos of bogs, bog plants, and bog surroundings laid in stacks upon the floor, hung in groupings on the walls. We talked about her project and upcoming gallery shows and further travels. By the time I left, I had a bad case of bog envy, wanderlust, and wetland withdrawal symptoms. The only remedy, of course, was to get outside and find some water as soon as possible.

So, a little later in the afternoon, I visited the Cowling Arboretum at Carleton for a short walk beside Spring Creek, the Cannon River, and catchment pond wetland near their junction. Cloudy and cool, but even still, plants were sending up first leaves and flowers were readying themselves for the next spate of sunshine and warm weather. Likewise, the sheer number of people passing me by on the trails indicated that many of the students and residents have turned the corner as well, crossed the threshold of belief, and seem convinced it is now spring—no matter what the actual weather. Or, I simply missed the sign announcing some kind of cross-country running event.

I get out of the way and wander off the main trails and have a look around. Parts of the forest floor are carpeted by the vibrant green leaves of the escaped cultivar, Siberian Squill. And some few of these plants are in bloom with handsome blue flowers. In addition to the squill, I find a single Hepatica flower, the first of the year for me. Soon there will hundreds more. Ahead of me in the woods, kicking among the leaf litter and the greening vegetation for something to eat, is a Song Sparrow, teaching me something about patience.

Earlier in the day, I photographed a small beetle brought home from yesterday's walk. A number of these bright metallic beetles roamed up and down the side of a tree along with a number of flies, probably all attracted to leaking sap. I'd seen these beetles on the very same tree the previous year and had tried but failed to photograph them due to the shiny reflections off the beetle wing cases, so I decided, this year, to bring one home and try to photograph it using a flash. Even this approach wasn't as successful as I'd hoped, but the images were good enough to help identify the beetle as a Metallic Flea Beetle, genus *Altica*, a member of the leaf beetle family Chrysomelidae. There are numerous related and difficult to separate species in this genus, most being host specific and most are quite small in size. While photographing the beetle I discover a surprising thing, the beetle jumps, it actually springs off the ground just like a grasshopper. Which, obvious after the fact, must be why they are called flea beetles.

Anemone acutiloba – Sharp-lobed Hepatica
Melospiza melodia – Song Sparrow
Altica sp. – Metallic Flea Beetles

April 6   56°

PICK-UP STICKS   35°

American entomologist Karl V. Krombein conducted a magisterial study of wasps and bees across a full decade from 1953 to 1964, collecting and examining some 3,400 trap nests. The results of this study were published in 1967 by the Smithsonian Press as *Trap-nesting Wasps and Bees: Life Histories, Nests, and Associates*, or, as I like to refer to it, the trap-nesting bible. With the aid of this book and the information it contains, I've conducted a limited amount of trap-nesting myself and have enjoyed it immensely.

Last year I added a bundle of hollow stems (all snipped to about five inches in length) at the base of our front yard bee block, to provide some additional nesting opportunities for the wasps and bees. The hollow stems, which are easily split open, work very well as natural trap nests. Seeing that a number of the stems had been occupied by midsummer, I placed one of them in a container so that I could examine the adults when they emerged from the trap. Usually, I photograph the wasps that emerge and then release them. In this instance, the wasps all emerged while we were away on vacation and all died before I could release them. Five Keyhole Wasps (genus *Trypoxylon*) emerged from the stem, two males and three females.

Over the winter, birds continuously ransacked this bundle of stems, eventually scattering them all to the ground where they were covered by snow and lay until spring. Today I played pick-up sticks, replacing them into their holder on the bee block. One of the fallen stems that I picked up was still plugged at the end. A closer look showed the stem to be cracked lengthwise so I decided to open it and have a look at the wasps inside. I was in for a surprise.

Upon opening the nest, what at first I took to be pupae turned out to be masses of parasitic wasp larvae that had devoured the overwintering *Trypoxylon* pupae. At some point last fall or this spring the nest had been compromised. The cracks in the stem had allowed the tiny parasitoid wasps to gain access to the nest cells. I remembered reading about such infestations by *Melittobia chalybii* in the Krombein book. Here is Krombein's description: "This eulophid parasite was a very serious pest in trap nests. It not only parasitized a number of nests in the field at several localities but also it caused serious secondary infestations in other nests in the laboratory.... *Melittobia* has limited powers of dispersal, thus accounting for its presence in traps placed in these particular situations [i.e. dead trees or structural lumber], rather than traps suspended from branches of living trees. Although *Melittobia* females are winged, they apparently do not fly at all but merely hop a few inches or walk about on the substrate."

More recent studies have shown that these parasitoids have a complicated life history involving some curious dimorphism: short-winged and long-winged females as well as blind, flightless males that don't resemble the females in the least. I've kept the larvae with the hope that they will pupate and emerge as adults and I'll be able to have a look at some of these morphological differences. As can be seen from the photographs of the one dead female wasp found in the nest, their body shape is remarkably flat, no doubt helpful in prying their way into host nests, the head being especially odd in having the shape of a flat disc.

Melittobia sp. – Chalcid Wasp
Trypoxylon sp. – Keyhole Wasp
Dahlica triquetrella – Narrow Lichen Case-bearer

April 7

PASQUEFLOWER AND PRAIRIE SMOKE

60°
—
33°

A couple of solid wildflower firsts today, Pasqueflower and Prairie Smoke, both in bloom at McKnight Prairie.

As for the pollinators of these early flowers, I thought I saw one bee and actually did see a number of tachinid flies, probably of the genus *Gonia*. These large, goonie-looking, white-faced flies are thought to be parasitoids of Owlet moths (Noctuidae).

As with so many endeavors, along with the good comes a little bad—while enjoying the spring flowers I find the first tick of the year, an American Dog Tick crawling on my shoulder.

Gonia sp. – Tachinid Fly
Formicidae – Ants
Ceuthophilinae – Camel and Cave Crickets
Pellenes sp. – Jumping Spider
Anemone patens – Eastern Pasqueflower
Geum triflorum – Prairie Smoke

73°
—
44°

April 8

MIGRANT DRAGONFLIES

The temperatures soared today, going above 70 degrees for the first time this year. And with the first heat of the year came the first dragonflies.

Nearly as large and conspicuous as a small bird, the Common Green Darner (*Anax junius*) is a methodical migrant, showing up no matter what the weather. The other spring migrant, the Variegated Meadowhawk (*Sympetrum corruptum*), seems to show up in great numbers when certain weather occurs, conditions such as great streaming winds from the south that aid in their journey, conveyed north along immense storm fronts.

Because of its size, and because it lives up to its name by being uncommonly common, this is the dragonfly most people speak of when they speak of dragonflies. Nearly any day after their arrival in the spring, Common Green Darners can be found patrolling the shorelines of wetlands, ponds, lakes, and rivers, pretty much anywhere there is water.

Following the earth around, keeping track of occurrences across the years yields phenological data. And while the dates don't always line up, the sequence of events remains somewhat fixed. This happens. Then this happens. It's satisfying to compile and track the arrivals and first appearances of favorite flowers and insects. Even more satisfying, however, is when that data can be added and compared to data collected by others in earlier decades. Richard B. Primack in his recent book, *Walden Warming: Climate Change Comes to Thoreau's Woods* set out to do this for Concord, Massachusetts, comparing his data to that recorded by Thoreau. For Rice County, the only phenological data compiled and available (that I'm aware of) is that of Orwin Rustad, *A Journal of Natural Events in Southeastern Minnesota*. Among his numerous records—ice-out on area lakes, first blooms of many wildflowers and trees, spring bird arrivals—are just a few concerning insects—Mourning Cloaks, Monarch Butterflies, and Cicadas, but none for dragonflies. So, I'll add my observations of the arrival dates for Common Green Darners and Variegated Meadowhawks here to get the record started.

	CGD	VM
2008	May 5	no record
2009	April 28	no record
2010	April 15	April 15
2011	April 11	April 11
2012	March 27	April 5
2013	April 30	no record
2014	no record	April 10
2015	April 1	April 2
2016	April 14	no record
2017	April 8	no record

Anax junius – Common Green Darner

72°
—
53°

April 9

MOTHING

In 2014, I did a lot of mothing. At that time, I used a disarmed bug-zapper which could be hung near the top of our wide, white garage door. It worked great.

Last year, I purchased a moth trap from BioQuip, hoping to attract and capture more micro-moths. This set up consists of an halo-shaped, UV light mounted above a stainless-steel funnel that empties into a five-gallon bucket. So far I haven't had the same kind of results as with the bug-zapper light.

Because the new light is powered by a car battery, it's somewhat more portable. And I'm looking forward to trying it out in some remote spots.

Mangora placida – Tuft-legged Orbweaver
Stenolophus comma – Ground Beetle
Amara sp. – Sun Beetles
Enoclerus rosmarus – Checkered Beetle
Enoplognatha sp. – Cobweb Spider
Ophioninae – Short-tailed Ichneumonid Wasps
Olethreutinae – Olethreutine Leafroller Moths
Orthosia hibisci – Speckled Green Fruitworm Moth
Phigalia denticulata – Toothed Phigalia Moth
Cerastis tenebrifera – Reddish Speckled Dart
Copivaleria grotei – Grote's Sallow
Ophion – Ophion Wasps
Trochosa terricola – Ground Wolf Spider

April 10   53°/33°

INSIDE-OUT

Here's a spring poem in honor of the Bloodroot in blossom today:

INSIDE OUT

My eyes follow, fierce,
through window glass
the flaw of the winter crow
as it flies from the yard.

The darkness spilled
and wiped up;
now that it's April,
I'm ready for spring.

Vision turned inward
I watch a man I once knew
row a boat, a black boat
out to sea.

Vision turned outward
I shake leaves from my fingertips,
and hold the door
for flowers.

Sanguinaria canadensis – Bloodroot

April 11

LOWLY GREENS

53° / 32°

Dandelions are certainly one of the most successful and most recognized plants in the world. From the lowly greens to bright yellow flowers to the magical globe of drift-away seeds.

I'm thinking of the Greek poet Yannis Ritsos, specifically his long poem 'The Smoke-blackened Pot.' The pot referred to in this poem, a simple clay pot used over open fires in the mountains of Greece to boil greens, *khorta*. One of the greens frequently used is *redikia*, dandelion greens. The pot, in its simplicity, in its functionality, becomes symbolic for the preparation of revolution, the mind of the revolutionary cooking up a better future. From lowly greens to high ideals.

Taraxacum officinale – Common Dandelion

April 12

SUBTERRANEAN APHID FARMING

57°
—
40°

An overcast, middling sort of day, ending in rain.

I've observed ants tending their flocks of aphids and leafhoppers on stems and leaves. I was more than a little surprised, however, to discover ants and aphids beneath a rock in our flower garden. A quick internet search indicated these were ants of the genus *Lasius* and the aphids, root aphids.

These ants have been given the name Citronella Ants. When disturbed (as when someone turns over the rock they are nesting under) some species in this genus emit an odor of citronella. Another common name, Fuzzy Ants, derives from the meaning of the Latin name, *Lasius*. Curiously, in 1939, a ruling by the International Commission on Zoological Nomenclature ended a 130-year controversy over whether the name "lasius" could be used for this genus of ants or for a genus of bees. Two decades after the decision in favor of the ants, E. O. Wilson's doctoral thesis laid the groundwork for the currently accepted taxonomy.

Once the ants I'd disturbed were exposed, they quickly dispersed but then soon returned to the aphids, eventually carrying them off to safety further underground in their nest. I wondered about the details of this farming: Is one ant assigned to one aphid? Is one ant in charge of several aphids? How is the food gathered from the aphids shared? Someone somewhere probably has the answers. I simply closed up the mystery by putting the rock back in place.

Eriosomatinae – Woolly Aphids And Gall-making Aphids
Lasius sp. – Citronella Ants

55°
—
46°

April 13

BUMBLE BEE PUZZLES

Dutchman's Breeches, certainly one of the more distinctively shaped spring ephemerals, is a flower I've known since childhood and is practically unmistakable. However, I still find the common name to be rather odd, though of course I can see the resemblance to some long-ago Dutchman's laundry, pairs of puffy pants hung upside-down on a clothesline. There are other common names, some good, some bad: white hearts, eardrops, butterfly banners, kitten breeches and staggerweed, the last having to do with the toxicity of the leaves. Knowing, now, that this curiously shaped flower requires a specialist pollinator, the long-tongued bumble bee *Bombus bimaculatus* or Two-spotted Bumble bee, maybe they could be called Bumble bee Puzzles or Bumble bee Trinkets?

Like other spring ephemerals, Dutchman's Breeches bloom early, soon after the snow has left the forest floor. The plants also disappear quickly, both flower and leaf, as the forest canopy fills in, spending the entire summer as dormant bulbs.

The seeds of Dutchman's Breeches and their dispersal are interesting as well. The seeds each have an edible appendage called an elaiosome. Ants gather these seeds. After they eat the elaiosome they discard the seed in their nest debris where some eventually germinate. This process of seed distribution by ants is called myrmecochory. Don't ask me to pronounce that.

Dicentra cucullaria – Dutchman's Breeches

April 14   66°

FIRST BEES   48°

Which came first, the flowers or the bees? This year it was the flowers. But both have evolved, lock and key, together; the history of flowering plants is mirrored by the history of pollinators.

Lisa had the day off from teaching, so we were able to hike together today. Before we opened the car doors to get out, a Red Admiral grazed the windshield and landed in the parking lot. This yearly migrant, surges north in the spring, arriving just in time to lay eggs on its host plant, Stinging Nettle. We'd chosen wisely in visiting the Cannon River Wilderness Area. As we walked down the stairs and entered the forested valley, a variety of spring ephemerals greeted us. First Bloodroot, its curled, knobby-fingered leaves beneath half-opened flowers giving it the look of a chalice raised by an Arthurian knight's armored hand. Then Hepatica in numerous shades between white and violet, leafless, tall-stemmed flowers in tight groupings here and there across the forest floor. Lisa suggested they grew as living bouquets. More of a surrealist, the image that leapt to mind was of a group of Christmas carolers. Then Spring Beauty. Then False Rue Anemone and Dutchman's Breeches, only beginning to bloom. Reaching the trail's end at the oak savanna, Prairie Buttercup, small, egg-yolk-yellow flowers inches above the sand.

To the flowers go the bees. I had been concerned that the sustained cool weather was delaying the emergence of the early bees. For the last two weeks, flowers had been open but there had been no bees to speak of. Today, however, my concerns fell away. The bees were back. A single, tiny *Lasioglossum* clung torpidly to the petal of a Hepatica flower. *Colletes inaequalis* visited willow catkins and dozens of vigilant males flew low to the ground at the aggregate nesting sites on the open sandstone slopes of the savanna bluffs. Several different species of *Andrena* gathered pollen from Bloodroot flowers. Yes indeed, the bees were back. I was happy, and doubly so, being able to take this hike and share these encounters with my companion of some many years, so many springs. Lisa and I were equal partners today like the bees and the flowers, or, the pleasing thought came to me, like the naturalists Edwin and Nellie Teale at Trail Wood.

Ranunculus rhomboideus – Prairie Buttercup
Carex pensylvanica – Pennsylvania Sedge
Anemone acutiloba – Sharp-lobed Hepatica
Claytonia virginica – Virginia Spring Beauty
Enemion biternatum – False Rue Anemone
Sanguinaria canadensis – Bloodroot
Asclera ruficollis – Red-necked False Blister Beetle
Vanessa atalanta – Red Admiral
Andrena dunningi – Dunning's Mining Bee
Melandrena – Mining Bee
Dialictus – Metallic Sweat Bees
Colletes inaequalis – Unequal Cellophane Bee
Polistes fuscatus – Dark Paper Wasp
Mangora placida – Tuft-legged Orbweaver
Kateretidae – Short-winged Flower Beetles
Brachycera – Brachyceran Flies

April 15

THE QUIET END

72°
—
52°

After overnight thunderstorms that continued until late morning, the skies cleared and the temperature rose above 70 degrees for only the second time this year. About mid-afternoon I walked to the St Olaf Natural Lands. Common Green Darners patrolled the catchment pond at the corner of campus. An American Lady fluttered from dandelion flower to dandelion flower.

In the woods, on my way to the wooded pond, I stopped at several groupings of Bloodroot. The wide-open flowers were attracting a lot of bees. Male Andrena Bees, often close to a dozen, flew pretty much nonstop around each grouping of flowers. Twice I noticed two bees tangled together at/in a flower. I don't know if they were two males being territorial, or if they were male and female making a clumsy attempt at mating. In addition to the Andrena bees, several small sweat bees (probably *Lasioglossum sp.*) visited the flowers, as well as a deep-red Nomad Bee (*Nomada sp.*).

Heading home, I cut through the woods at the north end of Lashbrook Park. I'm well off trail when I happen upon the remains of a Raccoon, a small circle of fur and bones. I have a sense that the raccoon, sick or old, curled up here last fall, maybe even at the start of the first snowfall, closed its eyes and never opened them again, low elderberry thickets providing shelter and peace. No, the death had to have happened earlier for the carrion beetles to have done their job so completely. The bones have been only slightly scattered, picked at no doubt by crows. A shambles, both dignified and somber. Here, at my feet, is the quiet end of a life.

Naphrys pulex – Flea Jumper
Sminthurinus henshawi – Globular Springtail
Procyon lotor – Common Raccoon
Andrena dunningi – Dunning's Mining Bee
Dialictus – Metallic Sweat Bees
Melandrena – Mining Bees
Nomada sp. – Nomad Bees

April 16

STARTER DRAGONFLY

66°
—
47°

Not overly warm, but beautifully sunny. We spent much of the holiday with relatives in St Louis Park. While there, Stella, my young niece, and I did some insect netting. I caught a Cabbage White butterfly. She caught a Small Carpenter Bee. We both attempted to net dragonflies without success. Definitely a bit slow with the net after nearly half a year of winter.

After we put down the nets, a Common Green Darner was found on the lawn, perhaps slightly injured. My niece's eyes widened at the sight of this huge insect. At first she was wary even to touch it while I held it, but eventually, she found the courage to hold it herself. Each time I've had the opportunity to introduce children to dragonflies an infectious mix of pride and curiosity wells to the surface of their beings as they connect with these wonderfully mysterious animals. The Common Green Darner is an excellent dragonfly to start with, a good starter dragonfly.

After the dragonfly, Lida spotted another notable insect, a Two-spotted Bumble bee. This large bumble bee seemed trapped in the grass of the yard. Perhaps she had only just now vacated her winter hibernaculum and was still a bit groggy and cool. Slowly she found her way out of the grass, then flew away, crossing the creek toward summer. This was the first bumble bee I've seen this year.

Anax junius – Common Green Darner
Halictus confusus – Confusing Furrow Bee
Zadontomerus – Small Carpenter Bee
Bombus bimaculatus – Two-spotted Bumble Bee
Nomada sp. – Nomad Bees

59°
—
45°

April 17

INUNDATED

I knew this day would come. After months of winter with rarely a trace of an insect outdoors among the snowdrifts, suddenly there are dozens and dozens of insect species flying about. I'm inundated. Just a few days into spring and I can't keep up. From now on each day's record reflects a choice and a focus, a few select observations excluding hundreds of other possible subjects.

For instance, today I saw several warbler-like birds in the trees but didn't run for the binoculars or telephoto lens. I saw three different kinds of butterflies in the yard but didn't give chase. Why? I was intent upon the mining bees and nomad bees that had shown up in the backyard. Then, while patiently waiting for these bees to land within range of my camera, I was further distracted and turned my attention from them for a while because the first syrphid fly of the season landed near my ear on the trunk of the ash tree.

How fitting that the very day my copy of *The Secret Life of Flies* by Erica McAlister arrived in the mail, a large syrphid fly arrived as well. This fly, *Brachypalpus oarus*, is a species I've never encountered before. Jeff Skevington who is working on the *Field Guide to Northeastern Nearctic Flower Flies* has given lively common names to the flies. This particular species gets the name Eastern Catkin Fly. Several sources suggest this fly is a bee mimic, resembling somewhat a large mason bee or a large mining bee. To me, however, it looks more like a robber fly, but that would make it a mimic of a mimic, a mimic once removed. In its maggothood, this fly would have lived in a rot hole of an old deciduous tree, grazing in some little pool of water high above the ground.

Brachypalpus oarus – Eastern Catkin Fly
Cerastium fontanum – Common Mouse-ear Chickweed
Nomada sp. – Nomad Bee
Andrena sp. – Mining Bee
Nomada sp. – Nomad Bee
Melandrena – Mining Bee

66°
—
48°

April 18

WILD GINGER

What a crazy flower this plant has. Hidden beneath the leaves, the flower opens at ground level, shaped somewhat like a tipped-over chalice with lipstick-stained rim. Apparently, very few pollinators have been observed visiting the flower and it's thought that it is largely self-pollinating.

Later, after the fruit splits and the seeds are discharged to the soil, ants scavenge the seeds, attracted to the oil-rich elaiosomes.

Chloropidae – Grass Flies
Scaptomyza – Fruit Fly
Prenolepis imparis – Small Honey Ant
Nabidae – Damsel Bugs
Philodromus sp. – Running Crab Spider
Asarum canadense – Canadian Wild Ginger
Armoracia rusticana – Horseradish

April 19

GOLD-EYE

51°
—
45°

To St Olaf Natural Lands. Cloudy with rain expected late. And cool, the temperature not climbing out of the 40s.

Near the edge of the catchment pond at the corner of campus, I noticed a tiny, black speck on a golden-yellow Dandelion flower. Using my camera to magnify the insect, the speck resolved to be a fly. It had long, many-segmented antennae, so one of the lower flies, the Nematocera (which translates literally as "thread-horn"), the group of flies that includes midges, mosquitoes, crane flies and gnats. The fly on the Dandelion, just a few millimeters in length, was from the family Sciaridae, the Dark-winged Fungus Gnats.

Because of the cool temperature and lack of sunshine, very few insects were active. There were, as if in exchange, some interesting birds. The first encountered, almost as rare as finding a gold coin, was the secretive Le Conte's Sparrow. This small, rufous sparrow flew from the grass to a leafless thicket where I was able to have a long look at it. Audubon lists Le Conte's Sparrow as climate endangered. The next encounters took place at the wooded pond. Along the shore, numerous Yellow-rumped Warblers flitted and fed in the overhanging branches. And a Swamp Sparrow worked the water's edge near my feet.

Turning for home, I decided to check on the wild plum blossoms. The blossoms were close to blooming. Probably they will open in the next day or two, certainly if the weather warms and the sun shines. An unusual clump of orange lichen on one of the plum branches caught my attention. As best I can determine, this is Gold-eye Lichen, quite a rare find if so.

Teloschistes chrysophthalmus – Golden-eye Lichen
Melospiza georgiana – Swamp Sparrow
Setophaga coronata – Yellow-rumped Warbler
Acer negundo – Boxelder Maple
Regulus calendula – Ruby-crowned Kinglet
Ammodramus leconteii – Leconte's Sparrow
Sciaridae – Dark-winged Fungus Gnats

April 20   52°

SPIDER IN THE POCKET   40°

On this day in 1854, Henry David Thoreau observed "A willow coming out fairly, with honey-bees humming on it, in a warm nook,—the most forward I have noticed, for the cold weather has held them in check. And now different kinds of bees and flies about them. What a sunny sight and summer sound!"

Here in Northfield, half a continent west of Concord, Massachusetts, and more than a century and a half on, the surprise is that the year progresses in so like a manner as Thoreau's. Unfortunately, no similar, early records exist for this part of Minnesota. No baseline inventory of species. No detailed phenological data. The best we can do is begin to assemble that information. Recently Richard Primack set about collecting data for comparison to Thoreau's data. He summarized his findings in the book *Walden Warming*. The ice on Walden Pond goes out earlier, wildflowers bloom earlier, migrant insects and birds arrive earlier. Everything seems to happen earlier than in Thoreau's day. The change in climate is indisputable.

Contrariwise, the cold continues to hold the willows and early native bees in check (even while it loses ground in the long run). Today is just as overcast, though not quite as rainy, and slightly colder. So I was happy to have a spider to photograph. Yesterday, while at the St Olaf Natural Lands, I looked over some recently cut down trees in one of the road ditches. Under a loose piece of bark, I discovered a spider inside a loose web. I collected it in a vial, put it in my pocket and brought it home to photograph. The arrangement of eyes and the tubular spinnerets put it in the superfamily Gnaphosoidea, the Ground Spiders.

Haplodrassus signifer – Ensign Ground Hunter

April 21   63°

"A SUN ITSELF IN THE GRASS"   37°

I really enjoy Dandelions. I like the flower itself and the insects attracted to it. Nor am I alone in my adoration of this common flower. Here are a couple of good tributes.

"... dandelions, perhaps the first, yesterday. This flower makes a great show, — a sun itself in the grass. How emphatic it is!" – Henry David Thoreau, from *Journals* (May 1, 1854)

DANDELIONS

The little children of the poor
is what I have named them.
They have come to town once again
to spread their brilliant color
on the neglected lawns of folks
who are half asleep, still rubbing
old man winter out of their eyes.
They very much want to belong,
to be liked for who they are.
But like the weak and poor everywhere
they are looked upon as a bad dream,
a nuisance, not wanted, stepped on,
cursed, kicked, and in the way.
I'm talking about my friends,
the yellow dandelions of this world,
who have made themselves at home
and are all being happy together
in the long, thick, wet spring grass.

Their clear-eyed, shiny, dew-washed faces
are full of hope for a good life.
But, sadly, their lives will end soon,
for here comes the loud lawnmowers,
the ugly shoes, the rough voices,
the smug, self-righteous attitudes.
It seems nobody loves them,
and few tolerate them if they
have the bad luck to flaw
a too fussy suburban lawn.
Only a child here and there
or a thoughtful schoolgirl now and then
will stop to prick a few of them
to take home to a sick mother,
or perhaps a caring teacher,
in some quiet country town.

– Dave Etter, from *Dandelions* (Red Dragonfly Press, 2010)

Polistes fuscatus – Dark Paper Wasp
Thlaspi arvense – Field Penny-cress
Melandrena – Mining Bee
Andrena sp. – Mining Bee
Viola sororia – Common Blue Violet
Taraxacum officinale – Common Dandelion
Equisetum arvense – Field Horsetail
Phormia regina – Black Blow Fly
Polygonia comma – Eastern Comma
Anax junius – Common Green Darner
Pherbellia parallela – Marsh Fly

April 22

69°

SCIOMYZIDS

41°

Marsh or Snail-killing Flies (family Sciomyzidae) are very common around wetlands. According to *Flies: The Natural History and Diversity of Diptera* by Stephen A. Marshall "most sciomyzids occur in moist to aquatic environments, where they develop as parasitoids or predators of aquatic or semi-aquatic snails or sphaeriid clams such as fingernail clams." According to Bugguide.net over two hundred species in twenty genera exist in North America.

The Marsh Flies I most often encounter are those of the genus *Sepedon*. This is a very distinctly-shaped fly, forward-thrusting antennae, a long face and grasshopper-like rear legs. Most often, this fly perches upside-down on grass stems and the stalks of other vegetation near water.

Over the last two days, I've encountered two additional kinds of Marsh Fly. Yesterday, the tiny *Pherbellia sp.* was observed on a dandelion flower next to a catchment pond. Today, the spot-winged *Dictya* was observed on a blade of grass along a small, temporary rivulet.

Eustala sp. – Orbweaver Spider
Melandrena – Mining Bee
Nomada sp. – Nomad Bee
Nomada sp. – Nomad Bee
Tipula sp. – Giant Crane Fly
Tetanocerini – Marsh Fly

April 23

EARLY DAMSELFLY

A lot happened in twenty-four hours. Bellwort and buttercup bloomed. The first damselfly of the year emerged.

I noticed Andrena bees flying around the Bellwort and saw one bee crawl into one of the long, yellow flowers. Photographing the Small-flowered Buttercup, I noticed a small ant rummaging around the flower. This little ant looks to be an Acorn Ant. Some of the species nest in acorns, hence the name. Over the years, I've checked hundreds of acorns hoping to find these ants, and now, suddenly, here one was, as easy as could be, feeding on this flower.

The early damselfly was an Eastern Forktail, a very common and colorful pond damsel. Often the first non-migrant to be observed in the spring. This particular damselfly was the earliest recorded for Rice County by nine days. (The previous early date was May 2, 2016.) The males have green and black striped thorax and blue-tipped abdomen. The females are orange and black when young but turn powdery blue as they mature. Occasionally the females mimic the coloration of the males and these damselflies are referred to as "andromorphs."

This damselfly had only just emerged and was still teneral. The term "teneral" refers to the condition of the adult immediately after emergence when the insect is still soft and vulnerable. The origin is Latin, "tener" which means tender. The exoskeleton will harden in a matter of hours, but the mature colors will take longer to form.

I've always admired the following description of damselflies by Swedish Nobel Laureate Harry Martinson:

"But even more fragile than the dragonflies are the damselflies with their almost needle-thin abdomens. You get the impression that they could at any moment make a tinkling sound and break like a crystal filament. Their eyes, thorax, wings, yes, their entire being is like the subtle lines of refined poems, so fragile that they would fall apart during the act of writing themselves down. These poems can only be read directly while they exist in the present moment and mood."

– Harry Martinson, from 'Views from a Tuft of Grass'

Ischnura verticalis – Eastern Forktail
Limoniidae – Limoniid Crane Flies
Melandrena – Mining Bee
Uvularia grandiflora – Largeflower Bellwort
Temnothorax – Acorn Ants
Ranunculus abortivus – Small-flowered Buttercup
Andrena miserabilis – Miserable Mining Bee
Harpalinae – Ground Beetle
Hyla sp. – Holarctic Tree Frogs
Tipula sp. – Crane Fly
Orthonama obstipata – Gem Moth

72°
—
49°

April 24

COROLLARY

If you find one dragonfly odds are you will find more. This is a fundamental proposition of dragonfly observation (and most other insects as well). A corollary of that proposition: if you encounter a teneral dragonfly (one that has just emerged) odds are the nymph of that species can be found in the water closest at hand.

Having had the privilege to assist Bob Dubois, author of *Damselflies of the North Woods* in the field, I learned this corollary first hand. We were surveying dragonflies at the Agassiz National Wildlife Refuge near Thief River Falls, Minnesota. Another member of our survey team found a teneral Black Meadowhawk. Bob immediately set about dredging several nearby boggy puddles and before long had found several full-grown nymphs. Ever since that day, when I've had the opportunity, I've made good use of this corollary.

Nymphs that are full-grown are easy to rear, usually requiring only a few days or weeks before the adults emerge. Rearing young nymphs requires a whole different level of responsibility as some species can take years to become full-grown. Because the identification of nymphs to species is difficult and there are no guidebooks yet, rearing a dragonfly nymph is a good way to identify it. Lately, I've become more and more interested in photographing the nymphs in addition to rearing them. And in this way help work toward a visual guide to the nymphs.

Having seen the teneral Eastern Forktail yesterday, I decided to look for nymphs today. At the catchment pond, it took a single pass through the silty shallows and cattail debris with the Tennessen racket (a racquetball racket with the strings replaced by steel mesh netting designed by Ken Tennessen) to find a number of damselfly nymphs. A few more swiffs through the bottom mud brought up a number of large skimmer dragonfly nymphs as well. Several large skimmer species have been observed at this pond. The most prevalent and the earliest to emerge is the Common Whitetail (*Plathemis lydia*). The nymphs closest to emergence, large in size and with full-sized wing pads, are probably this species. Checking the information on the nymph in *The Odonata of Canada and Alaska: Volume 3* by E. M. Walker and P. S. Corbet, one morphological difference between the Common Whitetail nymph and the nymphs of the other large skimmers, such as *Libellula pulchella* and *Libellula luctuosa* is the absence of a dorsal hook on segment 6 of the abdomen. I can't help but notice that the eyes are striped, knowing that the adult eyes are striped as well.

Plathemis lydia – Common Whitetail
Ischnura verticalis – Eastern Forktail

62°
—
47°

April 25

SIDE SPIKES

Finding a number of dragonfly and damselfly nymphs yesterday only planted the idea of looking for more. So today, I went down to the Cannon River and looked for more nymphs. Getting from the parking lot to the river bank took a while as the rampant spring ephemerals along the trail distracted me. Slowly, flower by flower I made my way.

The first swoop through the river silt yielded an Armored Mayfly nymph. These are wonderfully bizarre insects. They resemble tiny aquatic armadillos. Or, perhaps a better comparison, they look like tiny aquatic versions of the armored dinosaur *Ankylosaurus*, both protected by conspicuous side spikes.

A few scoops further down the shore, a large dragonfly nymph appeared. Recently molted, the nymph's long, pointed abdomen jutted out of a clump of muddy-hued river silt, an eye-catching pale green. Had it not recently molted, its color would have been equal to the silt, and it would have been much more difficult to spot. Compared to the mayfly nymph (around 7mm in length), the dragonfly nymph was gigantic (around 40 mm). I recognized it immediately as a clubtail dragonfly nymph, exactly what I'd hoped to find.

Clubtails or gomphids (a colloquial shortening of the family name Gomphidae) were a great discovery for me when I began studying dragonflies. To begin with, I was shocked that I'd never known of them before. And surely I had been where they had been but had simply never noticed them. Once I made their acquaintance they quickly they quickly took a place among my favorites. The appearance of the adults in late May and early June is much anticipated.

What species of clubtail was it? I didn't know. Seven species are known from this stretch of the river. The most pragmatic method of identifying it to species is to rear the nymph to adulthood and I planned to do that. But I'd also attempt to use available keys to try to identify the nymph. Currently, there are no published guides to the living nymphs and very few online photographs for comparison (a word of caution when working with online photos—many online photos are misidentified). There are several treatises that contain information about the clubtail nymphs. The two that I have on hand are *The Odonata of Canada and Alaska*, a three-volume masterpiece by Edmund M. Walker (with an assist by Philip S. Corbet on volume three), and *Dragonflies of North America* by James G. Needham, Minter Westfall, and Michael May. The keys in both of these books lead to the genus *Stylurus*.

Dragonflies of North America has the further key to the species of this genus. Keying out an insect can be satisfying when it works, or simply frustrating when it doesn't. There are difficulties in terminology. And technical difficulties, such as the need for high power microscopes or dissection. It takes practice, patience, and perseverance to master any given key. In this instance, the key is simplified because all but two of the species are not known from this watershed. It is only necessary to differentiate between the two species *Stylurus amnicola* and *Stylurus notatus*. This requires a bit of measurement and the calculation of some simple ratios. According to the key, the ratio of the dorsal length of abdominal segment 9 to the dorsal length of segment 8 should be greater than 1.5 for *Stylurus notatus*. For *Stylurus amnicola*, the ration of the length of abdominal segment 9 to the width of abdominal segment 9 should be equal or less than 0.75. The first ratio (9L/8L) for this nymph equals 1.56—good for *Stylurus notatus*. The second ratio (9L/9W) equals 1.11—too large for *Stylurus amnicola*. So, if the keys are correct and my measurements are correct, this is *Stylurus notatus*, the Elusive Clubtail.

Interestingly, this species has not been officially recorded, either by voucher or by photograph, on the Cannon River. It has been observed flying over the river on numerous occasions but never captured. It's called "elusive" for a reason. If I can successfully rear the nymph, this will be the first record.

Ranunculus abortivus – Small-flowered Buttercup
Ranunculus hispidus – Bristly Buttercup
Viola pubescens – Downy Yellow Violet
Enemion biternatum – False Rue Anemone
Cardamine concatenata – Cut-leaved Toothwort
Erythronium albidum – White Fawnlily
Phlox divaricata – Blue Phlox
Toxomerus geminatus – Eastern Calligrapher
Tringa – Shanks, Tattlers, And Allies
Stylurus notatus – Elusive Clubtail
Baetisca – Armored Mayfly
Ephemeroptera – Mayflies

April 26   47°

MICRO-MOTHS   36°

I've become a bit of a micro-moth fanatic. Why? Well, there are several reasons. Because they are so often overlooked. Because of the variety and beauty of these minuscule leps. Because of the technical challenges of getting good images. In fact, photographing micro-moths has greatly improved my macro photography. My usual routine is to capture the moth in a vial, cool it in the refrigerator, and take it's photo while it's cold. It's good to keep in mind that the smaller the moth the quicker it warms up and the sooner it flies away; I've missed many photographs because of this.

As the name suggests, these are small moths. While the division between macro- and micro- moths is arbitrary, a separation based more upon magnitude than taxonomy. Even so, there are several families of moths that fall entirely into the category of micro-moth. The three largest families being Tortricidae, Pyralidae, and Crambidae. Most of the moths in these families range in size from 7 to 15 mm in length.

A wonderful introduction to the main families is provided by the British guide *Field Guide to the Micro-moths of Great Britain and Ireland* by Phil Sterling and Mark Parsons. The most useful guide to the species of eastern North America is the *Peterson Field Guide to Moths of Northeastern North America* by David Beadle and Seabrooke Leckie (easily my most thumbed-through guidebook). Online resources such as Moth Photographers Group and Bugguide.net are extremely useful as well.

The micro-moth photographed today, a Red-banded Leafroller Moth, was caught in my moth trap two nights ago. A member of the Tortricid moths, its hosts are thought to include apple, cherry, grape, and spruce, a rather wide variety of trees and shrubs.

Argyrotaenia velutinana – Red-banded Leafroller Moth
Setophaga coronata – Yellow-rumped Warbler
Spizella passerina – Chipping Sparrow

April 27

WALRUS MOUSTACHES

36°
—
32°

The Andrena Bees (Family Andrenidae, genus *Andrena*) are important spring pollinators. Some species are polylectic, visiting many different flowers, while many species are oligolectic, gathering pollen from a single family of flowering plants. All species are solitary and nest in the ground. The numbers in *The Bees of the World* by Charles D. Michener give an idea of the vast diversity of this genus: 1,443 species worldwide and 476 species in the western hemisphere with many species still to be described. The entomologist O. A. Stevens, who studied the Andrena Bees in North Dakota, found 53 species but suspected that there were probably close a hundred species total. No doubt Minnesota has similar numbers.

Like this pair of *Andrena carlini*, the Andrena Bees exhibit a high degree of sexual dimorphism; the males and females look nothing alike. This leads to confusion and complications in the attempts to identify the bees. It's helpful when a mating pair can be captured. This particular pair of bees was captured while perched on the leaf of Large-flowered Bellwort.

Female Andrena Bees can be recognized by their facial fovea, a kind of vertical eyebrow, wide and velvety. They also have well-developed scopea on their rear tibias for collecting pollen. Male Andrena Bees are noticeably smaller than the females. They have longer mandibles, lack the pollen-gathering scopea, and have adorable, walrus mustaches, probably their most recognizable trait.

Andrena carlini – Carlin's Mining Bee

51° / 35°

April 28

A BIRD OR A BEE?

Early in the week, I saw something strange, so out of the ordinary that it didn't seem real at first. Replaying the incident in my head, it took on an animated, fictionalized, almost cartoonish quality.

For many years, the small birdhouse has hung under the eave of the garage. I've looked at it out the kitchen window or watched it from the screened porch and caught glimpses of its occupants. The resident birds have usually been House Wrens. But occasionally Black-capped Chickadees have nested there as well.

This time, however, as I looked at the birdhouse, a large bumble bee flew directly into the entrance. I was so surprised that I shook my head as if to shake off a hallucination. But then I remembered an account given by Edwin Way Teale of finding a bumble bee nest in an old bird nest, repurposed by the insect. While I watched the birdhouse, the bee didn't depart. Will it be successful in establishing a nest? Being seven feet off the ground seems a little arboreal considering most bumble bees nest at ground level or below. So, I'm keeping an eye on the birdhouse now, to see what enters or leaves next. Will it be it a bird or a bee?

Plutella xylostella – Diamondback Moth
Caulophyllum thalictroides – Blue Cohosh
Mertensia virginica – Virginia Bluebells
Bombus bimaculatus – Two-spotted Bumble Bee
Rhamphomyia – Dance Fly

April 29

WRITING ADVICE

56°
—
38°

When reading, because I write, the passages and descriptions that touch upon methods of writing always catch my attention. My eyebrow raises when I happen across even the smallest account—honest asides, unintentional insights, even prosaic details about ink or pencils, about typewriters or dictionaries. Some authors say far more than needed. "The lady doth protest too much, methinks," as the queen states in *Hamlet*. Some authors are quietly reticent. These I prefer. For when they do provide a hint as to their methods, the words seem all the more authentic and momentous, a treasured insight into the craft of writing.

Henry David Thoreau is a writer whose work I read very carefully. Embedded in his voluminous natural history journals (ten volumes in the edition I have) are small nuggets of writing advice and thoughts about revision. Here are a few examples from the year 1854:

March 28: "Got first proof of *Walden*."

March 31: "In criticising [sic] your writing, trust your fine instinct. There are many things which we come very near questioning, but do not question. When I have sent off my manuscripts to the printer, certain objectionable sentences or expressions are sure to obtrude themselves on my attention with force, though I had not consciously suspected them before. My critical instinct then at once breaks the ice and comes to the surface."

April 8: "I find that I can criticise [sic] my composition best when I stand at a little distance from it,—when I do not see it, for instance. I make a little chapter of contents which enables me to recall it page by page to my mind, and judge it more impartially when my manuscript is out of the way. The distraction of surveying enables me rapidly to take new points of view. A day or two surveying is equal to a journey."

April 20: "I find some advantage in describing the experience of a day on the day following. At this distance it is more ideal, like the landscape seen with the head inverted, or reflections in water."

I began my own natural history journaling in 2008, nine years ago. My preferred method is to write about observations early the following morning (of course this isn't always practical). After a good night's sleep, after a cup of coffee and an hour of reading (usually poetry), I find I'm best prepared to write. I have, at that moment, the energy and words needed to describe the previous day's observations. You can be sure my eyebrow raised sharply when I came across the April 20th passage by Thoreau, stating that he felt "some advantage" in this approach.

Lamium maculatum – Spotted Deadnettle
Agromyzidae – Leaf-miner Flies
Parasteatoda sp. – Orbweaver Spider
Enoplognatha sp. – Orbweaver Spider

April 30

APRIL'S END

49°
—
38°

The poet T. S. Eliot opened his long poem, *The Wasteland*, with these memorable lines: "April is the cruelest month, breeding / Lilacs out of the dead land, mixing / Memory and desire, stirring / Dull roots with spring rain." I've always considered February to be the cruelest month, but this year April vied for that honor, with a long string of cool, cloudy, snowy, rainy, windy days. Actually, another line by T. S. Eliot comes to mind, this time from the poem 'East Coker' in *The Four Quartets*: "In my beginning is my end." April's end is equal to its beginning. The temperature is the same on the last day as it was at the start of the month. The rain is the same, verging on snow. The paucity of insects is the same. But the plants have done all right, blooming and leafing out, carrying on throughout the cool weather, with or without the needed troops of pollinators. For instance, the plum blossoms which opened earlier this week have only known temperatures in the 30s and 40s. The few times I checked on them, there were no bees (usually there are audible swarms), only a few Winter Ants and cold-hardy Maggot Flies. So good riddance to April and onward to May.

Sanguinaria canadensis – Bloodroot
Ichneumonidae – Ichneumonid Wasps

May

Common Nighthawk – May 25

May 1

MAY DAY

41°
—
35°

In my twenties and thirties, I often attended the May Day Parade in Minneapolis. A mash-up of workers issues, grassroots politics, and pagan spirit. I loved it. The parade always culminated with a ceremony enacted by dancers and giant puppets in which Winter is vanquished and the Tree of Life reborn.

In recent decades, having moved away from the big city to a smaller town, the 1st of May passes without much fanfare. Still, I look forward to this day each year, but it's become a quiet milestone, a much more personal event.

The Greek poet Yannis Ritsos, whose work I've admired for close to thirty years, was born May 1st, 1909. Here's a poem of his that I translated and printed for May Day in 2002 because of its parallels with the endless cycle of the seasons, Spring's poetic life written into existence each year by the creative earth.

DEATH-PLEDGE

He said: I believe in poetry, in death,
which is why I believe in Immortality.
I write down a line; I write down the world.
I'm created; the world is created.
From the tip of my finger flows a river.
The sky is seven times as blue. Such clarity
is the very first truth, my final wish.

Carex radiata – Bracted Sedge
Carex albursina – White Bear Sedge
Carex sprengelii – Longbeak Sedge

62°
—
38°

May 2

APRIL SHOWERS BRING MAY . . .

Hoverflies!

Yesterday, with slush falling from the sky, felt like a warm day in February. Today, the skies cleared and the temperature rebounded to near normal for this time of year, reaching sixty degrees late in the afternoon. The hoverflies were as happy as I was.

Four different species encountered on my walk to St Olaf: Narrow-headed Sun Fly (*Helophilus fasciatus*); Orange-spotted Drone Fly (*Eristalis anthophorina*); another *Eristalis* species; and a *Syrphus* species. All visiting the plum blossoms.

They seem to show off, almost as if to taunt the predators with their bold, wasp-mimicking colors. Flaunt it if you've got it.

Like fish swimming yet stationary in the current of a stream, like a hawk falling yet staying aloft, hoverflies hover, hang in the air. Sometimes I've mistaken them for spiders suspended on an invisible web, so fixed in space were they. The great Chilean poet Pablo Neruda was fascinated by this motionless motion, this running in place, defining it *inmovil movil*, fashioning it into a personal symbol, the flag flying above his home on Isla Negra, and, at the center of the flag flapping in the wind off the ocean, a fish fixed in perpetual motion.

Eristalis stipator – Yellow-shouldered Drone Fly
Syrphus – Flower Fly
Eristalis dimidiata – Black-shouldered Drone Fly
Calyptratae – Fly
Helophilus fasciatus – Narrow-headed Sun Fly
Conoderus auritus – Click Beetle
Antennaria neglecta – Field Pussytoes
Andrena sp. – Mining Bees
Zanthoxylum americanum – Prickly Ash
Nomada luteoloides – Nomad Bee
Eristalis anthophorina – Orange-spotted Drone Fly

May 3

66°

45°

MCKNIGHT PRAIRIE, LAKE BYLLESBY, AND THE ANDERSON CENTER

On this day in 1819, the Long Expedition set out from Pittsburgh to explore the Rocky Mountains. The only reason I know this is that I recently paged through Howard Ensign Evan's well-written account of the Long Expedition, an account that aimed at bringing the work of naturalist Thomas Say and painter Titian Peale to the fore. A faint echo, a slight reenactment of that notable excursion, I made my own small expedition to Red Wing, with stops at McKnight Prairie and Lake Byllesby County Park along the way.

Everywhere one looks leaves unfurl, flowers blossom—it's a free-for-all. Joining and following, sipping and nibbling, a clamjamfry of insects as if summoned to the plants by the plants themselves. To the early-blooming Pasque Flower and Prairie Smoke, observed on my earlier visit to McKnight Prairie, can now be added Birdfoot Violet, Carolina Draba, Kittentails, and Field Pussytoes. About the violets and pussytoes, especially, the bees and flies are busy.

Chymomyza amoena – Vinegar Fly	*Colletes* – Cellophane Bees
Myotis lucifugus – Little Brown Bat	*Draba reptans* – Carolina Draba
Halictus confusus – Confusing Furrow Bee	*Lycosidae* – Wolf Spiders
Pergidae – Pergid Sawflies	*Andrena vicina* – Neighborly Mining Bee
Nomada – Nomad Bees	*Rhaphidophoridae* – Camel Crickets
Formicidae – Ants	*Viola pedata* – Bird's Foot Violet
Vanessa virginiensis – American Lady	*Veronica bullii* – Kittentails
Bombylius major – Greater Bee Fly	*Araneus pratensis* – Openfield Orbweaver
Lasioglossum pilosum – Sweat Bee	*Sphecodes* – Blood Bees
Vanessa virginiensis – American Lady	
Cicindela scutellaris lecontei – Leconte's Tiger Beetle	

69°
—
44°

May 4

BIG WOODS STATE PARK

I locked myself out of the house. This turned out to be a therapeutic mistake. Forced to find something to do for three hours, I decided to visit Nerstrand Big Woods State Park and have a look at the wildflowers. It was certainly a good day for a visit as pretty much every wildflower was in bloom. Walking the trail down to the waterfall and back, I was able to track down twenty-six species. Some interesting bees and flies were encountered as well, including the specialist pollinator of Spring Beauty, *Andrena erigeniae*.

Thoreau, when he went a-botanizing, had in hand Gray's *Manual of Botany*. I, too, own a copy of this book, a scuffed, leather edition (revised 1899), but it's a rare occasion that I take it out of its glassed-in bookcase and use it. Instead, I tend to use modern guidebooks, filled with color photos. And more and more often I use the online guide: Minnesota Wildflowers. This online guide has the advantage of being focused geographically on Minnesota.

There's an interesting connection to Asa Gray among the day's finds. In 1870, Mary Hedges, a botany teacher in Faribault, discovered the Dwarf Trout Lily. Specimens she collected made their way to Asa Gray, who subsequently described the species.

Here's an excerpt of the note by Asa Gray describing the Minnesota Dwarf Trout Lily:

A NEW SPECIES OF *ERYTHRONIUM*
by Professor Asa Gray

Ordinarily it is hardly worth while to make a separate article for a single new species of plant, even when discovered in a district in which a new flowering plant is unexpected. But the species of *Erythronium* are so few, and the present one is so peculiar, and its habitat so closely bordering the region included in my Manual of the Botany of the Northern United States, that I need not apologize for bringing it at once to notice.

The specimens before me, accompanied by a colored drawing, are just received from Miss S. P. Darlington (a daughter of the late Dr. Darlington, long the Nestor of American botanists and one of the best of men), and were collected, at Faribault, Minnesota, by Mrs. Mary B. Hedges, the teacher of Botany in St. Mary's Hall, a school of which Miss Darlington is Principal.

The flower is much smaller than that of any other known species, being barely half an inch long; and its color, a bright pink or rose, like that of the European *Erythronium Dens-Canis*, reflects the meaning of the generic name (viz. red), which is lost to us in our two familiar Adder-tongues, one with yellow, the other with white, blossoms. The most singular peculiarity of the new species is found in the way in which the bulb propagates. In *E. Dens-Canis* new bulbs are produced directly from the side of the old one, on which they are sessile, so that the plant as it multiplies forms close clumps. In our *E. Americanum* long and slender offshoots, or subterranean runners, proceed from the base of the parent bulb and develop the new bulb at their distant apex. Our Western *E. albidum* does not differ in this respect. In the new species an offshoot springs from the ascending slender stem, or subterranean sheathed portion of the scape (which is commonly five or six inches long), remote from the parent bulb, usually about mid-way between it and the bases or apparent insertion of the pair of leaves: this lateral offshoot grows downward, sometimes lengthening as in the foregoing species, sometimes remaining short, and its apex dilates into the new bulb.

This peculiarity was noticed by Mrs. Hedges, the discoverer of this interesting plant, to whom great credit is due. Most lady botanists are content with what appears above the surface ; but she went to the root of the matter at once. I learn that *E. albidum* abounds in the same locality. *E. Americanum* is also found in the region, but is scarce.

It is not easy to find or frame a specific name which will clearly express the most remarkable characteristic of this new species. But I will venture to name it.

– from *The American Naturalist* (1871, 298-300)

Barbarea vulgaris – Garden Yellowrocket
Galium triflorum – Fragrant Bedstraw
Galium mollugo – Hedge Bedstraw
Phlox divaricata – Blue Phlox
Podophyllum peltatum – Mayapple
Viola pubescens – Downy Yellow Violet
Claytonia virginica – Virginia Spring Beauty
Viola sororia – Common Blue Violet
Geranium maculatum – Wild Geranium
Caltha palustris – Marsh Marigold
Asarum canadense – Canadian Wild Ginger
Erythronium albidum – White Fawn Lily
Erythronium propullans – Minnesota Dwarf Trout Lily
Allium tricoccum – Small White Leek
Arisaema triphyllum – Jack-in-the-pulpit
Ranunculus hispidus – Bristly Buttercup
Cardamine concatenata – Cut-leaved Toothwort
Dicentra cucullaria – Dutchman's Breeches
Uvularia grandiflora – Largeflower Bellwort
Enemion biternatum – False Rue Anemone
Sanguinaria canadensis – Bloodroot
Anemone acutiloba – Sharp-lobed Hepatica
Ranunculus abortivus – Small-flowered Buttercup
Thalictrum dioicum – Early Meadow-rue
Pedicularis canadensis – Canadian Wood Betony
Anemone quinquefolia – Wood Anemone
Andrena erigeniae – Spring Beauty Mining Bee
Equisetum pratense – Shady Horsetail
Cryptogramma stelleri – Steller's Rock-brake
Conocephalum salebrosum – Snakewort
Plagiomnium – Thyme-Moss
Athyrium filix-femina – Lady Fern
Carex blanda – Eastern Woodland Sedge
Nomada – Nomad Bees
Priocnemis minorata – Spider Wasp
Chalcosyrphus nemorum – Dusky-banded Forest Fly

78°
—
46°

May 5

FINALLY A CIMBICID

At the Anderson Center in Red Wing. I hiked through acres of dandelions to reach the trail that leads down the bluffs to the Cannon Valley Bike Trail and the floodplain of the Cannon River. Nearing the bluffs and the edge of the dandelion fields, I noticed a mass of honey bees on a branch of a Bur Oak, high above two bee hives located there. A queen must have left one of the hives. Strangely, from a distance away, this bulk of bees draped on the underside of a branch looked like a three-toed sloth hanging there.

In the pools and channels of the spring-fed seeps at the base the bluffs, watercress grew thick and tall and lush. I picked a couple leaves and relished the spicy taste of the crisp greens. On one plant, a fancy, golden snail with a white swash in its shell moved along a leaf. A large insect flew by and plopped down nearby. Four wings, large abdomen, clubbed antennae, I recognized it almost immediately as something long on my entomological wish list, a large Cimbicid sawfly (genus *Abia*), the first I'd ever seen alive (I'd found several dead Cimbicids walking gravel roads in the far north and I'd seen them in insect collections). The larvae of this particular sawfly feed on Honeysuckle plants.

What else is on my entomological wish list you might ask? Being there are so many insects, and so many interesting insects, such a list could be very long, so I'll limit myself to a few that spring to mind. I'd love to see an Owlfly, though I don't think they are found in Minnesota. I'd like to have a longer look at a Tarantula Hunter, having had the briefest glimpse of one some years ago in Texas. A Luna Moth, a Cecropia Moth, or a Rosy Maple Moth at the mothing light would satisfy some of the moth envy I've built up over the years, years of seeing other people's photos of these showy moths. The dragonfly list is long, not to mention it recharges over the winter months so that even the most common species, so familiar from years of encounters, move high up on my wishlist; by the end of June, I truly long to see the year's first White-faced Meadowhawk.

Morchella – Morels
Caltha palustris – Marsh Marigold
Succineidae – Amber Snails
Abia sp. – Cimbicid Sawfly
Tetrix sp. – Pygmy Grasshopper
Tetragnatha sp. – Long-jawed Orbweaver
Athyrium filix-femina – Lady Fern
Ranunculus abortivus – Small-flowered Buttercup
Coleomegilla maculata – Spotted Lady Beetle
Toxomerus geminatus – Eastern Calligrapher
Veronica arvensis – Corn Speedwell
Apis mellifera – Western Honey Bee

69°
—
44°

May 6

FLY WATCHING

Fly watching never caught on in Minnesota. There's no Hoverfly Society, no Syrphidae database, no distribution atlas. In the U.K., all these exist (see for instance *Britain's Hoverflies: A Field Guide* by Stuart Ball and Roger Morris).

In 1939, Horace S. Telford (1909 - 2004) published *The Syrphidae of Minnesota*. At the time he was a Ph. D. student at the University of Minnesota, in 1939. Recently, I was lucky enough to obtain a copy of this publication and it arrived in today's mail. According to the introduction, several thousand syrphid flies collected in Minnesota formed the basis of Telford's study. He provides county data and dates for 135 species from 52 genera.

What better way to honor the work of Horace S. Telford than to go fly watching. Meadow Sedgesitters were abundant at Hauberg Woods and at St Olaf Natural Lands. This species was known from 18 counties at the time of Telford's study with a flight period from May 20 to September 13. Three, possibly four, additional syrphid species were observed at St Olaf Natural Lands. These were the Margined Calligrapher (not included in Telford), the Narrow-headed Sun Fly (known from 12 counties with a flight period of May 14 - September 14), a Swamp Fly (*Lejops sp.*), and a different species of Sedgesitter (*Platycheirus sp.*).

Platycheirus quadratus – Meadow Sedgesitter   *Lejops sp.* – Swamp Fly
Leucorrhinia intacta – Dot-tailed Whiteface   *Ischnura verticalis* – Eastern Forktail
Helophilus fasciatus – Narrow-headed Sun Fly
Platycheirus quadratus – Meadow Sedgesitter
Toxomerus marginatus – Margined Calligrapher
Melanostoma mellinum – Western Roundtail

May 7

PINOCCHIO

67°
—
49°

A spectacular spring day. Continuing my hoverfly kick, I decided to try a different habitat, lotic instead of lentic. In the afternoon, I hiked the trails along Spring Creek in the upper Cowling Arboretum. As in the past, the most insect activity was found along the sandy banks around the semi-open plunge pool where the creek enters the arboretum, flowing from a large culvert beneath a city street.

Many damseflies needled their way through the vegetation. A variety of bees worked the Creeping Charlie and Yellow Rocket flowers. I gave the bees and the damselflies the cold shoulder; I wanted to see more hoverflies. The same four species observed yesterday around the ponds, I observed today along the creek, with one remarkable addition—the American Heineken Fly (*Rhingia nasica*). The only species in this genus in North America, this fly is instantly recognizable because of its Pinocchio-like nose. This odd-shaped shnoz houses a long tongue used for reaching nectar in deep flowers. The European species *Rhingia campestris* was given the common name Heineken Fly after the humorous adds by the beer company of that name, adds that stated 'Heineken refreshes the parts other beers cannot reach,' recognizing that this fly could reach parts of flowers that other hoverflies can not.

One of the frustrating hindrances encountered when reading old natural history books and monographs is that many of the scientific names have changed or are no longer recognized as valid due to subsequent changes and revisions. Luckily for fly researchers, there is a wonderful online resource—Systema

Dipterorum: The Biosystematic Database of World Diptera (www.diptera.com). Old scientific names can be searched to find current statuses and currently accepted names. For instance, Horace Telford in *The Syrphidae of Minnesota* (published nearly eighty years ago) described a new species of *Microdon*. The name he proposed for this species isn't listed at bugguide.net. Checking the Diptera database, one finds that Telford's species' name, *Microdon robusta.* is now considered a junior synonym of *Microdon tristis*. I wish such a database was available for other orders of insects as well, especially Hymenoptera.

Dictynidae – Meshweavers
Naphrys pulex – Flea Jumper
Dolomedes tenebrosus – Dark Fishing Spider
Rhingia nasica – American Heineken Fly
Toxomerus geminatus – Eastern Calligrapher
Melanostoma mellinum – Western Roundtail
Chalcosyrphus nemorum – Dusky-banded Forest Fly
Platycheirus quadratus – Meadow Sedgesitter
Rhingia nasica – American Heineken Fly

May 8

IN WATER

70°
—
46°

"We are slow to realize water, —the beauty and magic of it. It is interestingly strange to us forever." Henry David Thoreau, *Journals*, May 8 1854

"If there is magic on this planet, it is contained in water." Loren Eiseley, *The Immense Journey*

Boyd Sartell WMA. A large natural area in southwestern Rice County. This WMA has, in previous years, supported populations of Taiga Bluets and Silvery Blues. And both damselfly and butterfly have been on the wing at the same time, usually about this time of year, both in and around the series of man-made ponds.

For the past several springs I've attempted to collect Taiga Bluet nymphs, each time without success. Last year, in North Dakota, I did finally locate several Taiga Bluet nymphs (of course this was while I was attempting to locate Prairie Bluet nymphs!). Like other Eurasian Bluets (genus *Coenagrion*) nymphs, the head of the Taiga Bluet nymph is speckled behind the eyes with tiny black dots. This makes it easy to separate them from the hordes of American Bluet (genus *Enallagma*), Forktail (genus *Ischnura*), and Sedge Sprite (genus *Nehalennia*) nymphs present.

Now that I knew how to identify the Taiga Bluet nymphs, I thought I'd try again to locate them at this site, searching the shallows at various points around the ponds with a dip net. No such luck.

Just like Thoreau's pronouncement on water, the most common water inhabitants, such as the nymphs of Sedge Sprites (which turned up in spades this day) and the trilobite-esque Isopods, will remain "interestingly strange."

Having a goal in mind can be an obstacle, can diminish the appreciation of other accomplishments, for example, the Two-lined Swamp Fly and the floating flight of Phantom Crane Flies, both observed near the ponds.

Coenagrionidae – Narrow-winged Damselflies
Nehalennia irene – Sedge Sprite
Caecidotea – Isopod
Libellula sp. – King Skimmers
Aeshna sp. – Mosaic Darners
Bittacomorpha clavipes – Phantom Crane Fly
Lejops bilinearis – Two-lined Swamp Fly

Viburnum lentago – Nannyberry →
Tenthredinidae – Common Sawflies
Lamium amplexicaule – Henbit Deadnettle
Malus sp. – Apples
Formica sp. – Wood, Mound, And Field Ants
Helophilus fasciatus – Narrow-headed Sun Fly
Icterus galbula – Baltimore Oriole
Capsella bursa-pastoris – Shepherd's-purse
Lasioglossum sp. – Sweat Bee
Cheilosia sp. – Grass Skimmer

May 9   71°

INTERESTING WEEDS   52°

This spring, it seemed every user of iNaturalist had photos of Henbit Deadnettle except for me. I was beginning to feel a bit left out. That is until today. Walking across the practice field adjacent to the lower Cowling Arboretum I saw patches of what seemed tall, pinkish-looking Creeping Charlie. A closer look showed it to be Henbit Deadnettle. Perhaps the grass seed used on the field contained its seed? This plant is not widespread in Minnesota, with records from the metro counties around Minneapolis and Saint Paul.

Shepherd's Purse is another distinctive plant. And a global success, as it seems to have spread to every part of the earth. It's easily recognized by the cluster of small white flowers, the deeply toothed leaves of the basal rosette, and the purse-shaped seed pods from which the plant gets its common name. A member of the cress family (also known as the cabbage or mustard family), it is eaten in some parts of the world, used as an herbal, and used by science as a model organism.

One of the great things about plants? They stay put. They don't run off or fly away. If it can't be identified, it can be revisited at a later date. If it's a tree, it can be visited many years in a row. Early in April, I photographed an unusual bud on a small tree but couldn't figure out what kind of tree it was, until today. Walking near the tree today, I remembered to have a look at the tree once more. The leaf shape and the flower clusters made it possible to identify it this time: Nannyberry (*Viburnum lentago*), a native shrub.

[Observations on opposite page]

69°
—
53°

May 10

APPLE BLOSSOMS

William Butler Yeats, to many the greatest Irish poet, published 'The Song of Wandering Aengus' in his 1899 collection *The Wind in the Reeds*. I've loved this poem for a long time. In fact, it's one of the dozen or so poems I know by heart. The hazel wood, a fire in the mind, moth-like stars, a little silver trout that magically transforms into a girl with apple blossoms in her hair, and the growing old with wandering are all contained in this short beautiful poem. Ah yes, and don't forget "the silver apples of the moon, the golden apples of the sun." A poem of magic and myth that takes place along life's transect through the natural world.

THE SONG OF WANDERING AENGUS

I went out to the hazel wood,
Because a fire was in my head,
And cut and peeled a hazel wand,
And hooked a berry to a thread;
And when white moths were on the wing,
And moth-like stars were flickering out,
I dropped the berry in a stream
And caught a little silver trout.

When I had laid it on the floor
I went to blow the fire a-flame,
But something rustled on the floor,
And someone called me by my name:
It had become a glimmering girl
With apple blossom in her hair

Who called me by my name and ran
And faded through the brightening air.

Though I am old with wandering
Through hollow lands and hilly lands,
I will find out where she has gone,
And kiss her lips and take her hands;
And walk among long dappled grass,
And pluck till time and times are done,
The silver apples of the moon,
The golden apples of the sun.

– William Butler Yeats

Malus sp. – Apples

67°
—
47°

May 11

DISINTRICATE

Just before I left for the day's hike, an email from local birder and dragonfly aficionado Dan Tallman alerted me to shorebirds at a newly constructed catchment pond just outside of town. So, on my way to the Cannon River Wilderness Area, I stopped and had a look. Several pairs of Least Sandpipers waded the muddy shore. A Killdeer, several Mallards, and a pair of Canada Geese joined them. Across from the pond, earthmovers, backhoes, and dump trucks rumbled and stirred up great clouds of dust. A few shorebirds along with a new housing development.

With less than an hour at the wilderness area, the hike was more aerobic exercise than attentive sauntering. I busted through acres of woodland flowers to reach the open oak savanna only to find it had recently been burned. So for the few minutes I had to poke around, I headed upslope to the bluff tops to see what I could see and was soon out of the burn area. While admiring a grouping of three small Hoary Puccoon plants a mason wasp landed on the dirt and gathered up a mouthful of clay for her nest. This was a welcome encounter, the first mason wasp of the year.

After the wasp flew away, a weird little fly made an appearance at the tip of a dandelion leaf. A ruby-and-ivory-colored Thick-headed Fly, genus *Myopa*. I'd seen this species before. Known parasites of Andrena Bees, these flies have preposterous ovipositors that resemble hand-operated can openers. Using this contraption, they are able to pry apart segments of the host bee's abdomen and insert their egg.

❖ ❖ ❖

Louis MacNeice, in his preface to *The Poetry of W. B. Yeats*, claims that "all literary critics are falsifiers in that they try to disintricate the value or essence of a poem from the poem itself; they peel away the onion." True as this may be in the literary realm, there seems to be a parallel distinction to be leveled at the so-called "objectivity" of scientific articles. The cool, matter-of-fact style put forth in journals and textbooks, cuts out what it can of the messy emotions and subjective realities and contains blessed little passion for the world. The goal, as I see it, is to keep observations intricate. While gathering data, why not work to include indications of connections and complications? One's loyalties must remain with the senses and the living encounter. I suspect this is why we still read Darwin and Wallace with pleasure.

Calidris minutilla – Least Sandpiper
Ancistrocerus sp. – Mason Wasp
Lithospermum canescens – Hoary Puccoon
Myopa clausa – Thick-headed Fly
Enallagma sp. – Bluet
Bibio femoratus – March Fly
Moehringia lateriflora – Bluntleaf Sandwort
Aquilegia canadensis – Red Columbine
Chalcosyrphus nemorum – Dusky-banded Forest Fly

May 12

"TAKE LIFE EASY"

Reading and re-reading a lot of W. B. Yeats. This line from his early poem 'Down by the Sally Gardens' makes a good motto for the day: "She bid me take life easy, as the grass grows on the weirs." Hiked to St Olaf Natural Lands at noon. An easy walk. Hopelessly behind on flowers and leaves and birds and insects, I have to let it all slide by, take some deep breaths and enjoy the day, come what may. A good Samaritan has spent the morning (or last several mornings) pulling up a large patch of Garlic Mustard in the woods to the side of the trail. As I walked along I pulled out a few plants that had been missed.

Having encountered the first bluet damselfly yesterday, I thought it likely to find more today. No such luck. Dozens of Eastern Forktails, but no additional bluets...yet. Lots of sawflies on Virginia Waterleaf, Elder, and Highbush Cranberry. Photographed two jumping spiders.

A large, multi-legged thing moved on the gravel in front of me. Actually, a combination of two distinct creatures, though bound by fate—a pitch-black spider wasp and a dark brown spider. The wasp succeeded in moving its prey across the open trail very quickly, then struggled in the short trailside grass, dropping the spider multiple times while doing its best to pull the spider over and under and through numerous obstacles. I watched patiently for some minutes. Eventually, I lost sight of the wasp and its labors when it reached a thick stand of grass at the wood's edge.

Howard Ensign Evans in *A Taxonomic Study of the Nearctic Spider Wasps Belonging to the Tribe* Pompilini (1950), writes: "Of several genera of taxonomic importance, nothing whatever is known. The patient field observer may yet learn a wealth of facts about these wasps, and his efforts will be rewarded not only by the thanks of taxonomists and students of animal behavior, but by his acquaintance with the wasps themselves, for few insects are so fascinating in their activities as these intrepid hunters of spiders." I agree; there's little better than to take life easy and watch these hunters of spiders when you have the chance.

❖ ❖ ❖

First Monarch! Chokecherry in blossom.

Prunus virginiana – Chokecherry
Andrena sp. – Mining Bee
Dictynidae – Meshweavers
Abia sp. – Cimbicd Sawfly
Platycheirus quadratus – Meadow Sedgesitter
Salticidae – Jumping Spiders
Priocnemis minorata – Spider Wasp
Pelegrina proterva – Jumping Spider
Danaus plexippus – Monarch

80°
—
54°

May 13

FISHING OPENER

In Rochester for the day. During a break between Lida's volleyball games, Lisa and I drove to Quarry Hill Nature Center for a hike.

Being an hour south of Northfield, I was expecting to see some dragonflies. We hiked around the fish pond, across some meadows, and searched the wetland in the upper quarry. And while we didn't happen upon any dragonflies, we did find a few damselflies: several Eastern Forktails and one teneral Sedge Sprite, still translucent and white after its recent emergence.

We returned to the pond. Along the north shore where the sun warmed the shallows, Eastern Forktails were emerging en masse. An impressive number of American Toads dotted the shallows, waiting for teneral damselflies to fly near their mouths. The trail brought us to the fishing dock just as a young man reeled in a Yellow Bullhead. I walked up to admire the fish and he was kind enough to let me take a photo of it.

❖ ❖ ❖

Evening mothing in Northfield.

Ameiurus natalis – Yellow Bullhead
Tettigidea – Pygmy Grasshopper
Lithobates pipiens – Northern Leopard Frog
Ischnura verticalis – Eastern Forktail
Anaxyrus americanus – American Toad
Xysticus sp. – Ground Crab Spiders
Ischnura verticalis – Eastern Forktail
Platycheirus quadratus – Meadow Sedgesitter
Nehalennia irene – Sedge Sprite

May 14   84°

SYRINGE AND SYPHON   56°

A beautiful summer-like day in Rochester, Minnesota. Outside the gymnasium during a break from watching volleyball, I found a Giant Water Bug on the pavement in the parking lot. Crawling, then flapping, then crawling again, it struggled to make progress. I picked it up, took it's photo and set it off the pavement. A long time ago, as a graduate student, I kept one of these insects as a pet. Some years later, I commemorated that experience with a poem.

    GIANT WATER BUG

    A stocky student in chest-waders,
    he joked about nearly everything, but spoke
    seriously about snails, passionately describing
    their perfect habitats, their golden spirals.

    He collected in shallow bays and bogs
    raking a net under lily pads.
    At the end of the day, the snails
    sloshed in a five-gallon bucket.

    One day water scorpions and
    crayfish, another day a giant water bug
    traveling like an origami rowboat
    in the bucket with the snails.

    We gathered round. He incanted
    the Latin name, *Lethocerus americanus*.

The bug's mouth both syringe and syphon.
You can feed it goldfish, he said.

So we fed it goldfish in a fish tank
back at our university lab.
Until it escaped and flew like a bat
down clean, well-lighted hallways.

◆ ◆ ◆

Evening mothing.

Lethocerus americanus – American Giant Water Bug

May 15 70°

MICRO-BEETLES & FIRST-SPOTTED 4-SPOTTED 60°

Emptying out the moth trap in the morning, I found one moth I hadn't noted the night before, a Peppered Moth (*Biston betularia*), a holarctic species made famous in England after studies of its industrial melanism. In addition to this one large moth and several large June Beetles, there were numerous tiny insects, mostly flies and beetles. Just as there are micro-moths, it appears there are also micro-beetles, beetles that don't get larger than several millimeters in length.

In the afternoon, I walked to St Olaf. Noticed some flies along the way. And, at the small hub meadow, in addition to Eastern Forktails and a female Dot-tailed Whiteface, I spotted the first Four-spotted Skimmer of the year. A large, golden dragonfly with a black spot at the node of each of its four wings. Like the Salt & Pepper Moth, the Four-spotted Skimmer is a holarctic species; it can be found around the entire northern hemisphere: in Japan, in Siberia, in Europe...there's even a relict population, held over from the last glaciation, in the mountains of Morocco.

In Europe, this dragonfly is known to migrate. There have been several occasions in Northfield where an irruption is the only likely explanation for the sudden appearance of vast numbers of Four-spotted Skimmers. One year, in particular, stands out. I found dozens pretty much everywhere I walked. Even our backyard held a number of them. I have a photograph of two perched, holding our car's antenna. Another curious observation, and more recent, was seeing a Four-spotted Skimmer (through binoculars) take intermittent feeding flights from a perch high in our neighbor's white pines, at least forty feet off the ground.

Libellula quadrimaculata – Four-spotted Skimmer
Scathophaga stercoraria – Yellow Dung Fly
Ichneumonidae – Ichneumonid Wasps
Leucorrhinia intacta – Dot-tailed Whiteface
Onthophagus sp. – Dung Beetle
Ischnura verticalis – Eastern Forktail
Tenthredinidae – Common Sawflies
Andrena sp. – Mining Bee
Calyptratae – Calyptrate Fly
Rhingia nasica – American Heineken Fly
Barypeithes pellucidus – Hairy Spider Weevil
Veronica serpyllifolia – Thyme-leaved Speedwell

May 16   84°

ACROBAT ANTS   61°

In a gap between later afternoon storms, towering walls of dark clouds blocked out sky canyons, and the sunlight raced through them like flash floods for a few minutes.

I went out to the big pond at the St Olaf Natural Lands, thinking I'd find bluet damselflies, but no such luck. I did happen across several Common Whitetail dragonflies, the first encounter with this large, showy species this year. This male will become even showier in a week or two as its abdomen turns bright white.

The sun disappeared behind a blue-black thunderhead. Distant thunder rumbled in the west. Even the flowers dimmed. So I went no further, turned and headed back to the car. On the way, I stopped for a quick inspection of one of the University of Minnesota bee survey blocks. It was immediately obvious that this nest block had been compromised by ants as they swarmed in and out of several of the nest entrances. A mason wasp had been captured. I took a photo of the wasp and the ants. Looking at the photo a little while later, I noticed the ant's peculiar, heart-shaped abdomens and the prong of propodeal spines. These are Acrobat Ants, of the genus *Crematogaster*, which actually hunt wasps. They secure their prey by the legs, pulling the legs out in all directions so that the wasp is incapacitated, stretched spread-eagle by a mob of ants. A neat trick to catch a wasp.

Sumitrosis inaequalis – Leaf Beetle   *Crematogaster sp.*– Acrobat Ants
Plathemis lydia – Common Whitetail   *Ischnura verticalis* – Eastern Forktail
Augochlorella aurata – Metallic Sweat Bee   *Maevia inclemens* – Dimorphic Jumper

72°
—
54°

May 17

SLEEPY BEES

For the second day in a row, my afternoon hike was accompanied by the rumble of encroaching storms. Today there seemed no chance of sunlight, rather a slight brightening of the grey clouds. I packed a plastic bag for my camera as there was a good chance I'd not make it back home before the next round of rain arrived.

Muddy trails, drenched and nodding flowers, dripping leaves. Along the way to the wooded pond, hoverflies and a bumble bees perched and nectared amid the blooming Waterleaf.

At the pond, in a stand of grass, I caught sight of a sleeping cuckoo bee. This has been high up on my hope-to-find list for quite some time. Numerous insect photographers have posted photos of these sleepy bees, but until today I hadn't encountered one myself. Other bees and wasps sleep in this fashion, clamped stiffly to leaf or twig. Except for the rigid posture, it's difficult to tell when a bee is asleep; they cannot close their eyes as we do.

Sometimes, while reading, I nod off and wake awhile later to find the book I was holding slipped from my grasp. I wonder if these bees ever relax their grip and fall from their perch while still asleep?

Nomada sp. – Nomad Bees
Eristalis transversa – Transverse Flower Fly
Evarcha hoyi – Hoy's Jumper
Toxomerus geminatus – Eastern Calligrapher
Bombus impatiens – Common Eastern Bumble Bee

May 18   56°/47°

HOUSE WORK

Today's observations were all out the front window of our house. Would have preferred the usual hike, but I got stuck doing house work. Luckily there were plenty of front yard visitors today.

The wren, the rabbit, and the Hermit Thrush all are regulars, live here with us. The Rose-breasted Grosbeak, which my daughter spotted and pointed out to us, hadn't been seen in our yard until today.

Catharus guttatus – Hermit Thrush
Pheucticus ludovicianus – Rose-breasted Grosbeak
Troglodytes aedon – House Wren
Sylvilagus floridanus – Eastern Cottontail

57°
—
44°

May 19

THE METAMORPHOSIS OF PLANTS

A chill day with only intermittent and vague brightenings of the sky. After dropping off packages at the post office, I had time for a short hike at the lower arboretum. The Cannon River level was up after the rain, near the tops of its banks.

While standing at the edge of the inundated creek, close to its confluence with the river, I experienced one of those occasional challenges to visual perception, one of those occurrences where the mind tries to make sense of unusual visual data. An object flew, splashed into the water, then swam away beneath the surface. I'd caught a glimpse of yellow and a swimming motion as if it were a small turtle. What could it be? It wasn't a Kingfisher, as that bird is larger and blue and wouldn't have remained underwater. The best my mind could do was the laughable—a flying turtle! After standing there perplexed for a moment, I took a step forward and saw the fisherman on the opposite side of the creek who was reeling in the lure he'd just cast.

Reading Goethe again recently, this quote of his came to mind: "every new object, clearly seen, opens up a new organ of perception in us." What about objects unclearly seen, I wondered?

I searched the meadow that rings the retention pond, then the fallow rectangle in the woods where a group of tennis courts had been removed. Neither place held any dragonflies. Any day now the clubtails, swift, black-and-yellow river dragonflies, would begin to emerge and commandeer these meadows.

Crossing the tennis court field, a Song Sparrow scurried out from under one of my footsteps, startling me. I checked the thick grass at my feet to make sure I hadn't destroyed its nest. Luckily I'd missed it by a few inches, the beautiful blue-speckled eggs were intact (and without a brown imposter Cowbird egg).

Reading Goethe long ago (and for the first time) as a regimented engineering student, I enthusiastically embraced Goethe's alternate and naturalistic approach to the sciences. Some of the simple experiments recounted in his *Theory of Color* and *The Metamorphosis of Plants* changed how I looked at the world. "Researchers have been generally aware for some time that there is a hidden relationship among various external parts of the plant that develop one after the other and, as it were, one out of the other... The process by which one and the same organ appears in a variety of forms has been called the metamorphosis of plants." – Johann Wolfgang von Goethe, from *The Metamorphosis of Plants* (1790)

It occurred to me to photograph the leaves of the Small-flowered Buttercup. The sequence of leaf shapes from base to flower changes quite dramatically for this plant and were similar to some of the illustrations in Goethe.

Lophodytes cucullatus – Hooded Merganser
Hirundo rustica – Barn Swallow
Ranunculus abortivus – Small-flowered Buttercup
Melospiza melodia – Song Sparrow

51°
—
42°

May 20

IT CRAWLED OUT OF ITSELF

Cold and rainy. There's been so much rain, recently, that water has started seeping, rising up like a spring, through the porous, seventy-year-old concrete in our basement. An old Johnny Cash song I'd learned when young began looping through my head throughout the day: "How high's the water Momma?"

Fortuitously, on this rainy day, I could stay home and observe nature (checking off-and-on the water in the basement). A dragonfly emerged in the morning in my office: a Common Whitetail, the nymph collected nearly a month ago (on April 24). One of several skimmer dragonflies that has banding of the eyes, it's not as complex as the wood grain patterned eyes of the Filigree Skimmer, but still interesting.

Recently, I showed a neighbor several dragonfly nymphs. He stared at them in disbelief. Like caterpillar and butterfly, the difference between the nymph and the adult dragonfly defies immediate comprehension. This difference seems more remarkable than the difference between caterpillar and butterfly because dragonfly nymphs live in an entirely different world underwater. The transformation is diminished by the technical term "incomplete metamorphosis," as though there is something missing from its life. Occasionally, this three-stage life cycle—egg, nymph, adult—is referred to as simple or gradual metamorphoses. Complete or complex metamorphosis includes the additional pupa stage.

Plathemis lydia – Common Whitetail

May 21 — 52°/46°

BOMBUS IN THE BASEMENT

Last summer several Half-black Bumble Bees found there way into our basement. I'd catch them and release them when this happened. These bees nested in a covered up window well. I observed them throughout the summer, the smallish workers flying in and out of the nest. Even though the nest was just a few feet from where we parked our car and where my daughter and her friend often played volleyball, the bees didn't mind our presence and we didn't mind theirs.

Today, I found a Two-spotted Bumble Bee, dead on the floor of an underground parking garage. Perhaps the bee had found its way in by accident while searching for a nesting site. And then couldn't find its way back out.

Bombus bimaculatus – Two-spotted Bumble Bee

67°
—
45°

May 22

AND THEN THE SUN CAME OUT

After a week of rainy cool days, the skies cleared, the sun came out, and the temperature bounced back up into the 70s.

For the last several weeks, I've had my eye on the Black Raspberries, watching the flowers get closer and closer. The flowering of the raspberries is a major phenological milestone; is this not the first day of summer? Timed almost to the minute with their bloom, the first skipper butterflies appear. So today, when I saw open flowers on the Black Raspberries I began to anticipate seeing a skipper. After passing by nearly all the usual patches of raspberry along my route, I'd stopped looking for the butterfly. Then I saw something fly that at first I thought was a moth. When I followed it to where it landed I was delighted to see that it was a Hobomok Skipper.

More surprising than the skipper was a spreadwing damselfly, the first of the year. I'm fairly certain this is a Southern Spreadwing just by the date. Very similar to the Northern Spreadwing, though in Minnesota Northern Spreadwings only begin to emerge in July and usually in the northern half of the state. The other early spreadwings for Rice County are the Emerald Spreadwing and the Slender Spreadwing and they usually begin to emerge in June.

Another highlight of this hike was finding a tiny hoverfly of a genus I'd not encountered before—*Paragus*. About the size of the superabundant *Toxomerus* species only solid black in color (*Toxomerus* is boldly yellow and black). The common name proposed by Skevington et. al. in their Nearctic Syrphidae Checklist for this genus is Grass Skimmer. According to bugguide.net, these flies are difficult to identify to species, as dissection is often required.

Poanes hobomok – Hobomok Skipper
Ancistrocerus sp. – Mason Wasp
Bittacomorpha clavipes – Phantom Crane Fly
Cicindela sexguttata – Six-spotted Tiger Beetle
Lestes australis – Southern Spreadwing
Cupido comyntas – Eastern Tailed-blue
Eristalis stipator – Yellow-shouldered Drone Fly
Paragus sp. – Grass Skimmer
Hypselistes florens – Dwarf Spider
Ischnura verticalis – Eastern Forktail
Ceratina calcarata – Spurred Ceratina
Veronica arvensis – Corn Speedwell
Lucilia – Greenbottle Flies
Ischnura verticalis – Eastern Forktail
Dialictus – Metallic Sweat Bees
Rubus occidentalis – Black Raspberry
Symphyta – Sawflies, Horntails, And Wood Wasps
Oncopeltus fasciatus – Large Milkweed Bug
Osmorhiza claytonii – Hairy Sweet Cicely
Myopa clausa – Thick-headed Fly
Sphecodes sp. – Blood Bees
Helophilus fasciatus – Narrow-headed Sun Fly
Pelegrina proterva – Jumping Spider
Andrena sp. – Mining Bee
Ichneumonidae – Ichneumonid Wasps

61°
—
49°

May 23

THE MAPS WITHIN

"I go about and look at flowers and listen to the birds. There was a time when the beauty and the music were all within, and I sat and listened to my thoughts, and there was a song in them."
– Henry David Thoreau, from *Journals* (Tuesday, May 23, 1854)

"He walked happily…and through his deteriorating summer footwear he felt the ground with extraordinary sensitivity when he walked across an unpaved section… For a long time he had wanted to express somehow that it was in his feet that he had the feeling of Russia, that he could touch and recognize all of her with his soles, as a blind man feels with his palms."
– Vladimir Nabokov, from *The Gift*

"Thus, although we are mere sojourners on the surface of the planet, chained to a mere point in space, enduring but for a moment of time, the human mind is not only enabled to number worlds beyond the unassisted ken of mortal eye, but to trace the events of indefinite ages before the creation of our race, and is not even withheld from penetrating into the dark secrets of the ocean, or the interior of the solid globe; free, like the spirit which [Virgil] described as animating the universe: *[Deum nanque] ire per omnes / Terrasque tractusque maris, coelumque profundum.*"
– Charles Lyell, from *Principles of Geology*

Chelymorpha cassidea – Argus Tortoise Beetle
Ichneumonoidea – Ichneumonid And Braconid Wasps
Plathemis lydia – Common Whitetail

May 24   63°

SNAIL-KILLING FLY   46°

To St Olaf Natural Lands with Dan Tallman in search of a Southern Spreadwing. The day was a little too cool and too cloudy for the damselflies to be active, so, unfortunately, we couldn't locate the spreadwing Dan hoped to find. The one dragonfly we found, a Common Green Darner hanging in a thick stand of grass, was too cold to even whirr its wings in order to warm up and fly away.

One oddity we did happen upon was a Snail-killing Fly standing on top of a heap of dead slug at the tip of a leaf. Talk about living in squalor, the life cycle of some of these flies begins with the adult fly killing the host snail or slug, and the subsequent young larvae living in and feeding on the decomposing flesh. While these flies are very common around ponds, this was the first time I'd seen one with its host.

❖ ❖ ❖

I spent much of the day working on some rough translations of a Russian dragonfly book recently acquired, *Dragonflies of Siberia* by B. F. Belyshev (1974). Ever since reading Dostoyevsky's *The Brothers Karamazov* and Tolstoy's *Resurrection* in college, I've been an avid admirer of Russian novelists and poets. This led to years of exploration among Russian composers—Sergei Taneyev, Nikolai Myaskovsky, Dmitri Shostakovich, Mieczysław Weinberg, Galina Ustvolskaya, and others. More recently I've enjoyed the tradition of Russian chess, admiring and replaying games by masters such as Mikhail Chigorin, Alexander Alekhine, David Bronstein, Anatoly Karpov, and Garry Kasparov. Now, expanding my research on the dragonflies of the genus *Sympetrum*, I've discovered the work of Russian odonatologists. Unlike Russian literature and chess books, a scant amount of the odonatological work has been translated into English, so I find myself nudged to essay the language, even if in a limited way.

Poanes hobomok – Hobomok Skipper
Tragopogon dubius – Yellow Salsify
Ichneumonidae – Ichneumonid Wasps
Chaetopsis sp. – Banded-wing Flies
Tetanocerini – Marsh Flies
Polydrusus formosus – Green Immigrant Leaf Weevil
Anax junius – Common Green Darner

May 25

A LITTLE WANDERLUST

71°

50°

Yesterday evening, the idea for a road trip suddenly overtook me. There exist two species in the dragonfly genus *Williamsoni*, the Ebony Boghaunter and the Ringed Boghaunter, both of which have been observed this month at a county park south of Eau Claire, Wisconsin, a mere two and a half hours from Northfield. Having never seen either, why not spend a day driving and try to find them? Curiously, within an hour of hatching this plan, I received an invitation via text to join members of the Minnesota Dragonfly Society on a survey next week of some bogs in northern Minnesota in search of these very species. Problem solved!

But then I set to work on a couple of loose ends in the text I began translating yesterday, simply checking a couple place names. On July 22, 1960, B. Belyshev happened upon a mass emergence of Yellow-winged Darters (*Sympetrum flaveolum*) at a dry temporary pond, so dry in fact that digging a hole 1.5 meters deep he failed to find water. How had the nymphs survived these conditions? After searching the grass where the exuvia and emerging dragonflies were found and after searching through the dried silt in the pond bottom, no nymphs could be found. "Finally, attention was drawn to lumps of almost completely dead moss scattered across the bottom of the dried up pond. In these damp (due to the absorption of atmospheric moisture) but not wet, mossy clusters, nymphs of the Yellow-winged Darter were found. They were in excellent condition: mobile, energetic and able to run quickly over the dry ground, which is so very unlike their slow crawling along the bottom in an aquatic environment."

Early in this account, Belyshev gives the location as the Tunka Basin, a large geological area west of the southern end of Lake Baikal. But later he adds more specific detail about the habitat, that the ponds are located in "an old channel of the Irkut River near the town of Tibelti." Amazingly, using Google Maps, one can follow the Irkut River through the Tunka Basin to the town of Tibelti and even see the abandoned ox-bow adjacent to the town containing several ponds.

After all this virtual travel, a little wanderlust resurfaced and pulsed once more in my veins. I headed out to look for dragonflies. Johnny Cash's song 'Drive On' on queue in the cd player. The first bluet damselflies of the year were found at Circle Lake where teneral Tule Bluets perched in the lakeside grass.

Driving away from Circle Lake, I noticed a Common Nighthawk flying back-and-forth above the road. So I stopped and tried my best to get an in-flight photograph of this living boomerang of a bird.

A little further on down the road, at Boyd Sartell WMA, mature male bluets cruised the surface of the small ponds, giving me my first look at the "blue" of the bluet damselflies this year. These were Boreal Bluets. Two other odonatological firsts for the year: a Twelve-spotted Skimmer and an Eastern Pondhawk.

Enallagma boreale – Boreal Bluet
Leucorrhinia intacta – Dot-tailed Whiteface
Erythemis simplicicollis – Eastern Pondhawk
Nehalennia irene – Sedge Sprite
Enallagma boreale – Boreal Bluet
Libellula pulchella – Twelve-spotted Skimmer
Anax junius – Common Green Darner
Libellula quadrimaculata – Four-spotted Skimmer
Ischnura verticalis – Eastern Forktail
Enallagma carunculatum – Tule Bluet
Chordeiles minor – Common Nighthawk
Sanicula odorata – Clustered Black Snakeroot
Geranium maculatum – Wild Geranium
Erigeron philadelphicus – Philadelphia Fleabane
Maianthemum stellatum – Star-flowered Lily-of-the-valley
Salix ×fragilis – Crack Willow
Sphaerophoria philanthus – Black-footed Globetail
Augochlorini – Metallic Sweat Bees
Lejops lineatus – Long-nosed Swamp Fly
Hypselistes florens – Dwarf Spider
Hyla sp. – Holarctic Tree Frogs
Anaxyrus americanus – American Toad
Eristalis transversa – Transverse Flower Fly
Andrena sp. – Mining Bee
Dermacentor variabilis – American Dog Tick
Andrena vicina – Neighborly Mining Bee
Eurosta solidaginis – Goldenrod Gall Fly

80°
—
61°

May 26

RIVER BEND

To River Bend Nature Center with Dan Tallman. Our destination at River Bend is a large, floodplain pond toward the north end of the nature center. This pond, which I've visited other years, supports a source population of Horned Clubtails, one of seven stillwater gomphids of the genus *Arigomphus* in North America and the only one with a distinctly Northern Great Plains distribution. A. D. Whedon published the first account of this species in Minnesota, describing its emergence at a stagnant pond in 1913 near the Minnesota River in Mankato, approximately fifty miles from the pond at River Bend. Almost always the first clubtail to emerge or be observed in Rice County, I look forward to its arrival each year.

These are large dragonflies and its likely that the nymphs require several years to reach full size, though admittedly very little is known about them. Emergence takes place at the water's edge or on suitable floating platforms near shore; this species seems particularly adept at emerging from the surface of floating algae mats. The adult females sport a tall, bi-lobed, bright-yellow occipital plate that distinguishes it from other clubtails, while the horns for which it is named are much more difficult to observe, especially in the field. The male looks quite different, being more black than yellow and with a thinner abdomen, and is a fleet and furtive creature, very difficult to observe.

One of the great pleasures of being in the field with Dan is his ear for birds; he hears all the instruments in the symphony. On the hike to the pond and back, Dan singles out the song of a Red-eyed Vireo, its tireless call and response, a Redstart, a Yellow Warbler, the trill of a Tree Sparrow, and the defiant tree-top blast of a Great Crested Flycatcher. When I ask what the mnemonic for the rather harsh call of the flycatcher might be, he smiles and answers, "Squawk!"

Approaching the pond we find the year's first Horned Clubtail. I suspected we'd see many more at the pond, but we only saw one more and that one from afar as it emerged on a float of algae some distance from shore. Probably there were others but their presence was simply masked by the mass emergence of Dot-tailed Whitefaces. Every step we took around the entire circumference of the pond flushed into flight dozens of newly emerged dragonflies. Several male Twelve-spotted Skimmers performed their showy (if not pugnacious) flight displays over the water, battling with each passing Common Green Darner. A Belted Kingfisher persisted in doing head-first dives into the shallow water, somehow avoiding serious neck injury or becoming stuck in the bottom. From a distance, the pond appeared to be covered by a white haze, closer this haze dissipated into thousands of white flowers of the Water Crowfoot.

Armoracia rusticana – Horseradish
Phyciodes cocyta – Northern Crescent
Coleomegilla maculata – Spotted Lady Beetle
Stellaria aquatica – Water Chickweed
Labidomera clivicollis – Milkweed Leaf Beetle
Ellisia nyctelea – Aunt Lucy
Ranunculus aquatilis – Common Water-crowfoot
Chrysemys picta – Painted Turtle
Dolomedes triton – Six-spotted Fishing Spider
Pyrausta orphisalis – Orange Mint Moth
Symphyta – Sawflies, Horntails, And Wood Wasps
Belostoma sp.– Water Bug
Arigomphus cornutus – Horned Clubtail

72°
55°

May 27
MOTHING: DOR-BUGS AND NIGHT BUTTERFLIES

The new moth light attracts less than the old, but that's OK by me. I captured and kept a half dozen different kinds for photographing. Moths can be a real challenge to identify and having six to work through makes for a reasonable task. Luckily I recognize all but one, so the task of labeling photos is made that much easier.

Large, bumbling scarab beetles were the big attraction at the light last evening. Thoreau referred to them as "dor-bugs." I was taught to call them Junebugs. Now, I know these names are a catch-all for a variety of species and that some large, fossorial wasps seek out their subterranean white grubs. But I can't shake the sound of them bouncing off the window screens late at night.

The moths collected this May evening included a Waterlily Leafcutter Moth, an Isabella Tiger Moth, a Common Gluphisia, a Little White Lichen Moth, an American Angle Shades, and a Moonseed Moth. A respectable sampling of moth variety. The Waterlily Leafcutter is a from the Crambidae family and its caterpillars are aquatic. The Isabella Tiger Moth is the adult of the familiar Woolly Bear caterpillar. The Moonseed Moth is an uncommon and uncommonly colored moth the caterpillar of which feeds on its namesake plant, Moonseed. The Little White Lichen Moth and the Moonseed Moth are new species for this site.

Plusiodonta compressipalpis – Moonseed Moth	*Melolonthini* – June Beetles
Euplexia benesimilis – American Angle Shades	
Clemensia albata – Little White Lichen Moth	
Gluphisia septentrionis – Common Gluphisia Moth	
Pyrrharctia isabella – Isabella Tiger Moth	
Elophila obliteralis – Waterlily Leafcutter Moth	

May 28   73°

CRITCHICROTCHES   56°

Reading Thoreau's journal entry for this date for the year 1854, I was curious about the subject of the following sentence: "Critchicrotches have been edible some time in some places." How do you even pronounce that word? Elsewhere in his journals, he makes reference to critchicrotches in conjunction with "Sweet Flag." This I found more confusing, since the only plant I know by this name is the local wild iris. It turns out Thoreau is referring to the fruit of *Acorus calamus*. This wetland plant goes by dozens of common names (hence my initial confusion): sweet flag, calamus, beewort, bitter pepper root, gladdon, myrtle grass, myrtle sedge, pine root, rat root, sea sedge, sweet cane, sweet cinnamon, sweet grass, sweet myrtle, sweet root, sweet rush, and sweet sedge. A good example of the usefulness of scientific names.

I continue to be amazed at how often I can observe the same plant or animal on the exact same date as Thoreau observes it. Today's intersection was a dragonfly. Thoreau writes "I see exuviae or cases of insects on the stems of water plants above the surface. The large devil's-needles are revealed by the reflection in the water, when I cannot see them in the air, and at first mistake them for swallows." This is written in such a way that it seems Thoreau may not recognize the exuvia as belonging to the dragonflies he's observing. Given the time of year and the multitude of exuvia, this suggests the mass, springtime emergence of Baskettails and Emeralds. Visiting the Cowling Arboretum today, the first dragonflies I encountered were, you guessed it, Baskettails.

[Observations →]

Orussus minutus – Parasitic Wood Wasp
Epitheca cynosura – Common Baskettail
Laphria sp. – Bee-mimic Robber Flies
Packera paupercula – Balsam Ragwort
Ellisia nyctelea – Aunt Lucy
Osmorhiza longistylis – Aniseroot
Aphaenogaster sp. – Collared Ants
Thaumatomyia sp. – Grass Fly
Callopistromyia strigula – Picture-winged Fly
Ancistrocerus adiabatus – Pathless Mason Wasp
Mallota bautias – Bare-eyed Mimic
Cercopoidea – Spittlebugs
Crabronina – Square-headed Wasps
Erysimum cheiranthoides – Wormseed Wallflower
Atherix sp. – Watersnipe Fly

May 29

MOTHING NIGHT 4

65°
—
54°

Wind in the sails. On a night like this, the wide, white, static backdrop of the garage door has its advantage. A mothing sheet would billow and snap.

Not only do moths come in a variety of colors, but they also come in a variety of shapes and sizes. Some guidebooks contain a page of moth silhouettes as an aid to navigating among the many families. Some moths have a distinctive posture, holding their wings in a peculiar manner or perching in odd, angular ways. The diminutive Leaf Blotch Miner Moths (Gracillariidae) that rest propped on fully-extended forelegs. The Plume Moths (Pterophoridae) that resemble crane flies. Clearly, however, one of the most unusual moths to visit the mothing light has to be the Dark-spotted Palthis. The creased and curled wings give it a kind of paper airplane look. But the male's elongated and upturned labial palps give it the appearance of having a preposterous snout.

Canarsia ulmiarrosorella – Elm Leaftier Moth
Proteoteras aesculana – Maple Twig Borer Moth
Palthis angulalis – Dark-spotted Palthis Moth
Lascoria ambigualis – Ambiguous Moth
Lacinipolia renigera – Bristly Cutworm Moth
Palpita sp.– Palpita Moth
Doryctinae – Braconid Wasp

60°
—
53°

May 30

AMMOPHILA IN MAY

To St Olaf. Cloudy. Temperature in the 50s. I walked the usual loop: to the wooded pond by way of the catchment pond and the trail through the trees, past the pond then across Highway 23 and back home by sidewalk (more often, I return through Lashbrook Park, especially if the weather is a little better).

Not much going on because of the weather; I noticed the Goatsbeard flowers were shut, long and scrunched together like closed umbrellas. Found a few spiders on webs in the tall grass at the shore of the pond, —a couple of jumpers and a couple of small orbweavers. On the sandy trail on the hill just north of the pond, I saw the first Ammophila wasp of the year. I don't know which species (does anyone?).

These cutworm-hunting wasps—black-bodied, with thin, wire-like waists, and usually a spot or section of red coloration at the beginning of the abdomen—are fascinating to watch. The females excavate a nest in the sand, with different species excavating in slightly different ways. There are over sixty species in North America, few of which can be identified by photos alone—so sayeth BugGuide. The behaviors of these wasps were studied and written about by both J. Henri Fabre and Niko Tinbergen.

> *Ammophila sp.*– Thread-waisted Wasp
> *Trichopoda pennipes* – Swift Feather-legged Fly
> *Salticidae* – Jumping Spiders
> *Tetragnathidae* – Long-jawed Orbweavers
> *Araneae* – Spiders

May 31
ST LOUIS COUNTY DAY ONE

The contentious issue of copper mining in northern Minnesota focuses, for the most part, around the possible impact this mining might have on the Boundary Waters Canoe Area. Poly-Met, located south of the Laurentian Divide near the town of Hoyt Lakes, claims its operations will not have an impact on the protected areas of the BWCA which are located north of the divide. What needs to be addressed, however, is the possible impact on the Upper Saint Louis River Watershed, where Poly-Met is located, which is an important natural area in its own right.

I left Northfield for Hoyt Lakes at 9 A.M. to take part in some Minnesota Dragonfly Society survey work. Nearly 250 miles due north, the drive took four and a half hours. Ami Thompson, Curt Oien, Mitch Haag and his son Jason, the remainder of the dragonfly survey crew, had arrived before me and were already at work in the Partridge River east of town, searching for dragonfly nymphs. Under the tutelage of Bob DuBois and Ken Tennessen, expert odonatologists from Wisconsin, Mitch and Curt have become the leading experts on dragonfly nymphs in Minnesota. Ami is also well-known in the dragonfly circles for having created and published dragonfly curriculum and for her current Ph.D. thesis research on the Common Green Darner. Jason's claim to fame is his netting ability; young and athletic, his netting is dead-eye. And me? I was along for the ride, mostly, but could lend some general knowledge and help document the days in the field with camera and pen.

This particular Minnesota Dragonfly Society survey centered on the upper Saint Louis River and targeted the Extra-striped Snaketail (*Ophiogomphus anomalus*), possibly the state's rarest dragonfly. Its presence in the state is known from a few exuviae found along this same stretch of the upper Saint Louis River several decades ago. No adults have ever been photographed or captured in the state. A second dragonfly, almost equally as rare, the Ebony Boghaunter (*Williamsonia fletcheri*) was being targeted as well. This mysterious dragonfly breeds in bogs and peatlands and flies very early in the year. Known from less than a dozen, widely-scattered observations in Minnesota, it's thought to be more rarely observed than actually rare.

When I arrived at the Partridge River, I found the others on the north shore of the river, upstream of the highway bridge where I'd parked. They all had waders on and were working in the water. Because of heavy rains the previous week the river was running very high and the nymphing work was confined to the edges. They had swooped up a fair number and diversity of nymphs by the time I had arrived but were having difficulty, because of the high water, reaching the deeper, sandier habitats where the targeted Snaketail might be located. While they continued to search the Partridge River for nymphs using their aquatic nets, I had a look around the river banks and the nearby remnants of a CCC work camp, aerial net in hand, hoping to locate some adult dragonflies. The only dragonfly I encountered was a Common Green Darner. A number of Tri-colored Bumble Bees visited currant flowers in the work camp clearing. Near the road, at the forest edge, I happened upon Sessile Bellwort, a wildflower I'd never seen before, as well as Nodding Trillium.

We left the river and drove to Site 15, an open bog. The small paved road we were on ended abruptly at a railway crossing, at a blip on the map given the name Allen. Gravel drives to several private residences started here. We crossed the tracks and turned onto a trail that followed the tracks on their north side. The railroad siding, on which we drove and parked, was interesting and a little treacherous to walk on due to a thick layer of spilled taconite pellets. Countless boxcars filled with these iron-ore-rich pellets had left the Mesabi Iron Range for the harbor at Duluth on Lake Superior along these tracks. From the railroad tracks, it was a long hike out to the open water. Nothing about bogs is fast. Early in our trek, we encountered a teneral Delicate Emerald (*Somatochlora franklini*). Parts of the open bog were sparsely wooded with stunted Black Spruce and Tamarack. Underfoot, Bog Laurel was in bloom. Despite its beautiful vivid pink flowers, this plant is quite poisonous, even honey produced by bees that visit the flowers is poisonous. Here we hoped to find Ebony Boghaunters. Instead, we found an abundance of Hudsonian Whitefaces, another early-emerging, bog-o-phile dragonfly, though much more common. These dragonflies had been out for a while, fully mature with many breeding pairs. There was a chance we had already missed the flight of the Boghaunters. A single Boreal Bluet. Several Henry's Elfins.

Next we visited the St Louis River south of Aurora. An incredible variety and number of dragonfly nymphs were swooped up—River Cruisers, Dragonhunters, Spring Darners, Fawn Darners, Shadow Darners, Pronghorn Clubtails, Mustached Clubtails. No adult dragonflies were seen. Interestingly, a lot of Phantom Crane Flies were present along the riverbank.

After this, we did a little scouting along the river further to the west. At the next bridge to cross the river, we found a Four-spotted Skimmer, Taiga Bluets, and Eastern Forktails. In the river, Ocelated Emerald, Elusive Clubtail, Spring Darner, Pronghorn Clubtail, and Dragonhunter nymphs were found.

The only place serving food this evening in Hoyt Lakes we discovered was the golf course, located near where we began the day on the Partridge River. So we ate there. And started to make plans for the next day.

Macromiidae – Cruisers
Eristalis sp. – Drone Fly
Trillium cernuum – Nodding Trillium
Bombus ternarius – Tricolored Bumble Bee
Anemone quinquefolia – Wood Anemone
Erynnis sp. – Duskywings
Uvularia sessilifolia – Sessile Bellwort
Cicindela limbalis – Common Claybank Tiger Beetle
Vaccinium – Blueberries, Cranberries, And Allies
Callophrys augustinus – Brown Elfin
Larix laricina – Tamarack
Leucorrhinia hudsonica – Hudsonian Whiteface
Drosera rotundifolia – Round-leaved Sundew
Sarracenia purpurea – Purple Pitcher Plant
Erynnis brizo – Sleepy Duskywing
Kalmia polifolia – Swamp Laurel
Enallagma sp. – Bluets
Somatochlora franklini – Delicate Emerald
Faxonius virilis – Virile Crayfish
Enallagma annexum – Northern Bluet
Carabus nemoralis – Wood Ground-beetle
Hagenius brevistylus – Dragonhunter

June

Dragon Hunters – June 2

June 1 — 77°/37°
ST LOUIS COUNTY DAY TWO

Curt was up early, returning to the room about the time everyone else was waking up. He had obtained information about a good place to see dragonflies from someone who worked at the hotel.

    Mitch: "Was it the big lady?"
    Curt: "No, it was the lady who got up at 7."

No matter how skeptical such a suggestion might be viewed, if a local resident who's not expert at differentiating different species is noticing dragonflies, the location is probably a good site, with good habitat and good diversity. Almost always these tips are worth exploring.

The decision was made to spend the day on the Saint Louis River west of Aurora, canoeing a stretch of river and continuing their search for Extra-striped Snaketail nymphs. I opted out, feeling not quite up to a day of nymphing, thinking perhaps I could be of more assistance if I surveyed one of the other targeted locations and looked for the other species of interest, the Ebony Boghaunter. However, before heading off on my own, I helped shuttle a vehicle from the canoe landing to their exit point a few miles downstream. At the landing, no more than a precipitous drop from road to river, I took a few photos: Eastern Forktails, a fishing spider, and wild rose spike galls.

I drove to Bird Lake, about five miles east of Hoyt Lakes. Because it was still early in the day, I hiked a short distance along a sunlit trail on the east side of the lake, just to see what insects were flying. This was a promising start. Teneral bluets. Hudsonian Whitefaces. Lots of interesting syrphid flies and flowers. Red-tailed Forest Fly on Alder catkins. Elfin butterflies. Leatherleaf. Labrador Tea. Bog Laurel. Mason bees. Arctic Skippers. Obviously, it was warm enough to start the trek across the open stretch of bog separating Bird Lake from Lillian Lake to the south.

While I was putting on my waders, someone drove into the otherwise empty parking lot, parked his truck, and walked over to talk to me. He lived not too far away, in Embarrass, Minnesota. He was curious as to why I needed waders and asked what I was looking for. My answer, that I was going across the bog to Lillian Lake to look for dragonflies, received the usual you-have-to-be-kidding-me pause. "You know there's a trail out there? You don't have to wade," he finally replied. He brought me over to the map and pointed out the trail. Then he asked about the dragonfly I was searching for, so I described the small, black dragonfly which I'd never seen. "Hope you have good luck," he said with a smile and a laugh.

Very few know the happiness of crossing a quaking bog. The net of submerged roots sinks and rebounds with each step, somewhat like a trampoline except its oscillations are damped by the surrounding water, both above and below. Each crossing is a performance. Each performance is witnessed with stoic aplomb by a sparse audience of Black Spruce and Tamarack. No one else sees. Insular in nature. Thoreau suggested that lakes were the eyes of the landscapes. Bogs represent inward, in-turned, vision and secret thoughts.

At my knees (when they weren't underwater) a miniature world of cranberries and sundew, the occasional Elfin butterfly, and numerous white-faced dragons. Bogbean was in bloom. Cottongrass and a clean variety of sedges.

A Racquet-tailed Emerald appeared momentarily at the edge of Lillian Lake, hovering over the open water. Further along the shore, I noticed styrofoam floats marking the site of bait traps, minnows or leeches, but probably the latter. An indication that there is indeed an easier route to this lake than the one I just took. Wordsworth's lines about the leech trapper come to mind: "To these waters he had come to gather leeches, being old and poor" and "I'll think of the leech gatherer on the lonely moor." More Four-spotted Skimmers.

I traversed a number of transects, crossing back-and-forth across the bog, hoping to encounter the rare Ebony Boghaunter. It was not to be. Hours later, back at the car, I took off the waders and headed out on the Bird Lake hiking trail a second time. A cool-down lap. Except this time I hiked the entire two-mile loop. Silver-bordered Fritillaries and Arctic Skippers made the miles of mosquitoes and deer fly misery all worthwhile. Taiga Bluet at a small woodland wetland along the way. Saw two wildflowers along the way that I'd never seen before: Threeleaf Goldthread and Naked Bishop's Cap.

Met up with the others at the hotel and dined with them at a bar in Aurora. Photographed an Eastern Least Clubtail nymph that the others had collected from the St. Louis River.

[Observations →]

Lysimachia borealis – Starflower
Stylogomphus albistylus – Eastern Least Clubtail
Carterocephalus palaemon – Chequered Skipper
Celastrina ladon – Spring Azure
Callophrys augustinus – Brown Elfin
Bombus ternarius – Tricolored Bumble Bee
Leucorrhinia hudsonica – Hudsonian Whiteface
Ischnura verticalis – Eastern Forktail
Sericomyia militaris – Narrow-banded Pond Fly
Libellula quadrimaculata – Four-spotted Skimmer
Eriophorum vaginatum – Tussock Cottongrass
Eufidonia sp. – Powdered Geometer Moths
Geometrinae – Emerald Moths
Menyanthes trifoliata – Bogbean
Viola sp. – Violets
Pterophoridae – Plume Moths
Carterocephalus palaemon – Chequered Skipper
Osmia sp. – Mason Bees
Synhalonia – Long-horned Bee
Hemaris diffinis – Snowberry Clearwing
Mitella nuda – Naked Bishop's Cap
Coptis trifolia – Threeleaf Goldthread
Actaea rubra – Red Baneberry
Carex gracillima – Graceful Sedge
Boloria selene – Silver-bordered Fritillary
Coenagrion resolutum – Taiga Bluet
Syrphini – Flower Fly
Cornus canadensis – Canadian Bunchberry
Gyromitra esculenta – False Morel
Carex limosa – Mud Sedge
Picea mariana – Black Spruce
Cardamine parviflora – Sand Bittercress
Erynnis brizo – Sleepy Duskywing
Tetrigidae – Pygmy Grasshoppers
Calligrapha alni – Russet Alder Leaf Beetle
Calla palustris – Wild Calla
Nomada sp. – Nomad Bees
Phaneta sp. – Phaneta Moth
Aralia nudicaulis – Wild Sarsaparilla
Leucorrhinia hudsonica – Hudsonian Whiteface
Thomisidae – Crab Spiders
Rubus sp. – Brambles

June 02

ST LOUIS COUNTY DAY THREE

82°
—
37°

The manager, a confessed dragonfly aficionado, spoke with us at breakfast in the hotel. She gave us a few bags of chaga tea that her brother had collected.

All five of us, the entire survey crew, set out to canoe a stretch of the St Louis River east of Hoyt Lakes. Our goal was to make it from the landing to Butterball Lake and explore its boggy margins which looked interesting in the satellite images. About three miles as the crow flies.

A long stretch of rapids began just below the canoe access, so some swooping for nymphs was done before we embarked upriver. Several freshly emerged Ashy Clubtails made their first flights from the river. These were the first adult clubtails we'd seen.

Curt and I took his canoe, an old aluminum craft. Mitch and Jason and Ami boarded a yellow fiberglass canoe. Right at the landing, Mitch asked about a plant growing there. Curiously, it was the plant I'd read about in Thoreau's *Journals*, a plant I'd never seen before, American Sweet-flag.

We set out. The river was wide and slow, more like a slim lake than a river. When we stopped to explore a small creek that joined the river, I made my way through a margin of forest to an ATV trail and followed the trail around to a bridge. Along a sunny stretch of the trail, a skipper butterfly caught my attention and I took some time to photograph it. It wasn't a species I recognized but was able to identify it later as a Pepper and Salt Skipper, a fairly uncommon butterfly.

Paddling across a wide expanse of the river, I noticed a number of large insects flying above the water in straight lines as if crossing. At first, I assumed, because of their size and shape that they must be bumble bees. After catching one, my mistake was revealed, these were Elm Sawflies, impressively large Hymenopterans with bright-orange antennae and fierce, pinching mandibles. Thankfully they don't sting!

For a while, we followed a group of Trumpeter Swans upriver. We came within a mile or so of our destination, but eventually, the decision was made to turn back and save some time to visit a few other sites. Not many dragonflies, but a long and pleasurable paddle, reminiscent of days spent in the Boundary Waters.

I drove back to Northfield in the evening.

Epitheca canis – Beaverpond Baskettail
Cimbex americana – Elm Sawfly
Cicindela limbalis – Common Claybank Tiger Beetle
Amblyscirtes hegon – Pepper and Salt Skipper
Synhalonia – Long-horned Bee
Acorus americanus – Sweetflag
Populus balsamifera – Balsam Poplar
Araneae – Spiders
Phanogomphus lividus – Ashy Clubtail
Geranium bicknellii – Northern Cranesbill
Ephemeroptera – Mayflies
Carabus maeander – Ground Beetle
Myrica gale – Bog Myrtle
Helophilus fasciatus – Narrow-headed Sun Fly
Parhelophilus – Sun Fly
Boloria selene – Silver-bordered Fritillary

June 3

BUPRESTID BLUES

90°
—
68°

Lisa and I took a morning stroll through North Valley Park in Inver Grove Heights while Lida practiced sand volleyball. We saw quite a few frisbee golf players and a fair number of dragonflies—Twelve-spotted Skimmers, Common Whitetails, and a single Horned Clubtail. We also saw a couple of skipper butterflies and a ringlet.

Leaving the park, a shiny, blue-metallic beetle on a leaf caught my eye. This was a buprestid beetle (family Buprestidae), also known as jewel beetles. Its size and shape, e.g. being small and narrow, suggested it belonged to the speciose genus *Agrilus*, a genus that includes something near to two hundred species in North America. The blue coloration further suggested it was the adventive species *Agrilus cyanescens*, introduced to North America in the early 1900s and associated with honeysuckle.

While some few species of buprestid beetles are considered pests, the real buprestid blues began with the introduction of the Emerald Ash Borer (*Agrilus planipennis*). This small, bright-metallic-green beetle was first observed near Detroit, Michigan, in 2002, just fifteen years ago, and has since spread into more than fifteen states. Because it kills ash trees, great efforts have been made to impede its march across the continent and to monitor known populations. One of the interesting programs associated with the Emerald Ash Borer is the biosurveillance program undertaken by the University of Minnesota which watches populations of the buprestid-hunting wasp, *Cerceris fummipennis*, to see what beetles it's capturing. Having seen this blue beetle this morning, I now know it's time to start looking for wasp nests.

A while after our walk in the park, I had a chance to read a few poems from a new collection by the Norwegian poet, Olav Hauge and came across these lines which made me smile:

> There is so much to ponder in this world
> that one life is not enough.
> After you're done with your tasks,
> you can fry up some bacon
> and read Chinese poetry.

My introduction to this poet came many years ago through the translations of Robert Bly, then some years later the translations of Robert Hedin. Being from rural Minnesota and a farm kid, I took an immediate liking to Hauge's spare, rustic, taciturn poetic utterances. Then, only a few years ago, I had the pleasure of meeting and sharing some Russian piano music with Hauge's widow, the artist Bodil Cappelen, during her stay at the Anderson Center. So it's wonderful to have such a substantial newly translated collection of poems and journal entries as *Luminous Spaces* (White Pine Press, 2016; translated by Olav Grinde).

Agrilus cyanescens – Jewel Beetle
Arigomphus cornutus – Horned Clubtail

June 4

PINEAPPLE WEED

90°
—
63°

To St Olaf. Did the season's first wasp surveillance, walking the playing fields in search of *Cerceris fumipennis*. No wasp activity of any sort. A number of ant piles, but no ants. The ground underfoot seemed hard as concrete. Six-spotted Tiger Beetles flew and ran pretty much everywhere I walked—they held dominion.

After checking the playing fields for hunting wasps, I stopped at some nearby wasp trees, looking for parasitoid wasps. At the first, the Giant Ichneumon Wasps (*Megarhyssa macrurus*) were out, at least a half-dozen males and one female, all on a branch well over ten feet up. The next few trees were quiet. I did find one buprestid beetle (*Dicerca sp.*), a large one, peering out from a shadowy shelter beneath a fold of bark.

For such a fine day, very few insects were active. Maybe it was too warm, too early? In the late morning the temperature was already 80 degrees. So, on a blue-sky, summer day I lower my standards and photograph the lowly Pineapple Weed, the rank-and-file species of the Chamomile family, adept of sidewalk cracks.

Dicerca sp. – Jewel Beetle
Dialictus – Metallic Sweat Bees
Matricaria discoidea – Pineapple-weed
Crepis tectorum – Narrow-leaved Hawk's-beard

82°
—
61°

June 5

A GOMPHIDAE QUARTET

To the Cowling Arboretum. Sunny, temperature in the 80s. Yesterday was wasp patrol; today it's clubtail patrol. Eight days had passed since my last visit which turned up zero clubtails. Three days ago in northern Minnesota, clubtails were emerging from the St. Louis River—so I was confident that there must be clubtails along the Cannon River now, this far south.

I decided to walk the edge of the Carleton practice fields along the river before entering the arboretum proper. And as it turned out this fringe of natural land was plenty. Four species of clubtail were encountered in succession in just a couple hundred yards—a Gomphidae quartet.

Midland Clubtail (*Gomphurus fraternus*). This is the most common clubtail along this stretch of the Cannon River and, probably, all the farmland rivers across southern Minnesota (perhaps it's Roundup Ready?). One of the most exciting mysteries of my early dragonfly days was discovering and documenting the presence in Rice County of a rare subspecies of this dragonfly (*Gomphurus fraternus manitobensis*), its coloration more similar to the Plains Clubtail (*Gomphurus externus*). This Midland was of the standard variety. When mature, the adult males will have turquoise-blue eyes.

Cobra Clubtail (*Gomphurus vastus*). Probably the most beautifully proportioned of our clubtails. There's something extremely pleasing about the colors and shape of this dragonfly, the long, slender abdomen and extra-wide club, the bright yellow and velvet black of newly emerged individuals. The Cobra Clubtail is found on rivers statewide. When mature, the adult males have malachite-green eyes.

Rusty Snaketail (*Ophiogomphus rupinsulensis*). Seeing this clubtail, with its lime-green thorax and cream-and-rust-colored abdomen, was a surprise. The past several years I've seen this dragonfly here. Prior to these sightings, however, I'd found them only upstream near the riffles at the Cannon River Wilderness Area and on the Straight River, a tributary to the Cannon. I'd like to believe their resurgence here indicates an improvement in the river, but it may be nothing more than population fluctuation and/or happenstance. Never abundant, it's always a treat to find this dragonfly. While easy enough to identify, its behavior today was very un-gomphid like, perching high up on the tops of dried weed stalks, flying out and returning to the same perch, foraging like a skimmer. I watched and admired it for several minutes before moving on.

Horned Clubtail (*Arigomphus cornutus*). The clubtails in this genus, Arigomphus, breed in still water. The Horned Clubtail thrives in shallow permanent ponds, especially floodplain ponds. Of all the pond clubtails (the common name attributed to this genus) the Horned Clubtail has the most northerly and westerly distribution. The females, when feeding, fly just above the meadows with their abdomens hooked downward like Dragonhunters. The males look quite different than the females and are much more difficult to approach, departing for the treetops at the least disturbance.

Lestes inaequalis – Elegant Spreadwing
Erythemis simplicicollis – Eastern Pondhawk
Calopteryx maculata – Ebony Jewelwing
Argia tibialis – Blue-tipped Dancer
Gomphurus vastus – Cobra Clubtail
Arigomphus cornutus – Horned Clubtail
Ophiogomphus rupinsulensis – Rusty Snaketail
Gomphurus fraternus – Midland Clubtail

82°
—
58°

June 6

EXUVIAE

Today I returned to the Cannon River to look for emerging clubtails and search for exuviae. Unfortunately, a thorough search was out of the question as I had less than a half hour, so I focused on the sandy spit of land upstream of the tributary Spring Creek.

Exuviae are the cast of exoskeletons left behind after dragonflies emerge. Because they are solid, they retain the exact shape of the full-grown nymph, a kind of death mask for that previous stage of life. There is a hole in each exuvia located behind the head and between the wing pads where the adult dragonfly made its escape, literally crawling out of itself. The white threads often seen dangling from this exit hole are the tracheal tubes.

There are a number of good reasons to pay attention to exuviae. Presence of exuviae proves the existence of breeding populations. Daily monitoring and collection of exuviae can provide the data for emergence periods and population estimates. Often the exuviae can be identified to species and in some instances can provide records for very elusive adults.

Poanes hobomok – Hobomok Skipper
Campaea perlata – Pale Beauty
Bombus borealis – Northern Amber Bumble Bee
Bombus impatiens – Common Eastern Bumble Bee
Papilio polyxenes – Black Swallowtail
Leucorrhinia intacta – Dot-tailed Whiteface
Libellula pulchella – Twelve-spotted Skimmer
Enallagma hageni – Hagen's Bluet
Plathemis lydia – Common Whitetail
Gomphurus vastus – Cobra Clubtail
Cicindela repanda – Bronzed Tiger Beetle

June 7

88°

MISSISSIPPI WATERSNAKE

61°

To Inver Grove Heights. While Lida practiced beach volleyball, I explored the nearby Mississippi River, visiting the Rock Island Swing Bridge city park. What's left of the bridge for which the park is named resembles an ocean pier more than a bridge. It no longer spans the entire river but extends nearly to the center and provides an excellent view.

I'd hoped to see more clubtail dragonflies, perhaps even a species or two not found along the Cannon River. However, during the hour at the park, I found only two dragonflies: the first a teneral that flew from the water's edge and was immediately snapped out of the air by a bird; the second was a Cobra Clubtail near the edge of the parking lot.

Unexpectedly, the most interesting encounters were with a snake and a frog. The snake was a young Northern Watersnake swimming in the water at the edge of the river. When I approached for a closer photo the snake dove to the bottom and entangled itself at the base of a weed. The frog, found at a safe distance down the shore from the snake, was a Blanchard's Cricket Frog. The Minnesota Department of Natural Resources lists this species as endangered in Minnesota and mentions only two currently known breeding populations, so this was an exciting find.

Carduus nutans – Musk Thistle
Nerodia sipedon – Northern Water Snake
Acris blanchardi – Blanchard's Cricket Frog
Gomphurus vastus – Cobra Clubtail

85°
—
69°

June 8

SOME QUALITY TIME AT THE WASP TREE

To the Cowling Arboretum. After admiring the Cobra Clubtail tree, where a dozen or so male dragonflies perched on sunlit maple leaves on branches overhanging the river, I walked the trails through floodplain forest to the open prairie and back.

On the way back, I remembered to look at the wasp tree where I'd observed a parasitic wood wasp on May 28th. Noticing several of these same wasps, I waded through the raspberry brambles and took a seat on one of the fallen logs at the base of the wasp tree. The wiry spine-waisted ants went about their labors. A large metallic wood-boring beetle uncamouflaged itself, then ambled off toward the dark side of the tree. The wood wasps, black with ivory markings on their legs, landed on the bare wood of the dead tree and raced about like large, hyperactive ants, stopping and starting, changing directions often. They came and went, one after another, and often several at the same time. I settled in for some quality time at the wasp tree.

After a little while watching these wasps perform at eye-level, I noticed closer to the base of the tree, nearer to my feet, a larger wood wasp with a bright red abdomen—a second species!

While the wood wasps were the anchor for this meditative pause in the middle of my daily hike, the sails of the senses were filled by all the other activity here. Instead of proceeding through the world, I sat and witnessed the procession.

[Observations →]

Lichenophanes bicornis – Horned Powder-post Beetle
Stenoscelis brevis – Snout Beetle
Laphria sp. – Bumble Bee-mimic Robber Fly
Laphria thoracica – Bumble Bee-mimic Robber Fly
Orussus terminalis – Parasitic Wood Wasp
Orussus minutus – Parasitic Wood Wasp
Gomphurus vastus – Cobra Clubtail
Gomphurus fraternus – Midland Clubtail
Libellula luctuosa – Widow Skimmer

"Instead of calling on some scholar,
I paid many a visit to particular trees."
– Thoreau, from 'Walking'

86°
—
63°

June 9

PALE BLUE EYES

"Linger on your pale blue eyes."
– The Velvet Underground

Sometimes nature is cruel; it leads you on then leaves you. Other times it's constant, the love of your life. And, in reality, you have no choice but to enter into these relationships innocently and open-heartedly. Steadfastly alone, perhaps. And, even just once or twice in a lifetime is enough, magnificently a part.

Last year, the occupation of a small stormwater catchment pond by a number of Blue-eyed Darners for much of the summer, the nearest pond to our house, seemed too good to be true. To have such an uncommon dragonfly (uncommon for Minnesota at least) four blocks away was a boon; I stopped by ritually to watch them, each time expecting to find them gone. But they persisted until summer was over, well into autumn.

Most backyard swimming pools are larger than the shallow, silty basin of this catchment pond. And now, only three years after its construction, the dragonfly days are numbered as the open water succumbs to an invasion of cattails. My expectations don't rise above the forktails and whitetails that tolerate such compromised habitats. But as long as it holds water, I'll continue to stop by and have a look. And today, after I pushing through a fringe of slim cottonwoods and willows and parting the shoreline cattails, I found a small world a-whirl with Common Whitetails. And then, a great surprise, hovering, facing off my intrusion, a pair of pale blue eyes—the Blue-eyed Darners were back! Either they have found their way here from a nearby location where they are breeding or they are breeding here.

I continued on my hike, buoyantly. In the woods, a Pileated Woodpecker hatcheted away at a stump just off the trail. After it flew, I inspected the stump. All I could find were a few ants, though possibly the bird had found some larger food items like the larvae of beetles or wasps or flies. At the sand piles beside the baseball field, several wasps were present. *Oxybelus*, a tiny black wasp that hunts even tinier flies. And *Hoplisoides*, a sturdy black and yellow wasp that hunts treehoppers. The first I've seen of either species this year. And the first active sand wasp nest building of the year as well. I scouted the nearby playing fields for *Philanthus*, *Cerceris*, and *Sphecius*, but found none.

Hoplisoides sp. – Treehopper Hunter Wasp
Oxybelus sp. – Fly-stabber Wasp
Dryocopus pileatus – Pileated Woodpecker
Tenthredinidae – Common Sawflies
Ichneumonidae – Ichneumonid Wasps
Rhionaeschna multicolor – Blue-eyed Darner
Plathemis lydia – Common Whitetail
Basiaeschna janata – Springtime Darner
Ladona julia – Chalk-fronted Corporal
Epitheca – Baskettails

June 10

TRAP NEST

Inexplicably hot and windy. Clouds flower debris blew down from our neighbor's white pine trees, forming brown crunchy drifts in our driveway, several inches deep in places. Lisa did some yard work, enjoying the heat, accompanied by several Common Whitetail dragonflies. Lida and I ducked the heat as best we could indoors.

Inspecting the bee block in our front yard garden, I noticed one of the trap nests placed below the block was freshly sealed. The plug at the opening was dark green, a plaster of minced and masticated plant material. This alone indicated the nest belonged to a bee and not a mason wasp (though it's true that some bees use clay to partition and seal their nests as well). The wood grain of the trap nest wasn't as linear as it looked from the outside and I had some difficulty splitting it open. The diameter was 3/8" and the depth was 11.5 cm. Inside, after the opening plug and a small gap, where ten cells in series. With the exception of the deepest cell, the one furthest from the opening which seemed to contain a cocoon and pupated larva, each cell contained a nearly full-grown larva. The cell divisions were constructed of the same material as the entrance plug. Some portion of the pollen stores remained in the outermost cells and the larvae were not full-grown. After photographs, the nest was closed, the two halves put back together, taped to seal the cracks and placed in a container so that the adults, when they emerge in a month or so, can be observed and identified before being released.

❖ ❖ ❖

A second indoor observation. The moth had emerged from the pupa collected last week in northern Minnesota. That pupa had been found in the flowerhead of a Purple Pitcher Plant. A Verbena Bud Moth, a colorful Olethreutine leafroller moth that feeds on a large variety of bog and wetland plants.

Osmia sp. – Mason Bees
Plathemis lydia – Common Whitetail
Endothenia hebesana – Verbena Bud Moth

79°
—
61°

June 11

MASKED BEES

Morning thunderstorms. One of those days when it's brightest early in the morning then progressively darker as the storm looms in from the west. I could read on the porch at 6 A.M., by 8 A.M. I had to turn on lights to see the page. About an hour later the storm hit, fierce straight-line winds followed by heavy rain. And then it was over. All as programmatic as Beethoven's Sixth Symphony, The Pastoral.

With the Goat's Beard newly blossoming and a few stray Ground Elder in bloom as well, I didn't need to set foot outside our backyard to see plenty of insects. For a short while late in the day when the sun nearly came free of the clouds, bees arrived at the flowers in droves, accompanied by a few flies and beetles, even a butterfly or two. My attention fixed upon what appeared to be some very small, mosquito-sized wasps. Black and yellow, slim and sleekit, these wee hymenopterans were (as I eventually remembered) Masked Bees, of the genus Hylaeus.

Certainly, the coloration and the absence of pollen-gathering hairs make the Masked Bees the most wasp-like in appearance of all the bees. And yet they are bees and they need to gather pollen. These bees fill their crops with nectar and pollen and regurgitate the mix into the nest cell. Like other members of Colletidae, their provisions are protected by a cellophane-like lining to their cells. These interesting and diminutive bees nest in the narrow hollows of dead twigs and plant stems.

Myopinae – Thick-headed Fly
Nomada articulata – Nomad Bee
Celastrina ladon – Spring Azure
Toxomerus marginatus – Margined Calligrapher
Trigonarthris minnesotana – Flower Longhorn Beetle

Hylaeus sp. – Masked Bees
Sphecodes sp. – Blood Bees

June 12   83°

PATTERNEYES   66°

Last fall I discovered the Wavy Patterneye (*Orthonevra nitida*), a small flower fly with a dark, squiggly labyrinth traced over its eyes. At that time, I learned that some species in the same genus had much simpler patterns on their eyes. So today, when I saw a fly with a single, dark line across its eye, I suspected it might belong to this genus.

The genus name, *Orthonevra*, probably refers to some "straight" or "correct" feature in the wing venation rather than the straight lines and scribbles that decorate the eyes of these flies. The genus is divided into two groups: 1) the *Orthonevra belli* group with eyes that have both horizontal and vertical markings. 2) the *Orthonevra pictipennis* group with eyes that have a single horizontal line. Using the *Canadian Guide to Syrphidae*, the fly found in the backyard appears to be *Orthonevra pulchella* because it doesn't have clouded wing markings.

It appears that little is known of the life histories of these flies. And why do they have such interesting eyes? The secret lives of these flies remain secret.

❖ ❖ ❖

Observed the first fireflies of the year while walking the dog late, around 11 pm.

Orthoneura pulchella – Beautiful Patterneye
Andrena sp. – Mining Bee
Nomada sp. – Nomad Bees
Myopa sp. – Thick-headed Fly
Zodion sp. – Thick-headed Fly

91°
—
66°

June 13

BUMBLE BEE MIMIC

One of the most enjoyable natural history monographs that I've come across is *The Systematics of Lasiopogon* (Diptera: Asilidae) by Robert A. Cannings, curator of entomology at the Royal British Columbia Museum in Victoria, Canada (2002). While I've yet to encounter a robber fly from this genus, I learned a lot from this book, especially from the section about the biogeography of *Lasiopogon*, which has implications or parallels to other insects that share Holarctic distributions. Here's Cannings' description of the family Asilidae:

"The robber fly family (Diptera: Asilidae) contains more than 6,700 described species worldwide. Robber flies are predators that as adults pursue other insects (usually flying ones), seize them and kill them with paralyzing saliva injected through the hypopharynx (tongue). They then suck up the liquefied contents of the prey through the proboscis. The morphology of the adult fly, especially the prominent eyes, the mouthparts, and the raptorial legs, reflects this mode of prey capture and feeding. Robber flies usually hunt in open, well-lit areas and are most active in the warmest part of the day; overcast skies greatly curtail their activity. Different genera, and often different species within a genus, have different hunting behavior and preferences for perching sites."

For our area, Minnesota and Wisconsin, the most useful resource has to be the Robber Fly guide included on the Wisconsin Butterfly website (www.wisconsinbutterflies.org). The genus *Laphria* includes a number of very convincing bumble bee mimics, including the impressive *Laphria thoracica*, which I had the pleasure to observe today in our backyard. The give away is the way they perch on leaves; bumble bees just don't stand on leaves like this. Of course, another tell is seeing them fly out and then return to their perch with their prey stabbed to their face.

Laphria thoracica – Bumble Bee-mimic Robber Fly
Curculio – Nut And Acorn Weevils
Nysson lateralis – Nysson Wasp

85°

66°

June 14

MORE FLIES

A late morning walk at the Cowling Arboretum began with sunshine but quickly transposed to clouds and shadow. Still, a comfortable outing with plenty of insects. I was surprised to encounter a number of darners feeding above the small meadow. I couldn't get a good look at them, but they may have been Blue-eyed Darners. Lots of Ornate Snipe Flies on the vegetation near the creek. Valentine Ants. The first Stream Bluets. Eastern Forktails. The ubiquitous Common Whitetails. An Eastern Pondhawk. A tiny Big-headed Fly. A spider-wasp-looking sawfly.

In the backyard, the Goat's Beard blossoms continue to attract a spectrum of insects. New today were Carrot Wasps (*Gasteruption sp.*), Streaktails, and Black-nosed Grass Skimmers. The latter is a fly I've not encountered before. Several males were present, exhibiting territorial behavior, perching, hovering, chasing. The larvae of these tiny flies feed on aphids and are especially adept at capturing ant-defended aphids.

Walking the dog at the end of the day, I can't take my eyes off the magnificent clouds to the east, large cumulus clouds billowing up into thunderheads. Bright white summits, the lower slopes and vales capture the colors of the sunset. So close and so mountainous that I suddenly feel as though I'm walking the streets of a foothill town in the Rockies, our dusty midwest streets transposed and transported.

[Observations →]

Paragus haemorrhous – Black-nosed Grass Skimmer
Toxomerus marginatus – Margined Calligrapher
Allograpta obliqua – Oblique Stripetail
Mordellidae – Tumbling Flower Beetles
Gasteruption – Carrot Wasp
Platydracus – Rover Beetle
Pipunculidae – Big-headed Flies
Crematogaster – Acrobat Ants
Macrophya sp. – Common Sawfly
Chrysopilus ornatus – Ornate Snipe Fly
Condylostylus – Long-legged Fly

88° | June 15
67°

MCKNIGHT PRAIRIE

A windy day on the prairie. The sun and the sound, the calm fury of the place, forced my hand to write this small poem.

REMNANT

The prairie
what's left of it
whispers and sighs in opposition
to the surrounding fields and lawns
planned and managed
rectilinear and regimented
where the drift of poisonous
thought applied strictly
threatens taproots and deep water
with runoff and erosion
and lawn mowers
with cash flow.

The prairie
a slim reserve
a resistance alive with beetles and bees
writes a letter to the editor
against simplification
starts a petition for flowers
and antlions and for the time being
an accord with the lancing plows
allows a kind of silence to persist
a solace a sigh
a lingering whisper in the grass
shhh...it's going to be all right.

[Observations →]

Spiza americana – Dickcissel
Epeolus – Variegated Cuckoo Bees
Efferia albibarbis – White-bearded Robber Fly
Philodromus sp. – Running Crab Spider
Polites themistocles – Tawny-edged Skipper
Dianthus armeria – Deptford Pink
Anemone cylindrica – Cylindrical Thimbleweed
Anticlea elegans – Mountain Deathcamas
Astragalus crassicarpus – Ground-plum
Heuchera richardsonii – Prairie Alumroot
Vanessa virginiensis – American Lady
Enallagma exsulans – Stream Bluet
Hetaerina americana – American Rubyspot
Gomphurus vastus – Cobra Clubtail

86°
—
65°

June 16

ANT FLY

To St Olaf. It's now been over seven months since I've seen a meadowhawk. According to the Roman slave-turned-stoic-philosopher Epictetus, "It doesn't matter what the external thing is, the value we place on it subjugates us to another… where our heart is set, there our impediment lies." And as Diogenes the cynic famously said, "It's the privilege of the gods to want nothing, and of godlike men to want little." These two statements come together like the same poles of a magnet, you can feel them repel at absolute contact, but between the two is a powerful truth. My heart is set upon seeing a certain dragonfly, and yet I feel it's not asking a lot.

I try not to let this want impede my awareness of other insects, or diminish the pleasure of a walk without meadowhawks. As it happens, something much more uncommon than a red dragonfly was encountered: an Ant Fly (*Microdon sp.*).

These are beautiful flies. And strange. The larvae live in ant nests where they prey upon the larvae of the ants. Some species are generalists and occur in ant nests of several species, some are host specific. The Ant Fly larvae resemble scale insects, domed and intricately patterned. They move slowly in a slug-like manner. And, in fact, they were first described to science (mistakenly) as a kind of mollusk. The adults are plump, fuzzy, with a metallic luster to their hair. Unlike other flower flies, the Ant Fly neither hovers nor visits flowers. Phylogenetic research revealed that the Ant Flies were the first and early offshoot of the branch that includes all the remaining flower flies.

[Observations →]

Apantesis virguncula – Little Virgin Tiger Moth
Hyles lineata – White-lined Sphinx
Stratiomys obesa – Soldier Fly
Climaciella brunnea – Wasp Mantidfly
Bombus citrinus – Lemon Cuckoo Bumble Bee
Leucospis affinis – Chalcid Wasp
Lepyronia quadrangularis – Diamondback Spittlebug
Lestes rectangularis – Slender Spreadwing
Oxybelus sp. – Fly-stabber Wasp
Junonia coenia – Common Buckeye
Microdon sp. – Ant Fly
Cyanocitta cristata – Blue Jay
Sceliphron caementarium – Black-and-Yellow Mud Dauber Wasp

85° / 63°

June 17

EMERALD SPREADWING

After driving Lida to the airport for her flight to Florida with her volleyball team in the early early morning and before Lisa and I returned to the airport in the afternoon for our own flight, I made a small outing to the catchment ponds and prairie planting below the science building at St Olaf. Purple Coneflowers were in bloom and being visited by bees. And a Sachem Skipper nectared at one. In addition to the one Midland Clubtail found near the pond, I photographed a mature male Emerald Spreadwing (*Lestes dryas*).

This is one of my favorite species, and I look forward to seeing it each year. The mature males are stunningly colored, a bright, metallic green. The stockiest of our spreadwing damselfly species. Interestingly, this species can be found around the globe in the temperate regions, so not only Minnesota, but Alaska, Japan, Siberia, and Europe. Like the Four-spotted Skimmer, the Emerald Spreadwing occurs in Morocco as a glacial relict.

Emerald Spreadwings breed at temporary and seasonal wetlands. The eggs overwinter inside the plant stem where the female had inserted them. The eggs hatch in the spring and the nymphs develop quickly in the shallow snowmelt-filled wetlands. The nymphs are fascinating creatures in their own right, efficient predators that stalk and capture their prey with a sling-shot-like lower jaw. Having reared several of these nymphs, I've had the opportunity to observe them closely.

Lestes dryas – Emerald Spreadwing
Eumelissodes sp. – Long-horned Bee
Gomphurus fraternus – Midland Clubtail
Tetraopes tetrophthalmus – Red Milkweed Beetle
Agapostemon virescens – Bicolored Striped Sweat Bee
Atalopedes campestris – Sachem
Megachile sp. – Leafcutter Bee

June 18

FLORIDA: DAY ONE

90°
—
73°

We arrived at our rental house late last night, a few minutes after midnight, after eight hours of traveling. A flight from Minneapolis to Orlando. Then an hour drive in a rental car in the dark on unfamiliar toll roads to a rental house in a gated community in Loughman. The Sanders family from Farmington, our housemates for the week, arrived shortly after us.

❖ ❖ ❖

In the morning, eager to get a look at our new surroundings, I stepped out the front door and inspected the hedges and landscape plantings in front of the house where the sun was shining. Paper wasps, a couple bumble bees, and a Brown Anole on the trunk of a small tree—the latter was exotic enough to remind I was no longer in Minnesota, that and the palm trees.

A little while later Lisa and I walked to the catchment pond near the house. The weedy area around the catchment pond attracted a lot of wasps and butterflies, especially the small white flowers of the Tropical Mexican Clover that grew there in abundance. Being so far from home, nearly ever insect encountered was new to me. There was even a dragonfly that was new to me, a Little Blue Dragonlet (*Erythrodiplax minuscula*). More new dragonflies were to follow.

Later in the morning, Lisa and I drove to Allen David Broussard Catfish Creek Preserve State Park, about an hour south of where we were staying. We hiked the white sand trails as far as the second wetland pond (maybe a half mile). I warily approached the water of the first pond, hyper-vigilant for snakes or alligators. Dragonflies and damselflies were abundant, an enticement to the water's edge. A large skimmer with orange wings caught my attention and I worked to get close enough for a photo. It was a Purple Skimmer (*Libellula jesseana*), a Florida endemic, and quite abundant at this location. But so were the other dragonflies: a Golden-winged Skimmer was observed; four different species of Pennant dragonflies; and the astonishing, black-bodied, black-winged Variable Dancers. The conditions were very hot and very humid. To me, this was a wonderfully exotic wetland habitat—I could easily spend many days exploring this park. In the end, I felt fortunate to have spent only a few hours here.

On the way out I revisited a stand of Rosy Camphorweed along the shore of the first pond and photographed an impressive variety of wasps. Of the fifty-one observations submitted for the day, twenty-eight were species I'd never seen before.

❖ ❖ ❖

Rain began mid-afternoon. We drove to ESPN Wide World of Sports to purchase our passes to the upcoming volleyball tournament. Watched a little volleyball, some teams from our club whose tournament started before ours. Then drove back to our house, getting our first taste of Disney traffic.

[Observations →]

Serenoa repens – Saw Palmetto
Tillandsia usneoides – Spanish Moss
Antigone canadensis – Sandhill Crane
Cicindela – Common Tiger Beetles
Celithemis bertha – Red-veined Pennant
Celithemis amanda – Amanda's Pennant
Celithemis fasciata – Banded Pennant
Argia fumipennis – Variable Dancer
Tramea carolina – Carolina Saddlebags
Enallagma doubledayi – Atlantic Bluet
Ischnura ramburii – Rambur's Forktail
Libellula auripennis – Golden-winged Skimmer
Libellula jesseana – Purple Skimmer
Erythrodiplax minuscula – Little Blue Dragonlet
Sabatia grandiflora – Largeflower Rose Gentian
Eriocaulon decangulare – Ten-angled Pipewort
Bicyrtes sp. – Stink Bug Hunter Wasp
Campsomeris dorsata – Scoliid Wasp
Ammophila procera – Common Thread-waisted Wasp
Pluchea baccharis – Rosy Camphorweed
Latrodectus mactans – Southern Black Widow
Gasteracantha cancriformis – Spinybacked Orbweaver
Portulaca oleracea – Common Purslane
Richardia brasiliensis – Tropical Mexican Clover
Portulaca amilis – Paraguayan Purslane
Leptotes cassius – Cassius Blue
Tramea onusta – Red Saddlebags
Campsomeris sp. – Scoliid Wasp
Ischnura ramburii – Rambur's Forktail
Erythrodiplax minuscula – Little Blue Dragonlet
Pachydiplax longipennis – Blue Dasher
Perithemis tenera – Eastern Amberwing
Celithemis eponina – Halloween Pennant
Pontia protodice – Checkered White
Spragueia onagrus – Black-dotted Spragueia Moth
Bembix americana – American Sand Wasp
Hylephila phyleus – Fiery Skipper
Bombus impatiens – Common Eastern Bumble Bee

Quercus virginiana – Southern Live Oak
Toxophora – Bee Fly
Myrmeleontidae – Antlions
Sagittaria – Arrowheads
Sphex dorsalis – Sphecid Wasp
Philanthus ventilabris – Beewolf
Polistes bellicosus – Paper Wasp
Eumeninae – Potter Wasps
Prionyx parkeri – Sphecid Wasp
Scolia nobilitata – Scoliid Wasp

89°
—
73°

June 19

FLORIDA: DAY TWO

ESPN Wide World of Sports from 7 A.M. to 2 P.M. Lida's team won all three of their matches. During the first break in the action, Lisa and I walked the grounds out past the baseball fields and soccer fields. At one edge we found a fenced in catchment pond with a sign warning about the presence of alligators and snakes. Just as Lisa was commenting on the sign I said: "There's one!" And Lisa jumped back to the sidewalk thinking I meant alligator. But really I meant snake. A large Black Racer was in the grass right by her feet. A very beautiful, if possibly slightly sassy, snake.

During the second break, we walked to the car in the parking lot to retrieve some water that would soon be too warm to drink. After sitting in the car with the air conditioning on for a while, I talked Lisa into walking past the parking lot to an unfenced catchment pond with the hope of getting closer to the water and seeing some dragonflies. I approached cautiously prepared to be respectful of any large reptiles that might be there and would have the right of way. I was in the process of taking a photo of a Four-spotted Pennant when a security guard intervened and chased me from the water's edge. Probably the most well-protected dragonflies I've encountered.

After this, it rained.

◆ ◆ ◆

Florida is, so far, a difficult place to navigate, at least by road signs, signs which seem noncommittal, after the fact (too little too late) or simply nonexistent or unhelpful. For instance, the main parking lot for the state park visited the day before. We drove past it as there were no signs on or near the road. Luckily I'd looked at a map beforehand and knew it was north of the park's fire tower. So we turned around at the fire tower and drove back. At first impression, one feels it's a system based on intuition or an intuitive sense of destination, not logic.

66°
—
35°

Taxodium distichum – Baldcypress
Brachymesia gravida – Four-spotted Pennant
Erythemis simplicicollis – Eastern Pondhawk
Antigone canadensis – Sandhill Crane
Latrodectus mactans – Southern Black Widow
Mimosa strigillosa – Powder Puff
Coluber constrictor – North American Racer
Zelus longipes – Milkweed Assassin Bug

91°
—
75°

June 20

FLORIDA: DAY THREE

Orange County Convention Center from 7:30 A.M. to 4:00 P.M. We parked in the middle of a huge parking lot under a sign that read "Nature Lot." Inside the Convention Center, we gazed in disbelief at how immense this venue was, well over 100 volleyball courts in a single room. Lida's team lost to a Puerto Rican team, then beat teams from Minnesota and Indiana to move on in the tournament. An exciting day of volleyball. It was at a coffee cart here that I discovered Cuban café con leche, espresso and milk sweetened with specially-spiced cane sugar. During the breaks, Lisa and I stepped outside and inspected the landscaped grounds. Surprising few dragonflies were at the large pond to the north of the Convention Center, a couple of Four-spotted Pennants and Halloween Pennants. When leaving we saw a few Sandhill Cranes wandering the parking lot. Maybe they were wondering where the "nature" was as well?

◆ ◆ ◆

Back at the villa, I took a late afternoon amble, exploring some undeveloped acreage—weedy sand, with a few dune-like blowouts—literally where the sidewalk ends. I encountered a number of interesting flies and wasps, a Florida Scrub Lizard, and the only clubtail of the trip, a Tawny Sanddragon (*Progomphus alachuensis*).

"There are nothing but gifts on this poor, poor earth." – Czeslaw Milosz, 'The Separate Notebooks'

[Observations →]

Antigone canadensis – Sandhill Crane
Ischnura ramburii – Rambur's Forktail
Celithemis bertha – Red-veined Pennant
Progomphus alachuensis – Tawny Sanddragon
Enallagma doubledayi – Atlantic Bluet
Erythrodiplax minuscula – Little Blue Dragonlet
Brachymesia gravida – Four-spotted Pennant
Cerceris tolteca – Weevil Wasp
Larra bicolor – Sand Wasp
Emilia sonchifolia – Lilac Tasselflower
Phoenix sp. – Date Palms
Megaphorus sp. – Robber Fly
Proctacanthus sp. – Robber Fly
Opuntia sp. – Prickly Pears
Pogonomyrmex badius – Florida Harvester Ant
Efferia sp. – Robber Fly
Sceloporus woodi – Florida Scrub Lizard
Hemiargus ceraunus – Ceraunus Blue
Anolis sagrei – Brown Anole
Myrmeleontidae – Antlions
Geron sp. – Bee Fly
Stichopogon trifasciatus – Three-banded Robber Fly
Dasymutilla quadriguttata – Velvet Ant
Strymon melinus – Gray Hairstreak
Myzinum sp. – Thynnid Wasp
Exoprosopa fascipennis – Bee Fly

91°
—
75°

June 21

FLORIDA: DAY FOUR

Orange County Convention Center from 6:30 A.M. to 2:00 P.M. Lida's team lost two and won their third which put them in the finals bracket for a chance at thirteenth place.

Afternoon walk at the villa. Lisa and I walked up to the lake by the clubhouse. Very hot and humid. Photographed a Beachcomber Wasp (Microbembex monodonta) on Lilac Tasselflower (Emilia sonchifolia). Photographed a Spicebush Swallowtail at the shore of the pond. We walked the gravel road outside the villa for a short distance. Photographed a Dainty Sulphur butterfly in the road ditch.

Walking back we passed three resident Sandhill Cranes. We've seen them all week. Today they were foraging in the grass along the sidewalk, making me a little curious about what they might be eating. A mottled—white and grayish blue—Little Blue Heron stood at the edge of the catchment pond. Driving out to the grocery store, we saw two Black Vultures dining on spilled garbage.

Celithemis eponina – Halloween Pennant
Egretta caerulea – Little Blue Heron
Polypodiopsida – Ferns
Nathalis iole – Dainty Sulphur
Pyrginae – Spread-wing Skippers
Erythemis simplicicollis – Eastern Pondhawk
Papilio troilus – Spicebush Swallowtail
Hemiargus ceraunus – Ceraunus Blue
Brachymesia gravida – Four-spotted Pennant
Microbembex monodonta – Beachcomber Wasp

June 22   93°

FLORIDA: DAY FIVE   75°

The final day of volleyball, two matches at 11 and noon. This left a little time in the morning and a little time in the afternoon for a last look at some Florida nature. The morning observations consisted of what could be found right outside the house door and around the outside of the screened-in-pool at the back of the house. An antlion, two tree frogs, several spiders (including a male Southern Black Widow), and an interesting ant. The latter moved along the top of the fence behind the house, and at first, I believed it to be some kind of wingless wasp or velvet ant. Only later did I discover it was a Twig Ant.

Lisa and I spoke to a resident walking two dogs. He commented on the newly installed alligator warning signs around the catchment pond. Having spent some quality time at the water's edge over the last several days, it was difficult not to believe these hadn't been hurriedly installed for my edification. This man told us about the small alligator that had lived in the pond for several years, but that had been removed because it was getting to large and the residents were treating it as a kind of pet.

Bubulcus ibis – Cattle Egret
Anhinga anhinga – Anhinga
Rabidosa rabida – Rabid Wolf Spider
Prunus – Plums, Cherries, And Allies
Argia fumipennis atra – Black Dancer
Perithemis tenera – Eastern Amberwing
Pseudomyrmex gracilis – Graceful Twig Ant
Crambidae – Crambid Snout Moths
Osteopilus septentrionalis – Cuban Treefrog
Trimerotropis maritima – Seaside Grasshopper
Periplaneta australasiae – Australian Cockroach
Agapostemon splendens – Brown-winged Striped Sweat Bee

Syrphini – Flower Fly
Carabidae – Ground Beetles
Pholcidae – Cellar Spiders
Larrini – Sand Wasp
Myrmeleontidae – Antlions
Hyla squirella – Squirrel Tree Frog
Polistes fuscatus – Dark Paper Wasp
Anaxyrus terrestris – Southern Toad

74°
—
58°

June 23

APHID-HUNTING WASPS

A travel day, beginning with a 4 A.M. (ET) wake up and an hour-long, pre-dawn drive to the Orlando airport. By noon, nine hours after waking, we were back in Northfield. Time to unpack, open up the house, retrieve the dog, take a few deep breaths and have a look around the yard before we needed to drive back to the cities to pick up Lida, who was traveling with her volleyball team.

Things were different here. After heat indices above 100 degrees in Florida, the low clouds, low humidity, and temperatures in the mid-sixties, the weather literally provided a welcome breath of fresh air. Looking out the front door, a chipmunk greeted us instead of an anole lizard.

Last week, before we left on our travels, I had placed a handful of empty trap nests in the holder at the bottom of the front yard bee block. Inspecting these today I found one trap nest sealed off and another with a wasp stationed at the entrance, *Passaloecus sp.*, an aphid-hunting wasp. These tiny hunting wasps are recognizable by their white mandibles and by their use of pine resin in partitioning and sealing their nests. The chemical properties of pine resin make it a good choice for nest construction, but one wonders how the wasps manage to work with such sticky material?

Passaloecus sp. – Aphid-hunting Wasp
Tamias striatus – Eastern Chipmunk

June 24   66°

APERÇU   54°

In the early afternoon, I walked to the St Olaf Natural Lands after a week's absence. I was greeted, at the first black raspberry bushes by the first Meadowhawk of the year as well as the first ripe fruit of the year. The small, tawny dragonfly, a week or so away from its mature red color, flew high into the tree canopy when I moved closer. It's hard to explain to myself and to others my attachment to these dragonflies, but I was deeply happy to see one, the first in slightly more than seven months. I'd see two more on this walk.

Johann Wolfgang von Goethe used the word Aperçu for observations of nature or other phenomena which gave the observer a sudden insight to the hidden and complex workings of the world. "All at once one sees things anew, looks at the world with new eyes and experiences a turning point in one's life." (from *Goethe* by Rüdiger Safranski)

Something like this still happens when I observe Meadowhawk dragonflies. Having studied them across many years, the familiarity and feelings I experience come from an accumulation of many previous aperçus. I'm familiar with their appearance and their behaviors. I'm grateful for their existence.

Laphria – Bee-mimic Robber Flies   *Bombyliidae* – Bee Flies
Somula decora – Spotted Wood Fly   *Tramea onusta* – Red Saddlebags
Enallagma hageni – Hagen's Bluet   *Pachydiplax longipennis* – Blue Dasher
Lestes rectangularis – Slender Spreadwing   *Euphoria fulgida* – Emerald Euphoria
Eristalis transversa – Transverse Flower Fly   *Zelus sp*, – Assassin Bug
Sympetrum obtrusum – White-faced Meadowhawk
Cycloneda munda – Polished Lady Beetle

69°
—
54°

June 25

HACKBERRY GALLS

Cool and cloudy in the morning. We celebrated Father's Day in Apple Valley with Lisa's family. My brother-in-law, my father-in-law, and I played a round of golf at a local course. If you are wondering how well I did, don't. My golf game is cobbled together from loose, untrained swings, swings that give me plenty of opportunities to check the sandtraps for wasps, the water features for dragonflies—my game is a reach, not a reality.

❖ ❖ ❖

Back in Northfield, late in the day, I moseyed about the backyard, searching for the day's observation. An arching frond of leaves caught my eye. This was a seedling Hackberry tree. Examining the leaves, I noticed a cluster of galls different in shape from the more conical Hackberry Nipple Galls that are very common on the neighborhood trees.

Hackberry trees host a variety of gall-making insects. These particular galls, tear-drop-shaped, were most likely Hackberry Acorn Galls, caused by the gall midge *Celticecis celtiphylla*.

Celticecis celtiphyllia – Hackberry Acorn Gall Midge

June 26

EATING THE STING

73°
—
50°

One of my favorite books by a Minnesota poet when I was first beginning to read poetry was *Eating the Sting* by John Caddy. The title poem is built around the observation of a White-footed Deermouse eating a hornet and the possibilities involved in the transformation of poison and pain into something wide-eyed and energetic. I think of this poem whenever I see dragonflies eating wasps and bees.

For instance, today, a Midland Clubtail flew off toward a blossoming patch of Canadian Thistle and returned to a perch with a Honey Bee in its mouth. I hurried to get a photo, but before I could get a clear angle the dragonfly suddenly shuddered, whirring its wings and dropping the bee. I think it may have been stung.

One wonders how often dragonflies do get stung and if the sting poses any real threat to them. I've watched many meadowhawks dine upon small bees and wasps, the dragonflies showing no adverse effects whatsoever. And the large Arrowhead Spiketails routinely dine upon large paper wasps and hornets. Maybe it's like the old adage says: "What doesn't kill you makes you stronger."

Many dragonflies, especially clubtails, take on the coloration of wasps, which brings to mind yet another old saw: "You are what you eat!"

Argia apicalis – Blue-fronted Dancer
Libellula luctuosa – Widow Skimmer
Gomphurus fraternus – Midland Clubtail
Lysimachia nummularia – Creeping Jenny
Ophiogomphus rupinsulensis – Rusty Snaketail
Erythemis simplicicollis – Eastern Pondhawk

Potentilla – Cinquefoils
Pompilidae – Spider Wasps
Enallagma exsulans – Stream Bluet

77°
55°

June 27

BLACK AND GOLD

To St Olaf Natural Lands, then back via Lashbrook Park. White Indigo in full bloom spires above the prairie. The large, boldly-patterned Black-and-Gold Bumble Bee worked the flowers. I've encountered this pairing before, flower and bee. A welcome sight.

Together they become a small icon, a badge against the sky, a living assurance that despite the heavy-handed, over-management of this small patch of prairie, a little wild nature continues.

Burnt completely (with no un-burnt refugia) last spring, then mown in the fall (destroying overwintering thatch), when the prairie grows back both host plant and insect have survived, a testament to the hardiness of the White Indigo and the bumble bee, not the care of the human overseers.

Ectemnius – Square-headed Wasp
Parhelophilus sp. – Bog Fly
Dalea purpurea – Purple Prairie Clover
Bombus auricomus – Black-and-Gold Bumble Bee
Satyrium edwardsii – Edwards' Hairstreak
Machimus sp. – Robber Fly
Laphria flavicollis – Robber Fly
Phryma leptostachya – American Lopseed

June 28

A BLACKER BLUE

83°
—
61°

After morning thunderstorms, I walked around the big pond at the St Olaf Natural Lands during a break in the weather. Scads of toadlets scurried about the gravel trails. Blue-eyed Darners patrolled where the open prairie abutted the woods, flying low and slow over a recently mowed patch of prairie. Several Blue Mud Daubers sunned themselves on the wooden roof of a bee block, perhaps nesting in the hollow of the metal post supporting the block.

Most people notice Blue Mud Daubers (*Chalybion californicum*) when they're around, because they're large and active, flicking iridescent blue wings, hunting for spiders. They are very similar in appearance to Steel-blue Cricket Hunters (*Chlorion aerarium*) which hunt crickets. If you're not fortunate enough to witness the wasp with its preferred prey, the two species do differ somewhat in color. *Chlorion*, the cricket hunter, is greenish blue with a much more polished look. The body of *Chalybion* looks much hairier, and is more steely-blue, a blacker blue.

Chalybion californicum – Common Blue Mud Dauber Wasp
Labidomera clivicollis – Milkweed Leaf Beetle
Anaxyrus americanus – American Toad

June 29

THE BEEWOLVES ARE BACK

Being late in the day and only intermittently sunny, I didn't expect to find any wasps but thought I could at least check the ball field for any active nests. Right away I noticed several new nests, the entrance holes with the characteristic alluvial debris field kicked to one side. Prior to this, when I visited this site, there had only been a scattering of ant hills. A man walking two dogs waved to me and I waved back. I wondered how long he had watched me as I made my slow circuit of the bases. Perhaps he thought I had lost something?

The entrances to the first several nests were plugged; the wasp had sealed herself in for the night. However, at the next several nests, wasps were still actively digging. I waited until one of the wasps entered her nest, then hurried closer so that I could get a good look at the wasp when it came back out. And when it did emerge and resume kicking away the sand and small pebbles, I could see that it was a Corrugated Beewolf (*Philanthus gibbosus*), possibly the most common species in Minnesota, and about the smallest.

This wasp had waited underground—nearly a full year—and was now actively creating new subterranean tunnels and cells. During the next few days and weeks, she will begin hunting the small bees she uses to provision her nest.

Philanthus gibbosus – Corrugated Beewolf
Heriades – Resin Bees
Isodontia mexicana – Grass-carrying Wasp
Libellula pulchella – Twelve-spotted Skimmer

June 30   80°/62°

BIOSURVEILLANCE: WASP WATCHING 101

Wasp Watchers is a citizen scientist biosurveillance program hosted by the University of Minnesota. The objective is to monitor the nesting sites of the buprestid-hunting wasp, *Cerceris fumipennis*, for the invasive Emerald Ash Borer. Yesterday I received from the director of the program a list of ball fields in and around Northfield to survey. One of the sites is just a few blocks from our house, so today, late in the morning, Lisa and I walked the dog in that direction.

The playing field at Greenvale Elementary was quite weedy and unmown. I walked the perimeter looking for nests and found none. Turning my attention to the rest of the field I noticed several black wasps flying near the center of the field, *Cerceris fumipennis*.

I returned to the field in the afternoon with my net and camera and spent an hour, from 2 P.M. to 3 P.M., watching the wasps. During this time I found two nests. There were probably more nests because they were very difficult to locate among the weeds growing on the field. The two nests that I found did not have any beetles abandoned near the entrance. And I wasn't able to collect any beetles from wasps returning to their nests (I didn't see any return to their nests with prey either).

Rhionaeschna multicolor – Blue-eyed Darner
Aeshna sp. – Mosaic Darners
Cerceris fumipennis – Smokey-winged Beetle Bandit
Typocerus sp. – Flower Longhorn Beetle
Nemotelus kansensis – Soldier Fly
Philanthus gibbosus – Corrugated Beewolf
Polistes dominula – European Paper Wasp

July

The Grotto – Zion National Park – July 14

July 1

QUEEN ANT KIDNAPPER

77°
—
61°

Mid afternoon. Cowling Arboretum. Mostly cloudy at the beginning of the hike as I set out to look for Queen Ant Kidnapper wasps.

Because it was the weekend, quite a number of people fished the banks of the Cannon River. At one location an extended Hmong family fished—moms and dads, grandmothers and grandfathers, infants in car seat carriers, music, and coolers of food—obviously there for the duration. I said hello to several of the family members and kept walking, headed to the sunny upland trails to search for wasps.

On the gravel trail along the edge of the prairie, I saw no clubtail dragonflies (I've seen them here in the past) nor any wasps. The weather was warm enough but too overcast. On the sandy access road, where I'd discovered the Queen Ant Kidnapper last summer, numerous wasp nests were present, many looked freshly dug. A single wasp flew down the road ahead of me, disappeared, and did not return.

I crossed Highway 19 and visited the Carleton College baseball fields, thinking I might find another *Cerceris fumipennis* nesting site there. The fields, which I'd never visited before, were immaculately cared for, but they were also empty, fenced, and padlocked. The low strata of clouds drifted away and the sun came out. I turned around and revisited the wasp sites.

On the sandy access road, several wasps were now active, flying, digging, or simply peeking out of their nests. A Weevil

Wasp (*Cerceris clypeata*) peered cautiously from one mound-shaped nest, recognizable by its solid yellow facial markings. Several feet away, out of a tunnel-shaped nest, the digging cast out in front of the entrance, a Queen Ant Kidnapper made an appearance, the yellow markings of its face divided by two wide vertical stripes. This was the wasp I'd hoped to see. In the days and weeks ahead, this wasp will hunt and provision its nest with queen ants.

Pleased to have found the wasp I went searching for, I headed toward the car. However, I was soon delayed by more wasps. On the gravel trail along the edge of the prairie, several black wasps were flying. These turned out to be *Cerceris fumipennis*. I'd passed by these nests earlier; with their large entrance plugged with sand from within (when it was cloudy), I mistook them for anthills. I counted five active nests (there were probably more). I watched one wasp carrying a beetle drop directly into its nest entrance without landing and disappear. Another returning wasp appeared, flying slowly. I didn't have my net, but I'd read, just this morning, that you could tap the wasp with your hand while it was flying and it would drop the beetle it was carrying. I gave it a try. I followed the wasp and tapped it. It landed on the ground and then flew away. Where it had landed, I found a slim, metallic beetle. It worked!

Agrilus quadriguttatus – Jewel Beetle
Cerceris fumipennis – Smokey-winged Beetle Bandit
Aphilanthops frigidus – Queen Ant Kidnapper
Cerceris clypeata – Weevil Wasp
Calliopsis andreniformis – Beautiful Mining Bee
Perdita halictoides – Mining Bee

July 2   80°

THIRTEEN NESTS   65°

Late in the day. English Plantain in bloom several places along the trails. I saw a single White-faced Meadowhawk perched on the tip of a dead branch at eye level. How long before these first meadowhawks take on their mature red color?

After the previous day's successful biosurveillance, intercepting the hunting wasp and capturing its prey, I keenly anticipated more, but I wasn't able to make it into the field until late in the day, well after the peak hours of wasp activity, so I settled on some simple reconnaissance work. There had been zero Cerceris nests on the practice field when visited several days ago. Today, I numbered thirteen nests (literally scratching the number into the infield dirt adjacent the nests) and there were wasps flying.

As I surveyed the ball field, a small mystery presented itself—not all nests were the same. In fact, there seemed to be four distinct varieties (if indeed they all belonged to *Cerceris fumipennis*): 1) active nests that had a large mound of sand with a large opening in the middle; 2) closed nests that had a large mound of dirt with the opening plugged by the wasp with sand from within; 3) exit holes of the same diameter as the active nests but without any mounded sand around them; 4) small pushed up mounds of sand. I suspect the latter may be wasps preparing to emerge? One of the flying wasps, a female, landed and inspected one of these small mounds. It then flew and landed at another of the small mounds before flying off for good.

George and Elizabeth Peckham, who studied a variety of wasps near Milwaukee, Wisconsin, over a century ago, had this to say about our wasp: "Fumipennis, large and handsome, with a broad yellow band at the front of the abdomen, is another wasp that has no regard for the convenience of the people who are watching her. You may sit by her big open hole for hours without seeing her, and when she comes she drops in so suddenly that, unless you are very much on your guard, you are not sure even then what she is." (from *Wasps: Social and Solitary*, 1905)

Philanthus gibbosus – Corrugated Beewolf
Cerceris fumipennis – Smokey-winged Beetle Bandit
Plantago lanceolata – Ribwort Plantain

July 3

81°

AN HOUR IN THE SUN

63°

"To a student of insects high summer is no time for dreaming: it is a time for being afoot and alert. The world of insects is at its shrill crescendo, in a few short weeks to fade to a whisper and then to months of silence." – Howard Ensign Evans, from *Wasp Farm*

Now that several populations of *Cerceris fumipennis* had been located, the wasp work began.

One hour of wasp watching at Greenvale Park from 11:30 to 12:30. Wasps flew continually over the scruff of vegetation covering the ball field. Perhaps some of these flights were orientation flights? Many, however, seemed more searching, perhaps searching for locations to excavate a new nest? And there were more nests than before. During the hour, I netted just one wasp that was carrying a beetle. As I was leaving, I interrupted a Great Golden Digger Wasp excavating a nest at the edge of the ball diamond. I crouched in the grass near where it circled. My presence bothered it, but eventually the wasp returned to her nest and began digging again. Intermittently she'd stop digging and fly up and inspect once more this new large object present in her space, hovering very near my face and sometimes around the camera I was holding. The bright orange abdomen and legs, the golden hairs on face and thorax, and the inquisitive, knowing vigilance are certainly good reasons to admire this wasp.

A second hour of wasp watching from 1:30 to 2:30 at St Olaf yielded similar results: a single beetle. Most entertaining at this site were several sand wasps, probably *Bicyrtes*, that raced across the bare surface of the practice field so close to the ground and at such speed that they resembled race cars cruising the salt flats of the Great Basin.

Spectralia gracilipes – Jewel Beetle
Lasioglossum – Sweat Bee
Agrilus vittaticollis – Hawthorn Root Borer
Sphex ichneumoneus – Great Golden Digger Wasp
Cerceris fumipennis – Smokey-winged Beetle Bandit

July 4

86°
—
67°

BUPRESTID BONANZA

I visited the *Cerceris fumipennis* nesting site at the Cowling Arboretum from noon to 1 P.M. Clear, sunny, hot and humid—ideal wasp watching conditions (ideal for the wasps not the watcher!). I counted twenty-two nests. I found three abandoned beetles, beetles dropped and left outside the nests by the wasps. In addition to the abandoned beetles, I was able to collect four beetles by gently tapping the returning wasps with my hand. Seven beetles total, a Buprestid bonanza compared to the single beetle collected during my previous visit.

Two of the beetles, one abandoned (*Buprestis sp.*) and one captured (*Dicerca sp.*), were very large. If I hadn't seen the wasp flying with the one beetle, I wouldn't have believed it possible. This feat raises further question of how the wasps capture these giant, well-armored beetles to begin with.

A lot of other wasp and bee activity occurred all around me, which I had to neglect while keeping my vigil for returning Cerceris wasps. I did manage to photograph an incoming Ant Queen Kidnapping Wasp, though I wasn't quick enough to get a photo of the kidnapped ant queen that the wasp left momentarily at the entrance of her nest. By the time I looked at it and recognized what I was seeing, the wasp grabbed it and whisked it into her burrow, a large reddish ant with wings, maybe *Lassius sp.*?

Polygonia comma – Eastern Comma
Sphinx kalmiae – Laurel Sphinx
Climaciella brunnea – Wasp Mantidfly
Aphilanthops frigidus – Queen Ant Kidnapper
Agrilus quadriguttatus – Jewel Beetle
Symmorphus sp. – Mason Wasp
Buprestis consularis – Jewel Beetle
Dicerca sp. – Jewel Beetle

92°
—
73°

July 5

SIZZLED SYNAPSES

Another very hot and very humid day. Fortunately, throughout the morning hours, thin clouds shielded us from the blazing spearpoints of the sun. Even still, without direct sunlight, the temperature reached the upper 80s by late morning when I visited the *Cerceris fumipennis* nesting site at St Olaf. For a half hour, I paced about the gravel of the practice field, ready for the wasps to take flight and return with their captured beetles. Instead, they seemed content to wait for more favorable conditions, simply hanging out at the entrance to their nests, peering placidly at the bright, yet overcast sky. Several diminutive males made the rounds. Not much else happened. I could see blue sky near the horizon to the northwest (perhaps the wasps could as well?), so I decided to leave and return later if the clouds passed.

By 1:30 the sun was out, the sky solidly blue, the afternoon sweltering. Surely the wasps would be active now. I grabbed my net and several vials for beetles and resumed my vigil at the practice field. There was even less activity than before! The wasps had departed from the nest entrances and, at first, I took this to be a good sign, that they were off procuring beetles. But as the minutes ticked by no wasps returned. Heat mirages distorted the landscape. Sweat stung my eyes. My purpose, my patience thinned and melted away; a molten disenchantment pooled in its place. The connection between earth and sun glittered and sparked like sizzled synapses. I left the ball field beaten, shutout.

Cerceris fumipennis – Smokey-winged Beetle Bandit
Bicyrtes quadrifasciatus – Stink Bug Hunter   *Dicerca sp.* – Jewel Beetle
Geina periscelidactylus – Grape Plume Moth

July 6   94°/76°

NATURAL CAUSES

A dead Four-spotted Skimmer was found on our driveway this afternoon. Mysteriously, the dragonfly held a rather lifelike pose. It looked as though it had died while perched, dried in that position, and then fell to the ground, dislodged perhaps by a gust of wind. This brought to mind an old question: How often do dragonflies die of old age? I've wondered about this off and on. Many adult dragonflies, at least in the northern latitudes, live long enough to succumb to the frosts and freezes of autumn and avoid predation that way. Still, a death by natural causes and not a death caused by nature (i.e. bats, birds, frogs, fish, or other dragonflies) seems to be a rarity. In fact, I know of only one eye-witness account. Ken Tennessen, a fellow odonatologist and poet, once told me of a time he was watching a dragonfly and it just quietly fell from its perch, dead. "It has to happen on occasion," he added, after a quiet philosophical pause.

❖ ❖ ❖

I turned the moth light on for several hours, shutting it off a little after midnight. A quiet, moonlit night. No tree crickets. A lot of caddisflies, a nice-looking long-horned beetle, and a number of moths, several of which remain unidentified, unnamed. "It's just as well that salvation doesn't depend on a name," says José Saramago in his *Journey to Portugal*.

Chrysobothris sp. – Jewel Beetle   *Dorcaschema alternatum* – Small Mulberry Borer
Spectralia gracilipes – Jewel Beetle   *Maliattha synochitis* – Black-dotted Glyph
Oxybelus sp. – Fly-stabber Wasp   *Cymatodera inornata* – Checkered Beetle
Philanthus sanbornii – Sanborn's Beewolf   *Crambus agitatellus* – Crambus Moth
Aphilanthops frigidus – Queen Ant Kidnapper   *Gluphisia sp.* – Gluphisia Moth
Cerceris fumipennis – Smokey-winged Beetle Bandit
Libellula quadrimaculata – Four-spotted Skimmer   *Geometridae* – Geometry Moths

84° / 65°

July 7

GOLD-MARKED WASP

To St Olaf Natural Lands. 2:30 to 4:00 pm. Windy, partly cloudy, warm (near 80 degrees) but less humid than the past several days. On the sidewalk, only a few blocks from the house, I noticed a dead cicada. The insect had made it out of the ground but died before emerging. As yet, there are no cicada calls during the day. No Cicada Killer Wasps either. Surely, now that the cicadas were beginning to emerge, the wasps would soon follow.

At the natural lands, after several stops to pick and eat a few handfuls of black raspberries, I encountered an uncommon and rather fugitive wasp, the Gold-marked Wasp (*Eremnophila aureonotata*). This graceful, black, thread-waisted wasp with metallic markings on the face and thorax, haunts sunlit openings in deciduous forests where it hunts large caterpillars. Only a handful of times have I seen this wasp and, as usual, all I get is a glimpse. In nine years walking the St Olaf Natural Lands and observing the plants and animals along the trails, this is the first time I've seen this species here.

For some reason, the markings of these wasps, at least the ones I've encountered in Minnesota, are silver not gold. Deflation or wishful thinking?

Eremnophila aureonotata – Gold-marked Wasp
Pachydiplax longipennis – Blue Dasher
Sympetrum obtrusum – White-faced Meadowhawk
Theridion – Cobweb Spider
Chalcidoidea – Chalcidoid Wasps
Asclepias syriaca – Common Milkweed
Cerceris fumipennis – Smokey-winged Beetle Bandit
Chrysididae – Cuckoo Wasps
Geum canadense – White Avens
Laphria – Bee-mimic Robber Flies

July 8 — 85° / 62°

WESTERN RED DAMSEL

Lisa and I visited Koester Prairie subunit of Prairie Creek WMA specifically to look for Red Damsels. Last year, they had been abundant at the spring-fed seep that surfaces and flows in the center of this open prairie preserve. This was the first opportunity I'd had to visit this year and I was hoping it wasn't too late to find these small, spectacular damselflies.

Right away, we got off track by following a trail to an isolated hill at the northeast corner of the wildlife area. We presumed, mistakenly, that this hill would be connected to the rest of the land, but that wasn't the case. It was a stand-alone, prairie hill separated from the rest of the wildlife area by a field of soybeans. Before crossing the field we had a look around on the hill, finding some good native prairie plants. On the sunny, south-facing slope where there was a lot of open sandy ground, we encountered a good population of sand wasps. A good site to revisit at some point, when not on a mission for damselflies.

There were numerous sand wasps on the main part of the prairie as well. The male wasps bombarded us belligerently, all bluff. The buzz they made, a kind of high-pitched whine, sounded like horseflies on helium. I laughed them off, knowing they were harmless. But Lisa tired of the persistent pestering and started back to the car. I lingered a while longer to photograph the Red Damsels, which were, indeed, abundant once again. No surprise, as they are obligate spring and seep dwellers and this is one of the finest permanent seeps around.

During the decade I've observed this species in Minnesota, the Minnesota species has gone from being called the Eastern Red Damsel (*Amphiagrion saucium*), then the Midwestern Red Damsel (*Amphiagrion intermediate*), to most recently being recognized as the Western Red Damsel (*Amphiagrion abbreviatum*). Possibly laboratory genetic analysis will settle the species question. Perhaps, like the Band-winged Meadowhawk, the variously named species will be lumped together as a single species with east to west variation in color and morphology? It's all too close to call.

Amphiagrion abbreviatum – Western Red Damsel
Hyles lineata – White-lined Sphinx
Satyrium titus – Coral Hairstreak
Sphaerophoria contigua – Tufted Globetail
Mecaphesa – Crab Spider
Ratibida pinnata – Grey-headed Coneflower
Bembix americana – American Sand Wasp

July 9
GREEN-EYED WASP

85°
—
67°

Cowling Arboretum. 1:30 to 3:00. A routine walk to the sandy upland trails. Along the way, I happened upon the remains of a Widow Skimmer, just the wings. This reminded me of a passage from Vladimir Nabokov's *The Gift* that I'd read earlier in the day. "But on occasion you find four black-and-white wings with brick-coloured undersides scattered like playing cards over a forest footpath: the rest, eaten by an unknown bird."

A few steps further down the footpath, I encountered a hunting wasp, *Cerceris halone*, an Acorn Weevil hunter. Just a few active nests in a small spot of sandy soil along an otherwise grassy stretch of trail, but almost directly under an oak tree.

At the sandy sloping stretch of trail where the *Cerceris fummipennis* colony had been active a few weeks earlier, a Green-eyed Wasp (*Tachytes crassus*) made a brief appearance. Amber hairs on its thorax and legs, silvery white bands of hair on its abdomen, and green eyes give this sand wasp its unique and pleasing appearance. These wasps provision their nests with grasshoppers. I don't see them very often and have yet to see one digging a nest or dragging a grasshopper, so I follow this one around for a few minutes until it flies off.

Tachytes crassus – Green-eyed Wasp
Curculio – Nut And Acorn Weevils
Libellula luctuosa – Widow Skimmer
Silphium laciniatum – Compass Plant
Gomphurus fraternus – Midland Clubtail
Desmodium canadense – Showy Tick-trefoil
Bombus griseocollis – Brown-belted Bumble Bee
Odocoileus virginianus – White-tailed Deer
Augochlora pura – Pure Green Augochlora
Chrysididae – Cuckoo Wasps
Cerceris halone – Weevil Wasp
Sphecodes sp. – Blood Bees
Oxybelus sp. – Fly-stabber Wasp
Chalepini – Leaf Beetle
Forficula auricularia – Earwig

86°
—
69°

July 10

BACKYARD WASPS AND BEES

About 5 P.M. The sun, about to lower behind the roofline of the neighbor's house, illuminated half of the backyard Hazel tree. I watched the sunlit leaves for quite some time, waiting for the tiny wasps and bees to land, take a break from their blurry orbits, and sit still for the camera. Every time they did so, I wasn't quick enough to get a photo. Eventually, I just netted a couple and placed them in vials in the refrigerator so that I could take their photo later after they'd cooled down and were less prone to instantaneous flight.

I photographed a leafhopper nymph because it had the good grace of sitting still on a leaf in front of me. Looking at the image later, many times enlarged on the computer screen, I noticed what I hadn't noticed to start with...a thick, brown disc jutting from the side of the nymph's abdomen. This was the larva of a Dryinid Wasp.

Dryinid Wasps are rather bizarre parasitoids. I've never seen an adult. Someday I hope to notice the parasite while taking a photo (instead of hours later on the computer screen), capture it and rear it. That's probably the only way I'll be able to see the adult wasp.

Symmorphus canadensis – Mason Wasp
Heriades – Resin Bees
Euodynerus foraminatus – Mason Wasp
Dryinidae – Dryinid Wasp

July 11

HORSE FLY MIMIC?

87°
—
65°

A late morning walk to the St Olaf Natural Lands. I noticed a lot of conspicuous galls on some Sumac. These are caused by Sumac Gall Aphids, one of the only aphids known to cause galls; usually, galls are caused by flies and wasps.

Encountered, at the same location and possibly the same dead branch as four years ago, *Astata unicolor*. The large eyes and odd behavior of this wasp convinced me, at first, that it was a horse fly or a soldier fly. It even had the high-pitched whine of a fly when it flew. Perched about four feet off the ground on a dead Sumac branch, this wasp made rapid, aggressive flights, returning again and again to the same perch—a male being territorial. The females of this species, stinkbug hunters, differ markedly in appearance, having smaller eyes (not meeting in the middle) and a red abdomens.

At the *Cerceris fumipennis* site, the playing field was being groomed, a small vehicle driving around and around the infield, smoothing out the playing field, erasing all the wasp nests. I found a wasp with beetle sitting on the ground, the entrance to her nest now gone. I wondered how many of the wasps would re-establish their nests and entrances.

Cerceris fumipennis – Smokey-winged Beetle Bandit
Astata sp. – Square-headed Wasp
Euphyes vestris – Dun Skipper
Microcrambus elegans – Elegant Grass-veneer
Melaphis rhois – Sumac Gall Aphid
Typocerus sp. – Flower Longhorn Beetle
Rhionaeschna multicolor – Blue-eyed Darner

100°
—
77°

July 12

THOREAU

On this day two hundred years ago Henry David Thoreau was born.

For myself and my family, this was the first day of a week's vacation and a travel day. From Northfield to Minneapolis to Las Vegas to Zion National Park, such distance in so little time (1,500 miles in 12 hours) is disorienting. Thoreau, who certainly enjoyed travel, preferred walking and was critical of the material and moral costs involved in greater journeys. In Walden he writes, "We do not ride on the railroad; it rides upon us."

We enter Zion National Park just before sunset. At the visitor center, we walk the beginning of The Watchman Trail. Along the trail, which follows the Virgin River, we encounter a number of very large Desert Stink Beetles. When disturbed by our passing too close, the beetles performed their characteristic headstands, a warning posture like a skunk raising its tail. These beetles can spray a foul liquid at an attacker if necessary, but the warning pose is probably an adequate defense in most instances.

Agave sp. – Century Plants
Eleodes sp. – Desert Stink Beetles

July 13

RIVER-WALKING

We started early, knowing we needed to beat the afternoon heat which was forecast to be 110 degrees. At the outfitter store near the entrance to the park we rented riverwalking gear—walking sticks, a pair of neoprene hiking shoes, and a waterproof backpack—then entered the park to catch the park shuttle that would take us to The Narrows. Only a few minutes after 7 A.M. and already there was a significant queue of people at the shuttle stop, many, like us, holding tall wooden riverwalking staffs, as if we were headed to a conference of wizards.

We reached the entrance to The Narrows before 8 A.M. At the beginning, our hike upstream was mostly in shadow, only occasionally did the morning sun align with the high cliffs of the narrow canyon to illuminate our path and the river. In the water, against the current, all three of us were enjoying this most unusual hike.

At a small pool, separate from the main channel of the river, I pulled a submerged branch out of the still water. Grasping it tightly was a dragonfly nymph, a nearly-full-sized darner. I hope it was something exotic like a Riffle or Giant Darner.

An hour upstream, at a sunlit boulder field, we climbed out of the water and stopped for a drink, snacks, and a family photo. Because of the sun and an area of open ground, insects were active here. I took the camera out of the waterproof pack and chased the insects for a while. A large red dragonfly flew over the river and then perched on the canyon wall opposite from where we stood, too far away for a photo, but possibly a Red

Rock Skimmer, one of the species I'd hoped to see. On the rocks on our side of the river, I photographed Sooty Dancers, Taxiles Skippers, and large, pitch-black Bot Flies. (9:45)

We ventured for half an hour further before turning around. We stopped briefly at the same boulder field and I spent some time photographing some sand wasps which were now actively digging nests and provisioning. (11:00) At a bend downstream of the boulder field, I stopped for one last photo. Several Canyon Rubyspots perched on the rocks and debris midstream. This species was new to me. We exited the narrows just after noon, against a torrent of people just setting out.

❖ ❖ ❖

We visited the park again in the evening. Taking the shuttle to the Weeping Rock trail. A short climb on paved trails to some hanging gardens. Photographed Netleaf Hackberry growing along the trail. We stayed as long as we could, catching one of the last shuttles back.

Celtis reticulata – Netleaf Hackberry
Hetaerina vulnerata – Canyon Rubyspot
Bicyrtes capnoptera – Stink Bug Hunter Wasp
Corydalus texanus – Western Dobsonfly
Aeshnidae – Darners
Tamias umbrinus – Uinta Chipmunk
Cuterebra sp. – Rodent And Lagomorph Bot Flies
Poanes taxiles – Taxiles Skipper
Cirsium arizonicum – Arizona Thistle
Argia lugens – Sooty Dancer
Otospermophilus variegatus – Rock Squirrel
Hordnia atropunctata – Blue-green Sharpshooter
Argia sp. – Dancers

July 14

ANGEL'S LANDING

106°
—
77°

We arrived at the park early, starting our ascent to Angel's Landing just after 7 A.M. With this hike, we left the world of down, the riverwalking world of yesterday, for the world of up, an elevation change of 1,488 feet and a distance of slightly more than five miles, round trip.

The first two-thirds of the ascent is on concrete, constructed trails, switchbacks, and wiggles. Easy going. This brings you to a high plateau, a sandy shelf high on the mountain. From here to the top the hike changes character and exhilarates. The trail follows the cutting edge of a thin rocky blade, with sheer drops on both sides. Hikers have fallen and died here. Fewer since the installation of chains.

On pretty much every list of the best hikes in the United States, Angel's Landing is a kind of pilgrimage. We pushed on. Lida, more determined than I've ever seen her, helped herself to the top by helping a young woman from Salt Lake City forget her fears. We made the climb without incident, enjoying the strenuous trail, clinging rather tightly to the chains in many places. At the top, Lida asked us to take her photo; she posed where innumerable other photos have been taken before, back to us, proudly sitting on a rock at the very edge of the landing, the valley of Zion opening majestically below. It is an astonishing view and a cherished photo.

Also, while at the top, I noticed what looked like a spider wasp, then a robber fly, then a cimbicid sawfly. Eventually, I admitted I didn't know what it was—some kind of fly—large, glossy-black, with orange antennae. I figured it out later, a Midas Fly.

At the end of our descent, as the trail neared the shuttle stop, I explored the river in search of dragonflies. Already the temperature in the sun was formidable. After a robber fly and a skipper, I photographed a White-belted Ringtail perched on a stick in the river, the only dragonfly photographed during our entire Utah trip.

We were back to the shuttle stop by noon.

We returned to our hotel in Hurricane and stayed there, relieved to have survived Angel's Landing, tired out by the heat, willing to take it easy and rest up for our next day's adventure.

Quercus gambelii – Gambel Oak
Erpetogomphus compositus – White-belted Ringtail
Orthoptera – Grasshoppers, Crickets, And Katydids
Efferia sp. – Robber Fly
Phoradendron juniperinum – Juniper Mistletoe
Acer grandidentatum – Bigtooth Maple
Abies concolor – White Fir
Bombyliidae – Bee Flies
Uta stansburiana – Common Side-blotched Lizard
Mydas xanthopterus – Mydas Fly
Solidago sp. – Goldenrods
Quercus turbinella – Sonoran Scrub Oak

July 15
CANYONEERING

109°
—
80°

This day we accepted the third (and probably greatest) of Lida's challenges...canyoneering. While I welcomed the day of hiking in the river, then wondered whether I was fit enough to make the climb to the top of Angel's Landing, the thought of repelling down rock cliffs unnerved me.

Luckily (and by necessity!) this was a guided trip—Zion Rock and Mountain Guides. We just had to trust Spencer, our guide for the morning's adventure. After gearing-up, fitting harnesses and helmets, we rode in the outfitter's van to Kolob Terrace at the western edge of Zion National Park. Hiking into the rocky canyons from below we encountered two rattlesnakes, one large and one small. Both were slow to move, still torpid in the morning shadows. I was very happy to have seen them. Lida, not so much.

Each of the descents we made—from 30 feet, from 60 feet, from 90 feet—I had to go first. This was fine. In fact, I quickly began to enjoy kicking away from the rock wall, then sliding down through the air. One of the rappel sites was given the name Eye of the Tiger, and because of the several routes one could take down Spencer referred to it as an opportunity to "choose your own adventure." One of the last rappels was at Snake Canyon. The name alone made Lida nervous. She requested that I do a thorough check for rattlers at the bottom before she would come down. From here we down-climbed through a very narrow slot canyon, having to proceed sideways for much of the time and having to scuttle down a couple of drops and rock chimneys, controlling our descent by pushing against both sides of the narrow passage with our feet and arms. It was a comfortable feeling to walk out the bottom of the canyon, the day and the entire landscape opening ahead of us.

◆ ◆ ◆

Ate at Oscar's Cafe in Silverdale for the second time.

◆ ◆ ◆

Hiked the Emerald Pools trail in the late afternoon. Lida hiked ahead on her own and hung out on the rocks near the upper pools until we arrived. I dawdled on the way up, taking photos of wildflowers and insects. The most interesting encounter was a robber fly that wasn't a robber fly but a Flower-loving Fly. These misleadingly-named flies prefer not to visit flowers and are observed most often on the sand or rocks of the deserts across the southwest. While closely related to robber flies, the flower-loving flies are also near relatives of the mydas flies. The exact phylogenetic relationships are yet to be worked out.

Abies concolor – White Fir
Stanleya pinnata – Prince's Plume
Berberis repens – Creeping Mahonia
Argia lugens – Sooty Dancer
Apiocera sp. – Flower-loving Flies
Marchantiophyta – Liverworts
Aquilegia chrysantha – Golden Columbine
Hyla arenicolor – Canyon Tree Frog
Primula pauciflora – Dark-throated Shooting Star
Crotalus oreganus lutosus – Great Basin Rattlesnake
Oenothera cespitosa – Fragrant Evening Primrose
Copaeodes aurantiaca – Orange Skipperling
Pheucticus melanocephalus – Black-headed Grosbeak
Penstemon rostriflorus – Bridge Penstemon
Ephedra viridis – Green Ephedra
Bombyliidae – Bee Flies
Quercus x pauciloba – Hybrid Oak
Dieteria canescens – Hoary Aster
Argia plana – Springwater Dancer
Heterotheca – False Golden Asters

July 16   86°

THE DESERT SMELLS LIKE RAIN   50°

We checked out of our hotel in Hurricane. Stopped at the River Rock Roasting Company drive-thru in La Verkin, a converted box-car coffee stop that Lida liked. Then drove to Zion. We drove through the park on the Mt. Carmel Highway on our way to Bryce Canyon, stopping for a look at Checkerboard Mesa before leaving the park.

Stopped at a sunlit rest area in the mountains. I chased insects for a while, finding a number of butterflies and damselflies along the wild edges of the rest area. Lida spotted a handsome Tiger Moth perched in the picnic shelter.

Stopped at a pull-off alongside the Sevier River. Arroyo Bluet, Spotted Spreadwing, a colorful long-horned beetle on Rabbitbrush, and a brown snake that disappeared before I could get a photo.

We arrived at Bryce Lodge shortly after noon, had lunch, then drove out to Rainbow Point and hiked around the Bristlecone Pine Loop. Photographed a beautiful jumping spider (Phidippus tyrellii), several interesting robber flies (Cryptopogon sp.), and a very colorful Pine White butterfly. Heading back, we made a few additional stops, seeing a couple of Podalonia wasps at Fairview Overlook, a Variable Tiger Beetle at a turnout with trails connecting to Swamp Canyon.

We checked into our room at the Lodge Motel, ate dinner at Bryce Lodge, then hiked the Navaho Loop down through Wall Street up past Thor's Hammer. The entire ascent took place in the increasing dusk. As we approached the rim of the canyon we were greeted by bats and the night's first stars.

❖ ❖ ❖

"We meet at dawn among the spires and hoodoos:
tourist European and tourist Yankee
on a switchback trail, watching honey rust
on the ridges and rims of canyons ragged with rainbows.
You could get labyrinthitis down here,
where searching for the lost calf used to be
an occupational hazard. Imagine snow, a rope
stiff with cold and a limping mare, and beauty
loses to bitter cowpokes lonesome for town."
— Robert Edwards,
from 'The German Tourist at Bryce Canyon'

Cylindera terricola – Variable Tiger Beetle
Phidippus tyrrellii – Jumping Spider
Satyrium saepium – Hedgerow Hairstreak
Cercyonis oetus – Small Wood-nymph
Neophasia menapia – Pine White
Oarisma garita – Garita Skipperling
Colias sp. – Clouded Yellows
Podalonia – Cutworm Wasps
Frasera speciosa – Monument Plant
Satyrium sylvinus – Sylvan Hairstreak
Poanes taxiles – Taxiles Skipper
Enallagma praevarum – Arroyo Bluet
Acmaeodera – Yellow-marked Buprestids
Dialictus – Metallic Sweat Bees
Eristalis flavipes – Orange-legged Drone Fly
Arctia caja – Great Tiger Moth
Lestes congener – Spotted Spreadwing
Crossidius coralinus – Longhorned Beetle
Bombus huntii – Hunt's Bumble Bee

Cirsium – Thistles
Argia sp. – Dancers
Enallagma sp. – Bluets
Asilinae – Robber Fly
Plebejus melissa – Melissa Blue
Hesperiidae – Skippers
Cyrtopogon sp. – Robber Fly
Mecaphesa sp. – Crab Spider

July 17   108°

FALLING ROCK   80°

In the morning, seated on the balcony of our hotel room, surrounded by Ponderosa Pines, I read the short story 'Coyote and Rattlesnake' by Barry Lopez from his book *Desert Notes*. "They are like a boulder fallen off a mountain," Akasitat to Coyote. This gives another meaning to the falling rocks signs—I thought—*we* are the falling rocks! A landslide really. Where does the expansion, improvements, and development stop? What will it be like then?

Another image floats up from readings past. In the oddly optimistic Soviet poem, *Bratsk Station* by Yevgeny Yevtushenko, a speaking Egyptian pyramid (in dialog with a hydroelectric dam) refers to mankind's many construction projects as "pigsnouting around."

After leaving Bryce Canyon, we drove through the mountains to Cedar, Utah. At a stop along the way, I photographed a Western Branded Skipper and a flightless grasshopper. In Cedar, we visited the Southern Utah University campus. Lida spoke with the assistant volleyball coach.

Checked into Excaliber late in the day. After dining at the hotel's lavish buffet, we took a walk on the Vegas Strip. Wide-eyed, appalled, somewhat amazed by the spectacle of nightlife in Vegas, Lida summed it up with a single word: extra!

Pinus ponderosa – Ponderosa Pine
Hesperia colorado – Western Branded Skipper
Tipula sp. – Crane Fly
Bradynotes obesa – Slow Mountain Grasshopper
Satyrium saepium – Hedgerow Hairstreak

82°
—
66°

July 18

BACK HOME

Before flying home, we made a quick visit to UNLV. Early morning but already the temperature was near 100 degrees. Our impression of the campus was that it was compact, efficiently arranged. Lida enjoyed visiting the science buildings, but overall it felt too urban and too big.

Crossing campus, I noticed a number of Saddlebags and Rainpool Gliders flying above an open quadrangle, far more dragonflies than during the previous days of our trip. Photographed a couple birds and a newly emerged cicada.

Back in Minnesota by early afternoon. And back to Northfield before dark. Interesting as Utah had been, it felt good to be home.

Quiscalus mexicanus – Great-tailed Grackle
Diceroprocta apache – Citrus Cicada
Mimus polyglottos – Northern Mockingbird
– – –
Lampyridae – Fireflies
Toxomerus geminatus – Eastern Calligrapher

July 19   83°/62°

THE CICADA KILLERS TAKE THE FIELD

To St Olaf Natural Lands. Still a bewildering number of Blue-eyed Darners. Throughout the early part of the summer, these large dragonflies have been the most abundant species flying over the nearby fields. They seem to fly closer to the ground than our other darners; I'm often looking down at them as they fly by.

In the past, I've consistently found one or two sand wasps active on the sand pile near the baseball fields. Today a cloud of wasps enveloped the mound of sand. It looked as though someone had disturbed a hornet's nest, except these were all solitary sand wasps, all emerged at the same time, all vying for the same space, a single heap of sand amid acres of green playing fields. I knelt down at the edge of the sand, enjoying the sound and swirl of wasps, a ground-level murmuration. Individuals would drop out of flight for a few seconds, then resume their flight. The concentration of flying wasps thickened and thinned, always just a couple of inches above the sands. A turbulence of wasps. A weather of wasps.

On the nearby practice field, Cicada Killers, around eight or ten males, did what the sand wasps were doing only on a larger scale, chasing about the infield and the third base line. I found two active female nests near home plate. I stood around for a while hoping to witness an incoming female with a cicada. After a while, I noticed a wasp struggling to fly about twenty feet away. This turned out to be *Cerceris fumipennis* carrying a large Buprestid beetle. I watched a little while longer, long enough to realize the Cicada Killers now held dominion of the airspace. They chased everything: birds, other Cicada Killers, smaller wasps, and butterflies. And they even chased me as I moved about their space. The Cicada Killers had taken the field.

Walking home, I found several Band-winged Meadowhawks perched on high bare branches on a hilltop, a habit of this species. And further on, a bright red White-faced Meadowhawk, the first mature male meadowhawk of the year.

Astata unicolor – Square-headed Wasp
Danaus plexippus – Monarch
Sympetrum obtrusum – White-faced Meadowhawk
Sympetrum semicinctum – Band-winged Meadowhawk
Lactuca canadensis – Canada Wild Lettuce
Cerceris fumipennis – Smokey-winged Beetle Bandit
Sphecius speciosus – Eastern Cicada Killer
Bicyrtes quadrifasciatus – Stink Bug Hunter
Erythemis simplicicollis – Eastern Pondhawk
Asilinae – Robber Flies
Teucrium canadense – American Germander
Carya cordiformis – Bitternut Hickory
Pachydiplax longipennis – Blue Dasher

July 20   89°

COMPOST FLIES   68°

Our compost pile occupies a corner of our backyard garden. As most of the yard clippings and fallen leaves go to the city compost, our compost takes the organic waste we have from the kitchen, so a lot of coffee grounds and peelings of vegetables and fruit. Even at this small scale, the compost provides food and shelter for a lot of insects. Quite a variety of fly larvae take advantage of the decaying food. And today, when emptying the compost bucket, I found an impressive bounty of soldier flies. Long-legged, green-accentuated, sleek-bodied, these flies were flying pretty much nonstop just above the surface of the compost pile. Like most insect swarms, their movement was rather mesmerizing to watch.

The following quote comes from a short note in the journal *Psyche* by Philip Rau, author (along with his wife Nellie) of *Wasp Studies Afield* (1918). Here are a few passages in summation of our compost fly, in Rau's characteristic flouncy style, written around 1930, in or near Kirkwood, Missouri:

"...these attractive greenish-colored flies were seen hovering in courtship dance above garbage heaps on the rear of a lot... To say that they dance incessantly is not wholly true for individuals often leave the throng to rest on a tin-can or bottle or cantaloupe skin... indeed, it's a pretty sight to see a flock of these flies moving in a horizontal plane in more-or-less irregular circles and in figure eights just an inch or two above the mass of multi-colored refuse."

Ptecticus trivittatus – Compost Fly
Bombus bimaculatus – Two-spotted Bumble Bee
Syritta pipiens – Thick-legged Hoverfly

83°
—
69°

July 21

MIDSUMMER MIGRANT

I visited the Cowling Arboretum at about 4 P.M. Full sun following an early afternoon rain made the world a true-to-life sauna. Not just my body, but everything seemed to sweat: leaves, flowers, trees; rocks, rivers, sand. I felt the steamy dampness as if I'd just crawled out of the river and into an oven. A tropical mess.

It figures that the one dragonfly encountered, a Spot-winged Glider, was a midsummer migrant, a visitor from the tropical south. Each year Minnesota welcomes a half-dozen or so species of migrating dragonflies. In the spring, Common Green Darners and Variegated Meadowhawks arrive shortly after the ice goes off the local ponds, usually in April. The Saddlebags and Rainpool Gliders arrive months later in June and July.

Also of note, noticed as I stewed in the steamy heat, were the American Bellflowers in bloom, a beautiful blue. The bright yellow flowers of the towering Cup Plants and the streamside Fringed Loosestrife. And a tiny, green-eyed Robber Fly found poised and vigilant at the pointed tip of a leaf.

Chrysopsinae – Deer Flies
Sceliphron caementarium – Black-and-Yellow Mud Dauber Wasp
Pantala hymenaea – Spot-winged Glider
Silphium perfoliatum – Cup Plant
Lysimachia ciliata – Fringed Loosestrife
Campanulastrum americanum – Tall Bellflower
Asilinae – Robber Flies
Polygonia comma – Eastern Comma
Cerotainia macrocera – Robber Fly

July 22   91°/71°
VALENTINE ANTS, SQUASH BEES, AND THYNNID WASPS

I went for a late morning walk, intending both Greenvale Park and St Olaf Natural Lands, but with the temperature and humidity both at eighty and with no breeze the conditions were too miserable to go further than the few blocks to the school and back.

The *Cerceris fumipennis* nesting site appeared inactive. Yesterday's thunderstorm smoothed over all the previous nest mounds and I could find no signs of new activity. Perhaps it was too early in the day.

Several squash vines had escaped the confines of the community garden, their large yellow flowers visible from the playing field. During the years Lida attended school here, we had a plot in the garden—it was, even then, an oasis for all kinds of insects. I decided to walk over and see if there might be Squash Bees in the squash flowers. Sure enough, in each and every blossom, I found one or more of these handsome bees stationed deep inside. Too shadowy in the well of the squash blossoms to get a good photograph without a flash, I captured one of the bees and brought it home. Interestingly, these bees buzz very loudly when they fly, especially when confined to a small vial.

The range of this bee has no doubt expanded as the cultivation of cucurbit plants expanded.

The other observations for today were of a Thynnid Wasp (before recent taxonomical changes, a Tiphiid Wasp) captured on some Canadian Thistle blossoms along with a tiny ant that must have been on the flower (an innocent bystander) when I'd captured the wasp, either that or it was hitchhiking a flight on the wasp.

Male Myzinum wasps are the stretch limousines of the wasp world, their abdomens go on and on. And to extend the car metaphor, each has their own trailer hitch, an elegantly curved pseudostinger. To me, this curved spine resembles a farm tool I used when young—a bale hook. Of course, the pseudo-stinger lacks any elaborate pseudo-purpose beyond its true purpose of coupling male abdomen to female abdomen when mating.

The female Myzinum wasps look entirely different—so much so that many were originally described to science as separate species. They are much shorter, stouter, and limbed with limbs made for digging, assisting them in their search for underground beetle larvae, their hosts.

Peponapis pruinosa – Pruinose Squash Bee
Crematogaster sp. – Acrobat Ants
Myzinum quinquecinctum – Five-banded Thynnid Wasp

July 23
FIRST AUTUMN MEADOWHAWK

79°
—
64°

Autumn Meadowhawks used to be known by the common name Yellow-legged Meadowhawks (and in at least one truly arcane guidebook Yellow-legged Meadowfly). The old name spot-lights a defining character of this species, its yellow legs setting it apart from most other species in this genus that have black or black-and-yellow legs. A perfectly acceptable name, especially when flying this early in the season. The flight season, while beginning now at the end of July, extends most years past the first freezes and first snowfalls to mid-November. Often it is the last dragonfly seen in the year. Its presence at the height of autumn—its abundance, frost-hardiness, and crimson-red color as well—make the current name fit ever so much better, having the justice of the best word in the best place, as in a poem. This name also provides a link, an elaborate linguistic lace, to the dragonfly art and literature of Japan. For instance, this haiku by Kaya Shirao:

> The start of autumn
>   is decided
>   by the red dragonfly

I like the scientific name, *Sympetrum vicinum*, as well. Hermann Hagen, a renowned 19th Century taxonomist, who described the species and provided the Latin epithet, *vicinum*. Perhaps he had been impressed by the geographic nearness of the type specimens to his Harvard office. He may have even encountered the species at some local park (if he ever left his office). *Vicinum*, which translates as "neighborly", is fitting in the sense of "friendly" as well. Especially late in the season, when these dragonflies tend to perch on vertical, sunlit surfaces like tree trunks and benches, as well as the bodies of people standing at the water's edge where these dragonflies are present.

Trichodezia albovittata – White-striped Black
Pieris rapae – Cabbage White
Cerceris fumipennis – Smokey-winged Beetle Bandit
Sphecius speciosus – Eastern Cicada Killer
Coelioxys sp. – Cuckoo Leaf-cutter Bees
Megachile latimanus – Leaf-cutter Bee
Lethe anthedon – Northern Pearly-eye
Sympetrum vicinum – Autumn Meadowhawk
Trichopoda pennipes – Swift Feather-legged Fly
Melissodes bimaculatus – Two-spotted Long-horned Bee
Silene vulgaris – Bladder Campion

July 24   79°/60°

OUTSIDE WHILE INSIDE

One of the finest features of our small house is a large screen porch. A simple structure: four-by-four posts, screen instead of walls, and a gable roof. A perfect pleasure during the temperate months. And a wonderful place to sit and visit with guests, as today with Sid Gershgoren and Eugenia McGrath, who drove down from Minneapolis for the afternoon. A way to enjoy the outside while inside.

Sid, who lives in Berkeley, California, has written some of the most imaginative and expansive poetry in English. He's also the author of the marvelously curious and vastly entertaining *Extended Words: An Imaginary Dictionary* (Red Dragonfly Press, 2009), a 430-page magnum opus that romps and plays just beyond the edges of the real dictionaries. Even at the age of 80, Sid remains a prolific writer. He brought two new manuscripts for me to look at today, one being a collection entitled *A World-widening Catalog of Bumper Stickers*.

Throughout the afternoon, during the several hours we conversed, Half-black Bumble bees worked the yellow pom-pom-like blossoms of the St Johnswort just outside the screen.

Trypoxylon – Keyhole Wasp
Megachile pugnata pugnata – Leaf-cutter Bee
Bombyliidae – Bee Flies
Bombus vagans – Half-black Bumble Bee
Cerotainia macrocera – Robber Fly

85°
—
67°

July 25

LYRE-TIPPED SPREADWING

The morning was an epoch of thunderstorms, downpours, and rainlight. The afternoon, with high humidity and heat, was molten.

I took a short, ambling walk at the St Olaf Natural Lands to look for Ruby Meadowhawks and found none. Perhaps it is still too early in the season for the dragonflies to show up along the prairie trails? Perhaps the morning weather drove them to shelter? Perhaps the unstoppable foraging of muskrats in the pothole breeding site, creating a sudden amount of open water, has altered the conditions for the nymphs? Perhaps it's just a down year for meadowhawks in general? These questions / worries surfaced as I searched for dragonflies and didn't find them.

As always there were other things to admire along the way. Lots of Leopard Frogs. And lots of Lyre-tipped Spreadwings. The latter is a very common late-summer damselfly here. The paraprocts of the male, the bottom two appendages at the end of the abdomen, are curved and resemble the shape of an Ancient Greek lyre. You'd probably need a pair of eyes with superhero vision to see this character in the field unaided, but it's readily visible using close-focusing binoculars or with the magnification of a good macro lens on a camera.

Lestes unguiculatus – Lyre-tipped Spreadwing
Lithobates pipiens – Northern Leopard Frog
Eryngium yuccifolium – Rattlesnake Master

July 26
UNDERWINGS

84°
69°

Lisa alerted me to the presence of a large moth on the porch screen. One look confirmed my suspicion that it was an underwing moth. The forewings were cryptically patterned to resemble tree bark. The hindwings (underwings) were brightly colored red or yellow.

We have a sincere population of these moths. Each summer a few of them show up and take shelter during the day beneath the eaves of the house or garage. Inevitably, we kick one into flight while moving garbage bins or repositioning the grill. However, I can't remember ever seeing more than one moth at any given time.

Some people go a little bit batty for these moths. One doesn't need to delve very far into the literature about moth sugaring to encounter some of these characters. Among the most interesting is the entomologist Augustus Radcliffe Grote, whose article "Collecting Noctuidae by Lake Erie" has become a mothing classic.

A lot of melodrama is encapsulated in the common names: Widow Underwing; Once-married Underwing; Tearful Underwing; Sad; Sordid; Charming.

Catocala innubens – Betrothed Underwing

85°
—
66°

July 27

A FINE MADNESS OF WASPS

What if this day was the best day of the year? The apex of the naturalist's year? Not likely. Nor are the days so easily judged. Each has its proper charm. Each is paired to the passions, the internal weather we bring along day by day—a tempest in a teapot, a calm blank slate.

Outwardly, it's a beautiful day; after days of storms, suddenly there is cool air out of the north and blue skies. And I feel buoyant as I step outside and walk toward the woods and fields, carrying the least expectations in quite some time.

I noticed a Carder Bee on some streetside flowers. A Blue-eyed Darner flew by. A Cicada Killer flew past my feet just as I entered the trail into the woods. And then inside the shadowy woods, a spot of sunlight at the edge of the trail drew my attention. I paused and watched. In the leaf litter, tiny band-winged spider wasps flicked and darted above and below the old leaves. More than a single spider wasp in one place is unusual and worthy of investigation. As far as I can ascertain, there are three of these small spider wasps, taking turns in the sunshine, each momentarily making an appearance then disappearing back into the shadows. While I'm unraveling this curiosity, several other wasps show up as well. A cuckoo wasp, a Carrot Wasp, another black spider wasp (without the bands on the wings), and then the extraordinary kleptoparasitic spider wasp, *Ceropales maculata*. A fine madness of wasps all within a circle of sunlight not much larger than a basketball.

Continuing on my walk, I added several more wasps to the day's tally: more Cicada Killers, Four-belted Stinkbug Hunters, and a number of diminutive Fly-stabbing Wasps. Along the way, I encounter a Lyre-tipped Spreadwing and four species of Meadowhawk, Ruby, Autumn, White-faced, and Band-winged. A Green Heron at the wooded pond. A Punctured Tiger Beetle on the sunlit trail. A Peck's Skipper in the sunlit grass.

It was going to be difficult to top such a fine day.

❖ ❖ ❖

Sam Shepard died this day:

"He's actually inside an enclosed screen porch with bugs buzzing, birds chirping, all kinds of summer things going on, on the outside—butterflies, wasps, etc.—but it's very hard to tell from this distance exactly how old he is" –Sam Shepard, from *Spy of the First Person*

Butorides virescens – Green Heron	*Polites peckius* – Peck's Skipper
Cerotainia macrocera – Robber Fly	Chrysididae – Cuckoo Wasps
Oxybelus sp. – Fly-stabber Wasp	*Gasteruption sp.* – Carrot Wasp
Liatris pycnostachya – Prairie Blazing Star	Pompilidae – Spider Wasps
Ischnura verticalis – Eastern Forktail	*Ceropales maculata* – Spider Wasp
Sphecius speciosus – Eastern Cicada Killer	Pompilidae – Spider Wasps
Bicyrtes quadrifasciatus – Stink Bug Hunter	
Lestes unguiculatus – Lyre-tipped Spreadwing	
Cicindela punctulata – Punctured Tiger Beetle	
Sympetrum semicinctum – Band-winged Meadowhawk	
Sympetrum rubicundulum – Ruby Meadowhawk	
Sympetrum vicinum – Autumn Meadowhawk	
Sympetrum obtrusum – White-faced Meadowhawk	
Astata sp. – Square-headed Wasp	

84° / 65°

July 28

AN EIGHTH TYPE OF AMBIGUITY

In general, I feel I'm far from understanding what's going on in the world around me, from world events, to local events, to personal events. And then there's the natural world. As my knowledge of names and my experiences increase, my grasp of the distributions and behaviors is revealed to have immense gaps, lacunae that perhaps cannot be filled. Each day I learn of new plants and animals that I know next to nothing about. The gaps shouldn't be such a surprise given the world's abundance, one person's limits.

In 1930, William Empson published his classic study of poetry and language, *7 Types of Ambiguity*. Ambiguity in poetry has to do with gaps between reader and writer, missing information on one side or another, multiplicities in word meanings like a road with crazy intersections, and expectations of syntax, our habits of interpreting words on the page. A fascinating subject and one of the reasons I find salt and savor in poems, one of the sources of profundity and pleasure. Hardly a day has gone by, over the last thirty years, when I haven't read or re-read some poetry.

There is, I know now (and perhaps I've always known or at least suspected this) a mirror-like relationship between the natural world and literature. This sounds facile and too easy at first. But the hooks of *Arctium* catch more than our clothing. The eye of the Paper Wasp knows your face. And in the space between observing a species for the first time and the failure of attaching a name, comes all the rich oblivion of our planet.

I'm troubled, for instance, that I've not noticed this small robber fly prior to this summer. After noticing it for the first time, I've been seeing it nearly every time I venture out. Even our backyard. This raises the terrible question, *What else am I missing?* I know the answer—a boundless number of things. Today, I found it three times on twig tips. Twice the fly at the tip of the twig was partnered with a second fly, a fuzzier, more silver version of the first, only perched further back along the branch.

Cerotainia albipilosa – Robber Fly
Bombus griseocollis – Brown-belted Bumble Bee
Bicyrtes quadrifasciatus – Stink Bug Hunter
Sympetrum obtrusum – White-faced Meadowhawk
Melissodes sp. – Long-horned Bee
Phymata sp. – Jagged Ambush Bugs
Asilinae – Robber Fly

84°
—
64°

July 29

RIVERSIDE

To Inver Grove Heights. While Lida practiced beach volleyball, I explored the nearby Mississippi River, visiting the Rock Island Swing Bridge city park for the second time (the first time was June 7th).

The bank of the Mississippi River is a perfect place to look for weeds. Seeds spilled from barges or washed downstream from anywhere in the watershed can find their way to this muddy bank. "The calyx of death's bounty giving back / a scattered chapter," so sayeth the poet Hart Crane.

Doing a quick survey beneath the old railroad bridge, I discovered a number of flowers I was unfamiliar with and others I'd rarely encountered before. Allegheny Monkeyflower, Side-flowering Skullcap, Chain Speedwell, and Lanceleaf Frogfruit. The latter, I'd only encountered once before and that was near the Mexican Border, in Mission, Texas.

Perithemis tenera – Eastern Amberwing
Melissodes bimaculatus – Two-spotted Long-horned Bee
Elaphrus sp. – Marsh Ground Beetles
Veronica catenata – Chain Speedwell
Asclepias incarnata – Swamp Milkweed
Mimulus ringens – Allegheny Monkeyflower
Scutellaria lateriflora – Side-flowering Skullcap
Lythrum salicaria – Purple Loosestrife
Phyla lanceolata – Lanceleaf Fogfruit

July 30   86°

PARANTHIDIUM   64°

In our front yard, we have a five-foot by five-foot raised bed garden in which we've started a number of native perennials. This one tiny square of replanted prairie is thriving and so thick with blossoms that one has to question the remaining patches of traditional lawn. We begin to imagine the entire lot relinquished to flowers and tall grass.

In our garden, the Brown-eyed Susans are probably doing a little too well, dominating the space. But at present they share the area with Purple Coneflowers, Butterfly Weed, Bottle-brush Grass, Foxglove, and a single Common Milkweed.

Now that these flowers are at their height, I keep watch to see what insects might be visiting. The usual visitors are some common hoverflies, honeybees, and bumble bees—Brown-belted, Half-black, and Twin-spotted. Today, however, I noticed an unexpected visitor, a carder bee. I hurried in to get my camera and returned to find the bee still there. This turned out to be a rather uncommon native bee, *Paranthidium jugatorium*. A brightly-colored wasp-mimic member of the leaf-cutter bees (Megachilidae) that nests in the ground and lines its nest cells with plant resins.

Paranthidium jugatorium – Carder Bee
Enoplognatha ovata – Common Candy-striped Spider
Polites peckius – Peck's Skipper
Allograpta obliqua – Oblique Stripetail

85°
—
69°

July 31

A WAITING MOVE

In chess, a situation may arise where the best move is a non-move, something that doesn't attack the opponent, something that doesn't damage the position of the pieces, perhaps a simple pacing back and forth of king or queen or rook. This procedure is called a waiting move. Most often this stalling tactic is employed while waiting for the opponent to commit a pawn or piece to a certain plan and reveal something of their future plan.

This idea finds uses in our day to day lives as well. There are many examples. And today it occurred to me that a waiting move has application to wasp watching as well. Locating the nesting sites of the hunting wasps is the first step. Being present and witnessing the wasps returning with their prey requires a little luck and a lot of patience. Practice from earlier wasp-watching vigils paid off today, as I was able to witness three different kinds of hunting wasp return with their prey. First, at a sand pile, amid a swirl of Four-belted Stinkbug Hunters, one wasp out of the dozens present dropped out of flight and landed directly in front of me carrying a stinkbug. Second, while stalking back and forth across the gravel base lines and infield of a practice field, keeping an eye on the various active nests, one of the Buprestid Hunters landed near its nest with a large Buprestid beetle. Third, a Cicada Killer returned to its nest with a Dog-day Cicada.

The latter observation was something I'd been longing to see for some years. I suspected I'd been missing the return flight of these large hunting wasps because I visited their nesting sites too late in the day, apparently after their labors had been complete. Today, I had arrived quite a bit earlier, at 10:00 A.M.

Paranthidium jugatorium – Carder Bee
Chinavia sp. – Green Stink Bugs
Sphecius speciosus – Eastern Cicada Killer
Dicerca sp. – Jewel Beetle

August

St Olaf Natural Lands – August 6

August 1
POLYTYPIC SPECIES

89°
—
67°

Having grown up in northwest Minnesota I was familiar with White Admiral butterflies, with their large, white-margined black wings. After moving to Northfield, in the southeast corner of the state, I was delighted to discover the large iridescent-blue Red-spotted Purple butterflies, then dumbfounded near disbelief to learn that they were the same species.

Most species are monotypic, meaning there is only "one type," and no recognized subspecies. However, there exist species, like *Limenitis arthemis*, for which there are more than one type, species having two or more recognized subspecies. The southern subspecies of this polytypic species mimics the Pipevine Swallowtail, a large poisonous butterfly.

Curiously, another species in this genus, the monotypic Viceroy, is also a mimic, replicating the look of the Monarch, another large poisonous butterfly.

Limenitis arthemis astyanax – Red-spotted Purple

84°
—
67°

August 2

WILD CUCUMBER

There are plants and animals that remind me of home. Actually "remind" is not the right word, instead, it's better to say that these familiar plants and animals make me feel that I'm at home.

Certain landscapes are deep-seated, early acquisitions. The outland of memory. Lakes with sandy shores cobbled at the edges by round, glacial boulders, beds of bulrush, a background of terminal moraine hills covered in basswood and maple, opened sparsely by a smattering of small farms. That's the landscape I knew each day as a child in Otter Tail County. And inside that landscape, I began to learn the plants and animals that shared that place.

Contrasting with the domesticated cucumber which sprawled across the tilled earth of our farmstead garden, some questions about the wider world were latent in the knowledge of the wild cucumber. Its vines climbed to the top of other plants and bushes. Its blossoms spired heavenward and broadcast a scent of jasmine at the wood's edge. Its light-brown seedpods, like tiny, dried pufferfish, dangled from the dead vines and bare branches in the winter. The Wild Cucumber had dealings with the supermundane and gave this farmboy a hint of the complexity of paradise.

> *Catocala innubens* – Betrothed Underwing
> *Robinia pseudoacacia* – Black Locust
> *Ischnura verticalis* – Eastern Forktail
> *Echinocystis lobata* – Wild Cucumber

August 3   67°

MICRO-FISHING   54°

After reading several recent journal posts on micro-fishing, my first reaction was one of consternation with myself; How could I not have thought to pursue fishlings and minnows in this manner!? As a child turning stones, I marveled at the stitlings found hiding beneath some of the stones. During my teen years, I pursued game fish using ultralight tackle. More recently, over the last several years, I've enjoyed some random encounters with native minnows and stream lampreys, usually while searching for dragonfly nymphs. How did I not think to try fishing for them?

The idea is even embedded, in a way, in the opening stanza of one of my favorite poems, 'The Song of Wandering Aengus' by William Butler Yeats:

> I went out to the hazel wood,
> Because a fire was in my head,
> And cut and peeled a hazel wand,
> And hooked a berry to a thread;
> And when white moths were on the wing,
> And moth-like stars were flickering out,
> I dropped the berry in a stream
> And caught a little silver trout.

It made perfect sense when I read that the sport of micro-fishing is big in Japan, the culture that has given us origami, haiku, and bonsai trees.

Emblazoned with the notion of micro-fishing, I decided to give it a try. I sought out the smallest hook in the tackle box and crimped the barb to make it easier to release any small fish caught on it. And using a short length of low test monofilament I tied a quick snelled hook. Then headed for the local stream.

While not being able to follow up on the Yeats poem by using a berry for bait, I did, however, have some success with several unusual micro-baits. A Touch-me-not flower bud, resembling a small kernel of corn, caught a Bluegill. An Eastern Forktail damselfly landed a minnow-sized largemouth bass. A single Swamp Milkweed floweret enticed a Green Sunfish to the hook. I enjoyed catching these tiny game fish, but found myself wondering How can I catch something smaller? No doubt the mantra of many a micro-fisher.

Lepomis cyanellus – Green Sunfish
Lepomis macrochirus – Bluegill
Micropterus salmoides – Largemouth Bass

August 4

DOORWAY TO AUTUMN

78° / 54°

The calendar indicates an exact number of days until summer ends and autumn begins. It's easy to calculate. And yet, I begin to see fallen leaves, reds and browns, and the first goldenrod flowers, the first field thistle blossoms. The inroads to autumn. And yes, it may take an entire month or so to walk there. Yet, to look at this day's landscape and see only summer is to error.

"For a naturalist, the most productive pace is a snail's pace," Edwin Way Teale instructs us in *Journey into Summer*. It may take a book or several books. It may take a lifetime or several lifetimes. The turning of the weather, the tilting of the planet, is a subtle journey—repeated now for millions of years—each footpath, each flowerpath, different from the next.

So summer ends with a deep breath, with a sigh of relief after heavy labors. And goldenrod stands in bloom at the doorway to autumn.

Sympetrum sp. – Meadowhawks
Brickellia eupatorioides – False Boneset
Megachile latimanus – Leaf-cutter Bee
Libellula luctuosa – Widow Skimmer
Eumelissodes – Long-horned Bee
Plantae – Unidentified Flower
Ancyloxypha numitor – Least Skipper
Anax junius – Common Green Darner
Argiope aurantia – Yellow Garden Spider
Sympetrum obtrusum – White-faced Meadowhawk
Solidago canadensis – Canada Goldenrod

79°
—
57°

August 5

EINE KLEINE NACHTMUSIK

Worked on the bathroom ceiling. Went for no walk.

Photographed a few bees at the flowers around the house. A Brown-belted Bumble Bee on the frontyard coneflowers. A cuckoo bee (*Specodes sp.*) on the Butterfly Weed, which may be a new species for the property.

At dusk, I set up the moth light, hoping to lure down a tree cricket. No such luck. They stayed in the canopy and kept up their concert, a little night music.

Photographed a couple of the visitors at the moth light; even though there wasn't much activity given the recent cool days. Several American Idia moths, which are attractive and abundant leaf-litter moths. A Dark-banded Owlet, another leaf-litter moth. A Grass Veneer. And a couple mosquitoes. I noticed, on the ground beneath the light, several small crickets, early instars of Fall Field Crickets. They will add to the night music in the weeks ahead.

Even though I cannot see them, I know they are up there, the tree crickets. I hear them singing through the holes in the leaves.

Phalaenophana pyramusalis – Dark-banded Owlet
Idia americalis – American Idia Moth

August 6   76°
           ___
QUIDDITAS 62°

Net in hand, I went to the prairie to look deliberately for Ruby Meadowhawks. At the edge of the prairie, I netted a male White-faced Meadowhawk. Farther out on the prairie, several hundred yards from the last trees, I netted a male Ruby Meadowhawk. I've found them here every year for the last eight years.

This preference in habitat for open prairie contributes to their quidditas, that which makes it the kind of thing that it is. This is just one of a combination of characters and behaviors that makes the mature male Ruby Meadowhawks recognizable in the field. They are also noticeably larger than the closely related White-faced and Cherry-faced Meadowhawks. On the open plains west of the Mississippi River, the wings of the male Ruby Meadowhawks have extensive amounts of color in the basal halves of their wings. This makes them easy to identify, instantly recognizable. Though I must admit I often feel I'm alone in this opinion; most everyone else professes that meadowhawk identification is intractable, not worth the trouble.

I, too, battled an early bewilderment with the red dragonflies. The first Ruby Meadowhawk I caught I submitted to the Minnesota Odonata Survey Project as a Band-winged Meadowhawk (because of the color in the wings). The path forward wasn't clear. For instance, the guidebook I used at the time contained mistakes and misidentified images: of the three photographs on the Ruby Meadowhawk page none (I would later discover) were of Ruby Meadowhawks! Luckily I had a net, a microscope, and steady populations of red dragonflies nearby to study.

Lestes rectangularis – Slender Spreadwing
Sympetrum rubicundulum – Ruby Meadowhawk
Sympetrum obtrusum – White-faced Meadowhawk

August 7   82°
THE TANGLED BANK   57°

There had been for several years a sandy approach to a pool on Spring Creek, an artifact from a large flood. Visiting the spot today, I found it completely overgrown. Saplings, raspberry canes, waist-high grasses, and stout burdock plants obscured and blockaded the previous path. Here was an unhappy version of Darwin's "tangled bank." A weedy wall, more in the spirit of the threatening riverside vegetation depicted in Joseph Conrad's *The Heart of Darkness*, a kind of living deterrent, "as if Nature herself had tried to ward off intruders."

I fought through to the edge of the water. A bullfrog, the largest I've ever encountered, lurched off the back and splashed into the slimy backwaters as though someone had kicked a flat football into the water. I turned and crawled back through the green wall.

Of the rare occasions when I've felt dismay or distress outdoors, most involved a struggle to find a way through dense, tall vegetation. A hillside forest of Stinging Nettle where all the plants were seven to eight feet in height. The monotonous slog through seemingly endless cattail swamps. The shudder that comes with the sudden ascension of the knowledge that we are not where we want to be. An unwelcome claustrophobia under acres of open sky.

Danaus plexippus – Monarch
Vanessa cardui – Painted Lady
Crabronina – Square-headed Wasp
Sympetrum vicinum – Autumn Meadowhawk
Neoxabea bipunctata – Two-spotted Tree Cricket
Eumelissodes – Long-horned Bee
Cirsium discolor – Field Thistle
Agapostemon – Striped Sweat Bees

84°
—
62°

August 8

PRAIRIE DAYS

McKnight Prairie rarely disappoints. Nearly every visit to this prairie remnant a new plant or insect catches my attention. Today was no exception. Not only did I encounter a number of Band-winged Meadowhawks, but also a number of interesting plants were in bloom: Flodman's Thistle, Spotted Horsemint, and a couple of unknown smaller plants.

I stopped and watched for wasps at the blowout. This open, sandy area at the base of one of the bluffs hosts a variety of sand wasps and solitary bees. The most common sand wasp here is *Bembix americana*. But in addition to this abundant species, I saw two larger sand wasps each with a different pattern of markings on their abdomen. One appears to be *Bembix texana*, wider and longer than *Bembix americana*. The other was a surprise as well, *Stictia carolina*, a Horse Guard Wasp. This wasp gets its common name from hunting horse flies at on and around horses being tormented by the large biting flies. This wasp is much more common in the southern states and is an unexpected observation for Minnesota.

Bembix texana – Sand Wasp
Phymata – Jagged Ambush Bugs
Liatris aspera – Rough Blazing Star
Melanoplus – Spurthroat Grasshoppers
Rudbeckia hirta – Black-eyed Susan
Chamaecrista fasciculata – Partridge Pea
Froelichia floridana – Plains Snakecotton
Colletes aberrans – Aberrant Cellophane Bee
Bembix americana spinolae – American Sand Wasp
Sympetrum semicinctum – Band-winged Meadowhawk
Bombus griseocollis – Brown-belted Bumble Bee
Myzinum quinquecinctum – Five-banded Thynnid Wasp
Schistocerca lineata – Spotted Bird Grasshopper
Thomisidae – Crab Spiders
Monarda punctata – Spotted Horse Mint
Stictia carolina – Horse Guard Wasp
Cirsium flodmanii – Flodman's Thistle

August 9

THINK OF THE POEM

71°
—
63°

A rainy day. Walking the dog in the rain, peering out from under the rain coat's visor, listening to the rain in the trees, I found a Blue Jay feather in the grass. An unexpected splash of color.

In the mailbox is a hand-written postcard from Brian Gooding, a dragon hunter from Texas. Brian's handwriting is a little difficult to read, small letters smeared by the machines that sort the mail, but it's worth the effort. I used to write quite a few letters. Certainly, I regret having given up the practice of handwritten letters. Even still, some of my old correspondents humor me, knowing that an answer will be unlikely. Each time they do write, I vow to send something back, and pay off a small portion of my debt of words.

Today, after finding a handwritten card in the mailbox after finding a feather on the ground, a memory of my printing mentor Don Olsen comes to mind. Among the many things he taught me, he once pointed out that instructions on how to make a goose quill pen could be found in the Encyclopedia Britannica. And then I thought, daydreaming extravagantly, just imagine the letters and poems one might write using a pen fashioned from the Blue Jay feather!

Herman Melville, describes catching a goose and plucking a feather to make a pen in his story 'I and My Chimney.'

While I don't write many letters at the moment, I do still write nearly every day—pen on paper. This is, I suppose, an act of defiance in the quickening electronic age, but, also, a pleasure. I discovered recently, that Lida, my own daughter, cannot read my handwriting. This probably shouldn't have surprised me as much as it did since she never was taught to write longhand or cursive in school.

Cyanocitta cristata – Blue Jay

August 10   74°
SMARTWEED   61°

Our wedding anniversary. Photographed a small smartweed plant blooming in the cracks of the sidewalk on our street. These plants and their flowers have long intrigued me. Even as a child I was aware of them. On our farm, they were a mainstay in the shallow pasture ponds, pink blossoms brightening the muddy shallows that were stoached by the cattle. I still remember a peppery and spicy aroma at certain times of the year, probably around the times I was trying to catch frogs for fishing.

A day of phantom car problems. A dashboard alarm announced air-bag and drive-train problems that needed immediate service. This restricted the car to the "limp-home mode." After arrangements with car dealer and rental car, we drove the car to dealership to drop it off for repair, but now there was no trace of the previous problem. So we drove home, perplexed, relieved, holding our breath.

During a sun-shiny interlude in the late afternoon between spates of rain, I battled the mosquitoes on the dripping trails. At the catchment pond at the corner of the St Olaf Natural Lands, I found a fruit fly on Giant Ragweed. Like many of these flies, it had attractively patterned wings, in this instance something like the elaborate treads of high-end hiking shoes.

Culicidae – Mosquitoes
Euaresta festiva – Fruit Fly
Augochlorini – Metallic Sweat Bees
Persicaria – Knotweeds, Smartweeds, And Waterpeppers
Ambrosia trifida – Giant Ragweed

80°
—
58°

August 11

SPHEX

Because the mosquitoes controlled the shadowy, pondside trails, I walked along the service road that cuts across the natural lands. Not surprisingly, the ditches of this gravel roadway contain a variety of weeds, like every roadway everywhere, the disturbed soils become home to some of the most opportunistic plants in the world. Even if largely unwanted, the adventitious species lend a kind of hothouse diversity to our temperate roadways and urban walks.

When John Clare, a nineteenth-century farm worker and poet, wrote of the fields that brought him such enjoyment, he described the weeds in flower as "troubling the cornfields with destroying beauty."

Broad-winged Thistle. Spotted Knapweed. Common Hemp-nettle. At a stand of the latter weed, a plant of the mint family, a Golden Northern Bumble Bee was visiting the flowers. Like myself, the native bee was making use of the non-native flowers.

Several large sphex wasps were nectaring on native wildflowers that were blooming just beyond the road ditch. A Golden Digger Wasp on White Prairie Clover. A Great Black Wasp on Rattlesnake Master.

Cirsium vulgare – Bull Thistle
Galeopsis tetrahit – Common Hemp-nettle
Bombus fervidus – Golden Northern Bumble Bee
Carduus acanthoides – Broad-winged Thistle
Solidago rigida – Stiff-leaved Goldenrod
Pycnanthemum virginianum – Virginia Mountain Mint
Dolichovespula arenaria – Common Aerial Yellowjacket
Sympetrum obtrusum – White-faced Meadowhawk
Sphex ichneumoneus – Great Golden Digger Wasp
Andropogon gerardii – Big Bluestem
Sphex pensylvanicus – Great Black Wasp
Ambrosia artemisiifolia – Common Ragweed
Centaurea stoebe – Spotted Knapweed

August 12 80°

TREE CRICKETS 58°

These wonderful creatures count among my favorite insects. Five years ago I didn't know they existed. Now I delight at each opportunity to observe them. The tree crickets have been singing for a number of weeks, but I've yet to see one this year.

After midnight last night, early this morning, I captured a Two-spotted Tree Cricket from the outside of the screen of the kitchen window. After capture, the cricket spent the evening in the refrigerator and was photographed and released this morning. Tree crickets are long and sleek, orthogonally structured, somewhat ethereal in the wings.

❖ ❖ ❖

The front yard elm is yellowing, already leaves have fallen.

"It is already the yellowing year." – Thoreau

Neoxabea bipunctata – Two-spotted Tree Cricket
Gasteruption sp. – Carrot Wasp
Eristalis flavipes – Orange-legged Drone Fly
Cricetidae – Vole

72°
—
61°

August 13

LETTERED MOTH

The Lettered Habrosyne Moth or The Scribe, a member of the Drepanidae family of moths, is a beautifully patterned and colored moth. In shape, it resembles the owlet moths of family Noctuidae, but it's not an owlet moth, which is most likely why they are sometimes referred to as false owlet moths. This species ranges from coast to coast in a fairly slender band across southern Canada and the northern United States.

According to *Caterpillars of Eastern North America* by David L. Wagner, the caterpillars of the thyratirine moths feed on Ironwood, Birch, Black Raspberry and leaves of other members of the birch and rose families. The caterpillars construct shelters by tying together the edges of a single leaf or several leaves.

In 1894, British entomologist Augustus Radcliffe Grote collected his first Lettered Habrosyne on a collecting trip near Lake Erie and described it as "a lovely blossom, tossed loose from the spray of spring." Not sure what this overly "flowery" praise might mean beyond that simple fact that he thought it pretty.

Habrosyne scripta – Lettered Habrosyne Moth

August 14

GREAT INDIAN PLANTAIN

76°
—
62°

"Seeds, of all the living the most inscrutable, keep their future goals closed like a fist clenched around a secret object. A ladder of leaves suddenly springs from the soil as if predestined. And the flowers become better, more fanciful suns."?

Several years ago I gathered the seeds. They sat in a bag in our cupboard until last fall when I thought to rake them into the new garden. Just last week, Lisa asked me about some odd looking leaves in the garden, spade-shaped. It occurred to me that they might be the Great Indian Plantain which I'd sowed and then forgotten about.

Today I visited a population to compare leaves and see what was visiting the flowers. Right away I recognized the new leaves in the undergrowth. On the flowers, despite overcast skies, Honey Bees, Drone Flies (a Honey Bee mimic) and Great Black Wasps with iridescent blue-black wings.

Seeds, of all the living the most brilliant at capturing / containing the future; they travel on the wind, in the pincers of an ant, hitched to the fur of an animal, and arrive where hope is most needed. And in the open palm, a key, shiny, yet to be used.

Arnoglossum reniforme – Great Indian Plantain
Ambrosia trifida – Giant Ragweed
Ischnura verticalis – Eastern Forktail
Sphex pensylvanicus – Great Black Wasp
Eristalis tenax – Drone Fly

78°
—
59°

August 15

BIG YELLOW BUTTERFLY

Continuing a string of less than sunny days like a grim character that occasionally lets down their hard exterior yet not ever smiling.

Lida drove me to St Olaf. She had obtained her learner's permit shortly after her birthday, but this was my first time riding in the car as a passenger while she took the wheel. She drove very cautiously, possibly a little too cautious. Yet she did a good job navigating the campus hill and the curvy roads and speed bumps, all brand new for her.

At the natural lands, a big yellow butterfly, an Eastern Tiger Swallowtail, nectared on the blooms of Field Thistle. A Lemon Cuckoo Bumble Bee joined it. Nearby, on goldenrod, I noticed the thick-headed fly, Physocephala tibialis. This convincing Mason Wasp mimic is thought to parasitize not the wasps which the adult flies resemble but Two-spotted Bumble Bees.

Spinus tristis – American Goldfinch
Perithemis tenera – Eastern Amberwing
Dolichovespula arenaria – Common Aerial Yellowjacket
Isodontia mexicana – Grass-carrying Wasp
Bombus citrinus – Lemon Cuckoo Bumble Bee
Papilio glaucus – Eastern Tiger Swallowtail
Physocephala tibialis – Common Eastern Physocephala
Melissodes trinodis – Long-horned Bee
Noctua pronuba – Large Yellow Underwing

August 16

PROBING THE UNDERWORLD

78°
—
64°

Nature happens. For instance, today, a walk to the library became a natural history outing. Last year, on a similar visit, I'd encountered an active colony of Great Golden Digger Wasps with several dozen nest entrances in a space only several square feet in size just outside the library doors. This year a Pelecinid wasp clung to the stone wall of the library. There's no avoiding nature.

Unlike the previous year, I was without my camera, unprepared but not entirely without means to secure an observation. I'd yet to photograph this wasp this year, so I hurriedly made a paper triangle out of a sheet of paper I had along, then captured the wasp by hand and stored it away inside the triangular envelope. I entered the library, holding the papered insect as inconspicuously as possible. I hurried through the stacks to retrieve the desired book, the wasp making quiet scraping noises all the while. I checked out the book, exited the library, and walked home, photographing the wasp later in the day.

Pelecinid wasps—their body of black enamel with a long, trailing abdomen like the tail of a Chinese dragon—are harmless to people, despite a wild and fierce appearance. They are parasitoids of scarab beetle larvae, the white grubs that can sometimes be a pest to gardens. They use that exceptionally long abdomen to probe the underworld of the soil for their host and when it finds one it deposits an egg. The egg hatches and then the wasp larva consumes the beetle larva. I wonder if they parasitize Japanese Beetles, an invasive scarab?

Pelecinus polyturator – American Pelecinid Wasp
Neotibicen canicularis – Northern Dog-day Cicada

74°
—
63°

August 17

BIRD'S NEST FUNGI

Emptying the compost bucket, I noticed a Giant Swallowtail, the first in years. After a dash inside to grab the camera and a subsequent data-card malfunction, I watched the butterfly flounce its way through the neighbor's yard, clear a backyard fence and pass out of sight. A little dejected to not get the photo but pleased to have seen this uncommon butterfly in our backyard, I returned to my original task of dumping a week's accumulation of coffee grounds and moldy vegetable and fruit peelings.

On the ground near the garden fence I noticed something even more uncommon than the butterfly, and if not more uncommon at least something I hadn't seen before, Bird's Nest Fungi. These unlikely-shaped fungi, common among woody debris in forests and the suburban equivalent of wood chip mulch, create a scattering of tiny hollow cones, shaggy on the outside, opalescent on the inside. A spontaneous creation, these conical fruiting bodies, of the unseen mycelium connecting in the dark like brain cells.

Each fruiting body flared and fluted like a fancy glass vase. The "eggs" of each nest, after the recent heavy rains, were missing, dispersed by splashing water drops. The same method of spore dispersal evolved among the lichens and mosses as well. As a saprophytic fungi, the Bird's Nest Fungi furthers the decomposition of organic matter in soils.

Cyathus striatus – Fluted Bird's Nest Fungus

August 18 75°

THE FIRST STARS ARE OUT

60°

After several rainy days, I took a short hike as soon as the skies cleared, late in the afternoon. I noticed the first asters of the year, exordium of autumn. The several plants all had white flowers. The shape of the leaves suggested Panicled Aster (*Symphyotrichum lanceolatum*).

Rounding the catchment pond, I paused to photograph a Spotted Spreadwing. Taking a few more steps, I noticed a large fly perched low to the ground. I recognized the Japanese animae coloration, the cylindrical, pill-shaped body as that of a Bot Fly. I'd seen another one in Saskatchewan, north of the Churchill River, in July 2013, then a pitch-black one in a canyon in Arizona in October 2014, but hadn't ever seen one in Minnesota. This particular fly was a Lagomorph Bot Fly, parasitic of rabbits.

Fallopia convolvulus – Black-bindweed
Trypoxylon politum – Organ Pipe Mud Dauber
Cuterebra fontinella – Lagomorph Bot Fly
Bicyrtes quadrifasciatus – Stink Bug Hunter
Symphyotrichum lanceolatum – Panicled Aster
Sympetrum vicinum – Autumn Meadowhawk
Melanoplus differentialis – Differential Grasshopper
Agapostemon – Striped Sweat Bees
Lestes congener – Spotted Spreadwing
Eumelissodes – Long-horned Bee

80°
—
59°

August 19

GIANT SWALLOWTAIL

I stepped outside to photograph a Giant Swallowtail that Lisa had spotted on the front flowers.

The host plant for this magnificently large butterfly is prickly ash.

Prickly Ash reminds me of a conversation I once had with poet John Rezmerski in his kitchen in Mankato, Minnesota. While preparing one of the best tuna-fish sandwiches I've ever eaten, John expounded on a wide variety of topics—not a rarity with this gregarious polymath—landing for a while on spices. His favorite spice, Sichuan pepper, had become difficult to obtain because of citrus bans in place to protect American citrus crops. "I've wondered, since it's related to Prickly Ash, if the prickly ash berries might be used as a substitute?" This was all news to me, though I said I'd collect some berries if I had the chance.

Prickly Ash grew into impassable thickets in places among the glacial hills in western Minnesota where I grew up. And early on I discovered their musky, lemony scent.

And now this note has gone as far astray as the tatter-winged Giant Swallowtail that found its way north to our yard.

Papilio cresphontes – Eastern Giant Swallowtail
Eristalis transversa – Transverse Flower Fly
Geometridae – Geometer Moths
Leucauge venusta – Orchard Orbweaver

August 20

HELPER FLY

86°
—
66°

Final push to make the renovated bathroom operational before the inlaws visited in the afternoon. Didn't happen. Installed the toilet and then tried to install the vanity, but found I was missing a piece needed to connect the drain. Ah well... it's that much closer. While doing some work on the sawhorses in front of the garage, a strange little fly landed several times on the bright yellow level. Eventually, I went and found my camera and photographed the fly. It turned out to be a rather uncommon syrphid fly, a Lesser Bulb Fly, this being the first observation for Minnesota that I know of.

◆ ◆ ◆

Janet Nelson, an iNaturalist acquaintance from Bemidji, was in town so we met at the St Olaf Natural Lands for a walk around the big pond. The number of Monarch butterflies impressed both of us. Nearly every spear of Blazing Star had its own butterfly. Sometimes there were dozens flying in iconic clouds above the Big Bluestem prairie grass. Amidst all the Monarchs, we found a single dragonfly, a Halloween Pennant. An old individual, its colors sun-faded—I'd say grizzled if it wasn't so transparent and airy—a dragonfly on the verge of no longer existing.

Before departing, Janet handed me several small jars filled with dragonfly exuviae, which she'd collected from Lake Kabetogoma earlier in the year. I promised to identify what I could, when I could. [When examined later, the exuviae included Springtime Darners, Stream Cruisers, Chalk-fronted Corporals, and Baskettails, a lot of Baskettails.]

Celithemis eponina – Halloween Pennant
Eumerus funeralis – Lesser Bulb Fly

80°
—
66°

August 21

LONG-NECKED GROUND BEETLE

A solar eclipse. Unfortunately, the day was heavy with clouds with intermittent rain. The sky darkened and, occasionally, lightened, but we were granted no direct glimpse of the eclipse. If it had been visible it would have been near total, about 90%, Northfield being several hundred miles north of the path of totality.

❖ ❖ ❖

"Such is the force of its 'ecology' " – St.-John Perse

Such is the force of its ecology, for the Long-necked Ground Beetle to go about its life almost completely unnoticed—uncelebrated in the annals of science, and virtually unknown throughout its range in eastern North America. Of course, it's not along on this account; countless insect species persist unnoticed, and I'd venture, unappreciated.

My eyesight is no longer what it used to be. Books with small print are more and more difficult to read. In low light the text disintegrates, blurs, the individual letters obstructed as if some mean hand had thrown a scattering of sand across the page.

So, when I collected this beetle from the inside of our screen porch, placing it in a vial, its small size and shape registered in my brain as an ant. A few hours later, after it had cooled in the refrigerator, I brought it out for some macro photos before releasing it. Only then did I see that it was a beetle, curiously shaped, curiously colored.

❖ ❖ ❖

Lisa and I walked the dog during a majestic sunset. The entire western horizon had a fiery, incandescent glow, burning for quite a long time above the neighborhood houses. I was reminded of the old saw my father often repeated, "red skies at night, a sailor's delight."

Colliuris pensylvanica – Long-necked Ground Beetle

76°
—
60°

August 22

GREEN-STRIPED DARNER

After the damp air moved east late yesterday, the humidity dropped, the temperature cooled, and the skies cleared. Unsettled to calm. Sticky to comfortable. This day started off fine and stayed fine to the end.

Being such a fine day, I indulged in two walks: the first, an early walk at the soccer fields, the second to St Olaf Natural Lands in the afternoon. It was an eventful walk. Collected the first meadowhawk eggs of the year. Photographed a Viceroy, a Painted Lady, a Giant Swallowtail, and an Eastern Tiger Swallowtail. Then I happened upon a Green-striped Darner. Only the second I've ever seen. Amazingly, the first was observed ten years earlier (September 8, 2007), less than two hundred feet from where this one was observed.

Green-striped Darners are very similar to the much more abundant Canada Darners in appearance. Because they are not often encountered, they keep a kind of mystique.

Xysticus sp. – Ground Crab Spiders
Danaus plexippus – Monarch
Atalopedes campestris – Sachem
Aeshna verticalis – Green-striped Darner
Erythemis simplicicollis – Eastern Pondhawk
Papilio glaucus – Eastern Tiger Swallowtail
Papilio cresphontes – Eastern Giant Swallowtail
Sympetrum vicinum – Autumn Meadowhawk
Bombus griseocollis – Brown-belted Bumble Bee
Coreopsis tinctoria – Plains Coreopsis
Sympetrum obtrusum – White-faced Meadowhawk
Sympetrum semicinctum – Band-winged Meadowhawk
Erythemis simplicicollis – Eastern Pondhawk
Amblycorypha – Round-headed Katydids
Vanessa cardui – Painted Lady
Limenitis archippus – Viceroy
Panorpa sp. – Scorpion Fly

August 23   77°

OWLET CATERPILLAR   56°

Inspecting the flowers in the front yard, I noticed a caterpillar at the center of one of the Brown-eyed Susan flowers. A dark and drab and overall rather nondescript caterpillar. The number of prolegs and their positions suggested it may be an owlet caterpillar, family Noctuidae. This prompted a look through the large and lavish guidebook *Owlet Caterpillars of Eastern North America* by David Wagner and others (Princeton University Press, 2012), though without finding any species that looked remotely like this caterpillar.

*

Yesterday evening, a young man came to the door asking us to support Clean Water Action. I gave him some money to aid their fight against legislation afoot that aims to undermine current environmental protections and restrictions. Money to fight money. While there's no question as to where my loyalties fall, I often resent that money should be the measure of commitment to a cause. Having cared and worried and fretted about such issues for my entire adult life, I still hope the little I have left...words, a worn-out heart, a well-seasoned anger... helps.

Noctuidae – Cutworm Moths And Allies
Bombus griseocollis – Brown-belted Bumble Bee
Lasioglossum – Sweat Bee

73°
59°

August 24

THE HISTORY OF THE DECLINE AND FALL

From here on out the days will chart the history of the decline and fall of the temperate summer, that green empire built in June and July. What better than to commence this history than with the harvest of some Bottlebrush Grass seed. These long-awned seeds will be sown in a couple promising spots in our yard later this fall.

Elymus hystrix – Bottlebrush Grass
Metcalfa pruinosa – Citrus Flatid Planthopper
Parasteatoda sp. – Cobweb Spider

August 25

ELUSIVE CLUBTAIL

71°
57°

On the downslope of summer, beneath a swirl of low gray clouds, against a cool presaging wind, I entered the Cowling Arboretum. Some middle school boys fished the pool downstream of the bridge and weir. They whooped and cheered as one of the boys held up a respectable Northern Pike.

I sauntered through the small meadow that had, until a few years ago, been the site of several tennis courts. Today, all the edges of the rectangular clearing were bright yellow with wide swaths of blooming goldenrod. Even in the absence of sun, the flowers attracted a lot of insects: Honey Bees, Jagged Ambush Bugs, Goldenrod Soldier Beetles, Black Blister Beetles, Beewolves, and various other bees and flies. The Five-banded Thynnid Wasps (moved recently from the Tiphiidae family) were especially prevalent, especially the large females.

After the goldenrod, I ambled riverside, enjoying the cool weather. A sparse rain fell at one point and then stopped after a few minutes. I walked to the turtle pond with the intent of looking for Saffron-winged Meadowhawks, which I've occasionally encountered near that pond in past years. Walking along the river, knowing September to be less than a week away, I reflected, wistfully, on not having seen a Riverine Clubtail this year. I'd missed seeing Plains Clubtails as well. These regrets faded as I began to encounter a number of darners along the trail, and they vanished completely when I happened upon Great Black Wasps and Great Golden Digger Wasps nectaring on the white blooms of Whorled Milkweed.

After inspecting the turtle pond, witnessing a few massive carp surge and shake up the shallows, hearing numerous frogs yelp as they launched themselves into the water, and finding no meadowhawks, I nearly turned back, but decided to push on to the river. A lucky decision. As I stood on the edge of the riverbank a black dragonfly lifted from the river, glided easily beneath my arm, and disappeared behind me. I comprehended, mid-turn, as I followed its flight, that it was a clubtail. Most likely, I thought it probably kept flying and landed someplace far off or high up. I searched the nearby grass anyway and saw it perched just over ten feet away. This late in the year it was probably a Riverine Clubtail, the very dragonfly I was lamenting not having seen earlier. Very cautiously I inched closer. This dragonfly was not one I'd seen before. Darker and marked differently than the Riverine Clubtail I'd expected it to be, it came in a flash that this was an Elusive Clubtail.

My first encounter with this species was in downtown Sioux Falls, South Dakota, in September 2010. That dragonfly, which I managed to capture by hand, had landed on the back of a bronze statue lion. A not-so-elusive Elusive Clubtail, I joked at the time. Since then I've found nymphs and exuviae in several rivers, including the Cannon River. Twice I've seen tenerals make their first flight to perches high in the treetops. But I have not had another good look at an adult since seeing that first one in South Dakota...until today. I've long suspected its presence in the Cannon River, having found nymphs several times (though never successfully rearing them). Ten years stalking the banks of this river for dragonflies and this was the first Elusive Clubtail I've seen here.

[Observations →]

Stylurus notatus – Elusive Clubtail
Anax junius – Common Green Darner
Sympetrum semicinctum – Band-winged Meadowhawk
Sympetrum vicinum – Autumn Meadowhawk
Sympetrum obtrusum – White-faced Meadowhawk
Xanthotype – Crocus Geometer Moths
Sphex ichneumoneus – Great Golden Digger Wasp
Ardea herodias – Great Blue Heron
Cuscuta gronovii – Scaldweed
Cupido comyntas – Eastern Tailed-blue
Timulla sp. – Velvet Ant
Epicauta pennsylvanica – Black Blister Beetle
Astata sp. – Square-headed Wasp
Myzinum quinquecinctum – Five-banded Thynnid Wasp
Philanthus gibbosus – Corrugated Beewolf
Asclepias verticillata – Whorled Milkweed

67°
—
60°

August 26

OBSIDIAN ARCHER

A day of rain and mist. While working in the basement, I discovered a tiny, black cricket. Seeing or hearing these insects at the end of summer and throughout the fall, I often think of this small poem written by my neo-surrealist friend, Roger Parish:

DEAD CRICKET

You sprawl
like a crushed piano,
O obsidian archer!

For you,
the siege is over
and the song forgotten.
Never again
will your dark arrows
pierce my screens.

Eunemobius carolinus – Carolina Ground Cricket
Pyralis farinalis – Meal Moth

Platydracus sp. – Rove Beetle
Cyprinus – Typical Carps
Sphex ichneumoneus – Great Golden Digger Wasp
Orchelimum nigripes – Black-legged Meadow Katydid
Sympetrum obtrusum – White-faced Meadowhawk
Hyla versicolor – Gray Tree Frog
Sparganium sp. – Bur-reeds
Sagittaria latifolia – Broadleaf Arrowhead
Ancistrocerus sp. – Mason Wasp
Phyciodes tharos – Pearl Crescent

August 27   73°
"DANCING AMONG THE WILDFLOWERS"   63°

Lisa, Lida and I attended an art opening at the Minnesota Landscape Arboretum in Chaska, featuring the work of Jim Fletcher.

Jim Fletcher, a remarkable artist and storyteller, has been a longtime family friend. He taught at the Pelican Rapids High School where my father taught and where I went to school. As a child, I marveled at his creativity, his joie de vivre. When I decided to try my hand at writing, it was Jim who first encouraged me. A black-belt in several martial arts and inventor of his own set of complicated and rigorous forms, the same strength and control (one could also use the words confidence and style) presents itself, shines through, Jim's artwork. Vibrant and kinetic even when representing a landscape. The present show was commissioned by the Arboretum, a series of paintings of various Minnesota wildflowers. I was surprised by this new subject and amazed at what he'd done. These were not botanical drawings, nor photographic replicas, but wonderful, artful, representations, full of movement and life.

Before attending, we visited the wetlands on the south edge of the arboretum and took a quick stroll on the boardwalks there. After the opening, we walked through some of the gardens. In the Kitchen Garden, amongst a wide variety of herbs, we watched a Great Golden Digger Wasp fly in with a Black-legged Meadow Katydid and were lucky enough to see it whisked down into the nest by the wasp.

[Observations on previous page]

August 28

AMERICAN DAGGER

An American Dagger Moth caterpillar crawled across our back step. An impressive caterpillar, long yellow hairs with a couple of paired black tufts and a shiny, onyx-black head, well over an inch in length. The moth is large and impressive as well, carrying two black dagger marks on its white wings (an echo, almost, of the caterpillar's paired black tufts of hair).

No doubt the caterpillar was searching for a place to pupate and make its cocoon. After that, it will bide its time through the cold months and emerge next summer, nearly a year's wait. Such patience among the short-lived.

Acronicta americana – American Dagger
Graphocephala – Red-banded Leafhopper

August 29
BUTTONOLOGY

80°
—
58°

Darner hunting—it's that time of year. These large dragonflies can be found patrolling above trails and over open fields. Earlier in the year Common Green Darners and Blue-eyed Darners held these airspaces along with a few Saddlebags and Gliders. But now the mosaic darners, of the genus *Aeshna*, have arrived on the scene. Like all darners, because they seldom land, they pose a real problem to the collector of either specimen or image. It's chancy with net or camera. Today I was using the latter.

The most common mosaic darner most years, at least here in Northfield, is the Lance-tipped Darner. Its formidable name comes from the harmless appendages on the female. This somewhat misleading name must be preferred to a common name based upon the scientific name, *Aeshna constricta*, which would translate to something like "pinched" darner, the adjective referring to the notch in the front thoracic stripe. Along an open and sunlit stretch of trail, I found a darner patrolling back and forth, much of the time on a level about equal with my eyes but also making great loops up among the treetops but then returning to its usual back-and-forth patrol along the trail. Each time it passed by me I could see green thoracic stripes and blue, patterned abdomen. And I assumed it was a Lance-tipped Darner.

With the camera, I focused on the ground about six feet away then tilted the camera up. When the dragonfly flew by at about that distance, I pointed the camera and held down the shutter button, taking rapid-fire digital images, hoping one might be in focus. Later, I'd dig through the hundred or so images for a usable shot. There were no guarantees, but I'd had some luck using this technique. What I found, when I looked at the photos, was not a Lance-tipped Darner like I'd expected, but another Green-striped Darner (I'd seen one near here the previous week).

◆ ◆ ◆

Earlier in the day, I'd read the chapter on "buttonology" in Fredrik Sjoberg's memoir, *The Fly Trap*. A deprecatory term invented by August Strindberg to make fun of human beings who collect things. Think beer cans, Hummel figurines, thimbles, shot glasses, editions of Moby Dick, square nails, and…yes…insects, whether pinned specimens or digital photographs. Logging into my online storage site, I noticed it now held 11,301 photos. Absurd. Well within the range of satire. One winces at just what level of mania is involved here.

I add a couple more photos of a drone fly in Sjoberg honor.

Eristalis stipator – Yellow-shouldered Drone Fly
Andrena helianthi – Sunflower Mining Bee
Augochloropsis sp. – Green Metallic Sweat Bee
Eumenes fraternus – Fraternal Potter Wasp
Aeshna verticalis – Green-striped Darner
Lestes congener – Spotted Spreadwing

August 30   83°

NET WORK   62°

Because Green-striped Darners look remarkable like Canada Darners, some close-up photos were needed to confirm the identification of the dragonflies I'd been seeing. And the best way to get those images was to net one and photograph it in hand. So I set out a little before noon to do just that.

It's good to know I'm still capable of netting a dragonfly when I need to. Though certainly I don't get near as much practice as I once did. My netting abilities peaked some years back—back when I had better eyesight and my fast-twitch muscles were twitchier. The summers of 2010, 2011, and 2012, most days I was in the field, net in hand. I was at the top of my game. Now, when I do take out the net, I practice more of a zen-in-the-art-of-archery approach to netting—deep breaths, calm patience, and a practical acceptance of failure.

And the net itself has seen better days. Its segmented handle and collapsible rim allow the net to be taken apart and stored compactly. The rim, when it was new, was sturdy and circular. Today, when I unfold it, the shape resembles the wobbly edge of a wide-brimmed hat. And it jiggles like jello as I walk. Still, the net bag is in good shape and I'm confident I can capture many more dragonflies with it.

The darners were not on the trails where I'd seen them yesterday. Eventually, I found one flying over a small clearing. It disappeared before I could get very close. Then, not far downhill, several more could be seen flying over the grassy margin at the edge of a small pond. Recent rain had flooded much of the area making it difficult to get near the dragonflies, but stepping hummock to hummock I made it to the edge of the tall grass. Then I waited and watched. There were four darners nearby, though there were no doubt more scattered around the pond. Each hovered three to six feet above the flooded grass, guarding their own territories. These territories seemed somewhat arbitrary and fluid. The darners chased away every invader, most often smaller dragonflies or a passing wasp or bee. Each skirmish with another darner usually ended with some adjustments to the amount and location of the airspace being controlled and guarded. So it was only a matter of time and chance until one of the dragonflies stationed itself near where I stood. About twenty minutes passed before one hovered off my left shoulder. As it moved forward, one quick upward sweep of the net was all it took to capture the dragonfly, a Green-striped Darner.

Aeshna verticalis – Green-striped Darner
Orchelimum nigripes – Black-legged Meadow Katydid

August 31, 2017 — 71° / 56°

OCYPTAMUS

Found a wonderful fly in the backyard, an Eastern Band-winged Hoverfly (Ocyptamus fascipennis). It was nectaring on the tiny yellow Oxalis flowers, behaving like a mosquito-sized hummingbird.

One of the enemies of the Scale Insects (superfamily Coccoidea) is the entomophagous fungi Cordyceps. Apparently, another foe is the larvae of this fly. Many hoverfly larvae prey upon aphids, but the larvae of the Eastern Band-winged Hoverfly preys upon scale insects. Interestingly, both scale insects and aphids are often tended by ants.

❖ ❖ ❖

Volleyball match. Lida set the JV game against Faribault. Her team won in two sets. Bruce and Karen, just back from their trip to Las Vegas, came to the match.

Ocyptamus fascipennis – Eastern Band-winged Hover Fly

September

Creek Chub – Cowling Arboretum – September 27

September 1

GRIEF AMONG THE GOLDENRODS

72° — 53°

Woke to a phone call from Karen who was at the hospital. Bruce, Lisa's father, had had a heart attack. Lisa got out of bed and drove to meet her mother in Minneapolis. I stayed home with Lida.

I was walking through the goldenrod fields when Lisa called in the afternoon with news about her father. It wasn't good. Bruce was on life-support, heart stable, but in an induced coma. It wouldn't be until the next day before they would begin to know if he could recover. Lisa was crying on the phone. I was crying among the wildflowers.

Photographed a Black-tipped Darner, so suitably named for this black-tipped day. This beautiful dragonfly was completely unexpected and a new Rice County record.

> *Aeshna tuberculifera* – Black-tipped Darner
> *Spilomyia sayi* – Four-lined Hornet Fly
> *Toxomerus politus* – Maize Calligrapher
> *Helophilus fasciatus* – Narrow-headed Sun Fly
> *Eristalis transversa* – Transverse Flower Fly
> *Toxomerus geminatus* – Eastern Calligrapher
> *Chalcosyrphus nemorum* – Dusky-banded Forest Fly
> *Andrena helianthi* – Sunflower Mining Bee
> *Astata unicolor* – Squared-headed Wasp
> *Nomada sp.* – Nomad Bees
> *Aeshna umbrosa* – Shadow Darner
> *Bombus auricomus* – Black-and-Gold Bumble Bee
> *Bicyrtes quadrifasciatus* – Stink Bug Hunter
> *Dissosteira carolina* – Carolina Grasshopper
> *Eremnophila aureonotata* – Gold-marked Wasp
> *Sympetrum vicinum* – Autumn Meadowhawk
> *Dialictus* – Metallic Sweat Bees
> *Eris militaris* – Bronze Jumper
> *Schizomyia racemicola* – Goldenrod Gall Midge

82°
—
64°

September 2

AN OVERLOOKED CORNER

Bruce died in the afternoon.

Unlike the sadness that comes second-hand in a novel or in a movie, this sadness, the sadness of losing someone close, someone you love, hurts and never stops hurting.

❖ ❖ ❖

Earlier in the day, while waiting for this heartbreaking news, I took a walk. A cross-country meet at the St Olaf Natural Lands deterred me from that place. A Carleton football game made the lower Arboretum inaccessible because there was no parking. So I drove on, trying to decide on an alternative. When I turned onto Canada Avenue I remembered that there were a trailhead and parking lot here, access to a corner of the Arboretum, that I've overlooked. At one time, according to a sign near the entrance, there was a cabin here belonging to the Carleton Women's League.

The open countryside, not restored to native prairie but simply re-wilding farm fields, consisted of acres of Canada Goldenrod in bloom. Broken here and there by trees and fence lines and colorful ridges of Sumac beginning to turn toward autumn colors. As I set out along one trail a large dragonfly lifted from the edge, its wings shimmering copper in flight. I followed its flight and took note of where it landed, which allowed me to sneak up to it and get a photo, a teneral Red Saddlebags. Where had it come from, there being no open water nearby?

As the trail proceeded uphill, the ground underfoot became very sandy and there were suddenly dozens of large, green tiger beetles, Purple Tiger Beetles, taking flight and landing and quite a number of smaller, black tiger beetles, Punctured Tiger Beetles, scurrying away on long legs. After watching several beewolves burrowing at the edge of the trail, I turned and walked back. Along the way, I noticed a truly large sphinx moth caterpillar on some grapevine, prominent horn, showy eyespots, a Pandora Sphinx Moth caterpillar.

Tramea onusta – Red Saddlebags
Eumorpha pandorus – Pandora Sphinx
Geometridae – Geometer Moths
Schizomyia eupatoriflorae – Gall Midge
Oxybelus sp. – Fly-stabber Wasp
Cicindela purpurea – Purple Tiger Beetle
Philanthus gibbosus – Corrugated Beewolf
Cicindela punctulata – Punctured Tiger Beetle
Enallagma civile – Familiar Bluet
Danaus plexippus – Monarch
Chalcidoidea – Chalcidoid Wasps
Callopistromyia strigula – Picture-winged Fly

85°
—
60°

September 3

OAK APPLES AND MILKWEED CATERPILLARS

Took a walk in the afternoon around the big pond at the St Olaf Natural Lands.

Oak apple galls littered the trail beneath the Red Oaks out on the prairie, the galls falling like crab apples from the branches above. These windfall "fruit" contain Cynipid wasp larvae. Curious to see if the larvae were still inside the galls I broke one and found a hard center core about the size of a lemon seed. Inside this hard pit, split open with my fingernail, nestled a small white grub, the gall wasp larvae. These larvae will pupate and emerge at some point in the winter. They will be wingless and they will make their way to the nearby tree, crossing snow if need be, and they will climb to the new leaf buds to lay their eggs. Such are the steps in the life cycle of the Oak Apple Gall Wasp. At a time when the oaks begin to drop acorns, the galls fall as well.

Searching the milkweed plants for Oleander Aphids, I hoped to find and photograph the hoverfly larvae that prey upon these aphids. Instead, a number of other interesting insects turned up. Ants tending the aphids. Milkweed Leaf Beetles. Monarch caterpillars. Weevils. And, most numerous on this day, Milkweed Tussock Moth caterpillars.

Amphibolips cookii – Oak Apple Gall Wasp
Euchaetes egle – Milkweed Tussock Moth
Araneus trifolium – Shamrock Orbweaver
Cynipidae – Fuzzy Oak Gall Wasp

September 4

COLD FRONT

76° / 55°

In the late afternoon, while Lisa and Lida were out shopping for school supplies, a cold front blew through town. A low wall of black rain clouds suddenly darkened the sky to the northwest. Cool winds preceded the rain. I hurried to replace a few patio bricks I'd been raising and leveling, then reinstall the rain gutter downspout I'd removed.

As the front moved through and passed by, the temperature dropped from the 70s to the 50s, twenty degrees in a matter of minutes. A dramatic end to summer vacation.

❖ ❖ ❖

The circle of the seasons isn't circular, rather serrated and looped like the circumference of a flower or the circumnavigation of an island.

Solidago flexicaulis – Broad-leaved Goldenrod
Araneus thaddeus – Lattice Orbweaver

65°
—
51°

September 5

FISHING SPIDER

Visited the Cowling Arboretum. A large cottonwood tree had crashed down onto the soccer field, crushing the pipes and netting of a soccer goal. No doubt the high winds of yesterday's cold front had toppled it. The upper branches disintegrated upon impact with the ground, leaving a rubble of sticks and leaves, the wide and voluminous canopy of the tree collapsed into a thin outline of debris, all the space held in its great branches disappeared. It was a strange site. Most other kinds of trees when they fall hold their shape. Only the main trunk remained unbroken. It was immense, at least ten feet in circumference near the base, a toppled tower.

Not far from the downed tree, I noticed what looked to be the big, silky nest of tent caterpillars, except that it was too late in the year and not in the branches of a tree but around the top of a goldenrod plant. The entire top of the plant, flower and leaves, was encased in silk. In the center was a dark mass. I reached for the plant, intending to bend it closer. In doing so I almost placed my hand on top of a very large fishing spider. This was a spider nursery.

> "It's an enormous spider, impeded by the abdomen
> from following its head.
> And I've thought about its eyes
> and its numerous legs…
> And today what worry that traveler brought me!"
> – César Vallejo, from 'The Spider'

While I wouldn't go out of my way to handle a fishing spider, enormous spiders don't strike terror into my being. Instead, I rather like watching them...from a distance. Given how difficult it can be for us to find our way with two feet, how much more difficult must it be for the spider? Or, how much easier!?

Bidens cernua – Nodding Beggarticks
Dolomedes scriptus – Striped Fishing Spider

68°
—
49°

September 6

A PARADE OF PAINTED LADIES

I noticed the first Painted Ladies in the morning while walking the dog. Several under the neighbor's plum tree, sampling the fallen fruit. Then, most decoratively, in a garden bright with multi-colored chrysanthemums, Painted Ladies perched, one per flower, each in the center with wings closed. Later, while driving, dozens flew across the city street. And, on a short walk along a railroad track, clusters of butterflies burst into flight. The day became a tapestry of butterflies. A parade of Painted Ladies.

I wasn't alone. So many and widespread were the reports of these butterflies, they made the evening news.

Toxomerus geminatus – Eastern Calligrapher
Regulus calendula – Ruby-crowned Kinglet
Vanessa cardui – Painted Lady
Melanoplus differentialis – Differential Grasshopper
Accipiter cooperii – Cooper's Hawk
Pompilidae – Spider Wasps

September 7
PYGMY MOLE CRICKET

76°
—
49°

And my day was already going well. Earlier in the day, I'd found a poem written some years ago and forgotten inside a book. While hiking, I'd found some interesting hoverflies, some showy butterflies, several handsome Autumn Meadowhawks, and a Lance-tipped Darner.

Things got even better when I reached a sunlit, sandy spot along the river. A variety of plants grew along this open stretch of riverbank and a wide margin of mud followed the river, but there were also several areas of clear sand and one of those was fully in the sun. I settled in and watched the Bronzed Tiger Beetles, at least fifty raced here and there across the sand. There were other insects here as well. A small spider wasp. Many tiny Oxybelus wasps, hunters of tiny flies. Several larger sand wasps.

Then I noticed what looked to be a small beetle digging in the sand. About half the size of the tiger beetles, or smaller. I pointed the camera at it and took a few photos. Looking at the magnified images, I saw that it was a Pygmy Mole Cricket. A second one showed up not long afterward. I'd never encountered these fantastic creatures before.

Syrphus sp. – Flower Fly
Cicindela repanda – Bronzed Tiger Beetle
Oxybelus sp. – Fly-stabber Wasp
Crabro – Shield-handed Square-headed Wasps
Tetrigidae – Pygmy Grasshoppers
Neotridactylus apicialis – Larger Pygmy Mole Grasshopper
Vanessa cardui – Painted Lady

[Observations continued on next page →]

Cicindela repanda – Bronzed Tiger Beetle
Pompilidae – Spider Wasps
Elaphrus sp. – Marsh Ground Beetles
Eupatorium perfoliatum – Common Boneset
Sympetrum obtrusum – White-faced Meadowhawk
Sympetrum vicinum – Autumn Meadowhawk
Necrophila americana – American Carrion Beetle
Aeshna constricta – Lance-tipped Darner
Polygonia comma – Eastern Comma
Helophilus fasciatus – Narrow-headed Sun Fly
Dolichovespula maculata – Bald-faced Aerial Yellowjacket
Anax junius – Common Green Darner
Atalopedes campestris – Sachem
Sphaerophoria philanthus – Black-footed Globetail
Syritta pipiens – Thick-legged Hoverfly
Bidens frondosa – Devil's Beggarticks
Vanessa cardui – Painted Lady

Sympetrum corruptum – Variegated Meadowhawk
Sympetrum obtrusum – White-faced Meadowhawk
Paragus sp. – Grass Skimmer
Syrphus sp. – Flower Fly
Aeshna umbrosa – Shadow Darner
Chalcosyrphus nemorum – Dusky-banded Forest Fly
Sympetrum vicinum – Autumn Meadowhawk

September 8   68°

VARIEGATED MEADOWHAWKS   52°

I'd nearly given up on seeing a Variegated Meadowhawk this year. Usually, I see them in the spring when they arrive with the other regular migrant, the Common Green Darner. This year the Common Green Darners arrived on their own. Conditions were such that the Variegated Meadowhawks either didn't arrive here in the spring or they arrived in low numbers and I didn't find them.

Today, walking home after a hike to the St Olaf Natural Lands, I decided to cut through the restored prairie of Lashbrook Park. Dozens of White-faced Meadowhawks perched low along the edges of the mowed trails. Then I noticed a golden-colored meadowhawk flying above my head, glider-like, and I knew in an instant that it was a Variegated Meadowhawk. I waited for it to land, thinking this might be the one and only chance I'd get to observe this species this year. After it perched, I was able to get close enough to photograph it and see that it was a juvenile female, recently emerged.

Ten minutes later, leaving the park, I encountered a second Variegated Meadowhawk. This one was a male, but also a juvenile, yellow colored instead of red. This species grows incredibly fast, going from egg to nymph (through all dozen instars) to adult in about two months time. Therefore the eggs for the two dragonflies encountered today must have been oviposited in a nearby pond sometime around the beginning of July. The eggs of all the other species of meadowhawks in North America all overwinter as eggs, undergoing a period of diapause, a suspended development during the cold months.

[Observations on previous page]

76°
—
53°

September 9

HOVERFLY PUPARIUM

Two days ago, examining the leaves and stalk of a Swamp Milkweed plant, continuing to search among the bright orange Oleander Aphids for fly larvae, I noticed something dark under the tip of one leaf. It looked like a dried slug or a solidified drop of oil. A hoverfly puparium. Proof that the fly larvae had been there prior to my search.

When I photographed the puparium today, magnified by the camera and lens, what looked like an adult fly was somewhat visible inside. It looked close to emerging. I placed it back in the vial and set it on my desk, suspecting to find a fly in the vial in the days ahead. [see September 24]

Hetaerina americana – American Rubyspot
Sympetrum vicinum – Autumn Meadowhawk
Sympetrum obtrusum – White-faced Meadowhawk
Syritta pipiens – Thick-legged Hoverfly
Syrphidae – Hover Flies

September 10

GOLDWESPEN

81°
—
61°

A mason wasp and a cuckoo wasp, one on the inside of the window, one on the outside. The mason wasp probably escaped from one of the trap nests being reared in my office. The cuckoo wasp was searching for a wasp or bee nest in the exterior of our house. I captured both and took photos of each.

The Chrysididae are thickly armored wasps and bright as living jewels, as colorful as hummingbirds. Bright metallic greens and blues and reds are the common colors. These colors are not from pigments but result from the microscopic structure of the chitin exterior that refracts and reflects certain colors. Because these wasps are kleptoparasitic, not provisioning their own nests but stealing into the nests of other wasps and bees, they are often referred to as cuckoo wasps. But just as often, because of their striking iridescent colors, they are referred to as jewel wasps or ruby-tailed wasps or goldwespen.

For a rich sampling of this family, visit Chrysis.net, a website devoted to the photography and natural history of these fabulous wasps.

Ancistrocerus adiabatus – Pathless Mason Wasp
Chrysis sp. – Cuckoo Wasp

82°
—
60°

September 11

WASP-POWERED SPIDER-RAFT

In a mood to excavate a wasp nest, I decided to revisit the sandy river bank where I'd seen several square-headed wasps nesting just last week. The main trail at the Cowling Arboretum follows the Cannon River for a good distance. I stopped at one point and looked out over the river, hoping to see dragonflies. There were none. But I did notice something about the size of a fallen leaf moving on the surface of the water below me. I climbed down the bank for a closer look and suddenly realized what I was seeing—a large spider wasp ferrying a large fishing spider. I hurried to get a few photos before it made landfall. What a spectacle to witness a wasp-propelled spider creating a wake like a barge.

As soon as the wasp came ashore, it abandoned the spider (probably I'd made it nervous tromping along the river bank to keep up). The wasp landed a few feet away and groomed herself, but eventually made her way back to reclaim the spider. After dragging the spider a few inches, the wasp flew off, a fair distance this time, disappearing beyond a downed tree and rocks. But the wasp soon returned again. Perhaps she had gone ahead to inspect and ready the burrow or crevice in which it would place the spider.

Looking at the photos I took, comparing them to the single set of photos I could find online, I noticed that in both instances the wasp extended it flat front feet and used them like water skis as it towed the spider across the water.

Cicindela repanda – Bronzed Tiger Beetle
Anoplius depressipes – Spider Wasp
Dolomedes tenebrosus – Dark Fishing Spider
Oecanthus niveus – Narrow-winged Tree Cricket

Sphaerophoria sp. – Flower Fly
Marchantiales – Liverworts

September 12

AN IMPORTANT VISITOR

85°
—
63°

By the time I was free to go outside, the temperature was in the eighties and I decided to forgo the usual hike. Instead, I sat down beside the three Zigzag Goldenrod blooming in the front yard pollinator garden. During the twenty minutes watching the flowers, an interesting variety of bees, wasps, flies and beetles visited. A single, gluttonous Honey Bee nectared the whole time.

Before returning to work, inside, I wandered, camera in hand, to our backyard garden. What happened next is the stuff of legend. I noticed a bumble bee on aster flowers, black across the thorax like Black-and-gold and Golden Northern Bumble Bees, but it didn't have the right amount of yellow on the abdomen. Curious, I stepped inside the garden gate and began taking photos as best I could. The moment I had a good look at the abdomen, the orange patch surrounded by yellow . . . I knew. This was a Rusty-patched Bumble Bee. What a surprise to see it, and doubly surprising to have it visit our backyard. If the weedy margins of our property offend a few of our neighbors, at least it seems to please the insects.

Truly, I'd nearly abandoned all hope of seeing a Rusty-patched Bumble Bee, especially after watching Clay Bolt's brilliant documentary about his search for the bee, *A Ghost in the Making*. This bee finally received federal protection in January of this year, the original petition being filed in 2013 by the Xerxes Society. Once a common bee and important crop pollinator in the east and the Midwest, it is now absent from close to 90% of its original range. In 2016, the city of Northfield, where

I live, declared itself a pollinator-friendly community. The appearance of this spotlight species (there are hundreds of other important native pollinators) rings a small note of success and brings some well-needed hope.

Bombus affinis – Rusty-patched Bumble Bee

September 13

A SECOND LOOK

87°
64°

"Lighting strikes (not once but twice)" – The Clash

On my way to St Olaf, I stopped in at the Kjar residence to tell Arlene about yesterday's discovery. Her house, on the opposite corner of the block from ours, has a yard thick with flowers and a long-established and well-used vegetable garden. Dennis let me into the house through the back door, into a kitchen where cucumbers and tomatoes were being pickled and canned, evoking deeply embedded memories from my childhood and my grandparents' farmhouse, the very same smells and clutter of jars and pans.

Arlene and I stepped out the front door to look at the flowers. And there, on the very same kind of aster as in our yard, was another Rusty-patched Bumble Bee.

Bombus affinis – Rusty-patched Bumble Bee
Sphex ichneumoneus – Great Golden Digger Wasp
Symphyotrichum novae-angliae – New England Aster
Hylephila phyleus – Fiery Skipper
Symphyotrichum drummondii – Drummond's Aster

91°
—
66°

September 14

ASTER HOURS

Seeing as the Rusty-patched Bumble Bees have been visiting the asters in the backyard, these flowers have become the focus of my attention. And since I was spending some time watching them, it seemed worthwhile to try to document as many other visitors as possible. In addition to the several Rusty-patched Bumble Bees, numerous Half-black and Common Eastern Bumble Bees frequented the flowers. A Carder Bee was a surprise. Another surprise was the two spectacular hoverflies, the Eastern Hornet Fly and the Orange-legged Drone Fly. Not bad for an hour among the asters.

Augochlora pura – Pure Green Augochlora
Dialictus – Metallic Sweat Bees
Halictus ligatus – Ligated Furrow Bee
Eristalis dimidiata – Black-shouldered Drone Fly
Anthidium oblongatum – Carder Bee
Spilomyia longicornis – Eastern Hornet Fly
Eristalis flavipes – Orange-legged Drone Fly
Bombus impatiens – Common Eastern Bumble Bee
Bombus affinis – Rusty-patched Bumble Bee
Syrphus sp. – Flower Fly
Pseudopanurgus sp. – Mining Bee
Bombus impatiens – Common Eastern Bumble Bee

September 15   84°/70°

THE METHOD OF THEREFORE

How do we get from here to there? Surely not by car. The roads we need are not those you'll find on maps. There are other ways. Ask the Chickadee. Ask the Crow. Ask the migrating Kingbird.

Or: ask the Greek poet Odysseus Elytis: "A little dry mint rubbed between your fingers, sends you straight into the thought of the Ionians. Your caresses are the transposition of a soft music, with all its andante and allegro, onto the skin's surface. And that plant opposite you, which divides the area unequally yet correctly, is the invisible geometry that pervades the whole universe. Now that's a real freedom, one that has the overall power to suspend all your individual inhibitions and, each time you eat grilled sea bream, to make you eat a little of the Aegean with, in place of lemon, two or three tart verses by Archilochus." (from 'The Method of "Therefore" ')

Sometimes the best way forward is the easiest, only the trails have been made invisible and hard to see by habit. Sometimes our destination is so near at hand and so close we overreach it. When I chew on a goldenrod leaf I time-travel back to childhood and walk again the maple hills of Otter Tail County. The taste sends me straight from here to there.

Bombus affinis – Rusty-patched Bumble Bee
Sphaerophoria sp. – Flower Fly
Bittacomorpha clavipes – Phantom Crane Fly
Platytipula – Crane Fly
Labidomera clivicollis – Milkweed Leaf Beetle

80°
—
58°

September 16

SAP BEETLE

Our colossal front-yard elm leaks. Sap trickles slowly out of the tree in two places. The more days without rain, the more insects and birds these seeps attract, each a small arboreal oasis.

Today I noticed a small red beetle mired in the sap runs. It resembled the Flat Red Bark Beetles found previously on this tree, except it was smaller, narrower, and a darker vermillion color. Instead of family Cucujidae, it's of family Passandridae, a parasitic flat bark beetle. The very first sentence of the description for this family in Beetles of Eastern North America by Arthur V. Evans states "the feeding habits of adult passandrids are unknown." Maybe there's a slight discovery here? Apparently more is known about the larvae: "larval passandrids appear to be ectoparasites of other wood-boring insects, especially the pupae of longhorn beetles."

One of the photos shows an extra leg. As the beetle pictured has all six of its legs, more than one passandrid has visited this little oasis.

Coelioxys – Cuckoo Leaf-cutter Bees
Danaus plexippus – Monarch
Enallagma civile – Familiar Bluet
Lestes rectangularis – Slender Spreadwing
Bombus affinis – Rusty-patched Bumble Bee
Zadontomerus – Small Carpenter Bee
Syrphus sp. – Flower Fly
Catogenus rufus – Passandrid Beetle

September 17

THE AUTUMN WASTES...

67°
52°

To St Olaf Natural Lands, into the wilder, more barren days.

> "The autumn wastes are each day wilder;
> Cold in the river the blue sky stirs."
> – Tu Fu, from the poem 'Autumn Wastes'
> (A. C. Graham trans.)

> "It is enough
> To smell, to crumble the dark earth,
> While the robin sings over again
> Sad songs of Autumn mirth."
> – Edward Thomas, from 'Digging'

Syrphus sp. – Flower Fly
Eremnophila aureonotata – Gold-marked Wasp
Sympetrum vicinum – Autumn Meadowhawk
Bombus affinis – Rusty-patched Bumble Bee
Bombus citrinus – Lemon Cuckoo Bumble Bee
Anax junius – Common Green Darner
Agapostemon sp. – Striped Sweat Bees

59°
—
55°

September 18

THE EYES HAVE IT

The White-faced Meadowhawk eggs collected on August 22 have eyespots. Now that the embryos are fully-formed, they are ready for diapause. Diapause is a kind of preemptive stasis or strategic pause in development used to survive periods of adverse environmental conditions—in this case, winter.

These eggs will not complete their development and will not hatch without this period of winter diapause which is triggered by the cold temperatures of the season. If they are not cold for the right length of time the eggs will perish. This may be an important factor in determining the southern range limit of this species.

Now that the eggs have eyespots, it's time to place them in the refrigerator. When removed in April they will all hatch within hours of warming up.

Sympetrum obtrusum – White-faced Meadowhawk

September 19

LITTLE THINGS

80°
—
59°

I stood looking at the trunk of the large tree for a long time. If there were any onlookers, no doubt they thought it odd to see someone standing motionless, nose to the bark of a tree like a punished schoolboy made to stand in a corner. But this was no punishment, more a meditation, as I surveyed the intricacies of the bark, a kind of vertical landscape, for its inhabitants. Standing like a tree in front of a tree.

Tree bark is one of the wonders of the natural world. For large oaks, the bark is a formidable armor, textured like the scales of some long gone saurian (if you question this analogy look at the bark of the southwest Alligator Juniper). Poplar trunks can be smooth as vases. And the cherry tree with its annealed bronze allowing a glint in the sun like that off an ancient Greek shield.

In a mossy crevice of the bark of our backyard ash tree, I noticed a very small snail, later identified as a Quick Gloss Snail. As part of their courtship, these snails create and use love darts, a kind of ritual stabbing that happens before actual mating occurs. Then a curious black fly caught my attention. This little fly turned out to be a Woodlouse Fly, a species native to Europe and an endoparasite of sowbugs and woodlice.

Acraspis macrocarpae – Jewel Oak Gall Wasp
Enallagma civile – Familiar Bluet
Sympetrum obtrusum – White-faced Meadowhawk
Populus deltoides – Eastern Cottonwood
Sympetrum vicinum – Autumn Meadowhawk
Aphaenogaster sp. – Collared Ants
Zonitoides arboreus – Quick Gloss Snail
Melanophora roralis – Woodlouse Fly

75° / 58°

September 20

DOUBLE-BANDED SCOLIID

Walking back from the St Olaf Natural Lands, I took a slightly longer route than usual, following some sidewalks and paths rarely visited.

On some Yarrow flowers in a garden, I noticed the very distinctive Double-banded Scoliid wasp. This velvety-black wasp with black wings and a pair of ivory bands on its abdomen is a parasitoid of June beetles. It's the first time I've noticed this wasp in my neighborhood. Previously, I'd found them flying on the sandy slope of a small oak savanna at the Cannon River Wilderness Area.

Ammophila sp. – Thread-waisted Wasp
Scolia bicincta – Double-banded Scoliid Wasp
Syritta pipiens – Thick-legged Hoverfly
Colias eurytheme – Orange Sulphur
Phymata sp. – Jagged Ambush Bugs
Eristalis tenax – Drone Fly
Helophilus fasciatus – Narrow-headed Sun Fly
Colias philodice – Clouded Sulphur
Eristalis dimidiata – Black-shouldered Drone Fly
Epicauta pennsylvanica – Black Blister Beetle
Eupeodes sp. – Flower Fly
Sphecodes – Blood Bees
Vanessa atalanta – Red Admiral

September 21   84°

MY FIRST BRISTLETAIL   53°

Cowling Arboretum. Turning over a shard of limestone at the base of a small outcropping of rock, I saw what appeared to be a souped-up silverfish, larger, more darkly iridescent than powdery gray. I moved it into the light for a photo. It remained absolutely motionless until it didn't, moving, almost flowing, into the leaf litter where it disappeared. Only later did I learn that this creature was a Bristletail, a member of the Archaeognatha, an order of insects I'd never encountered before. And knew next to nothing about.

The Bristletails have an ancient lineage, with a fossil record going back to the Devonian, sharing that strange, long-ago world with the trilobites, horsetail forests, and proto-dragonflies. An exception among insects, Bristletails continue to molt as adults and can live up to four years.

Further along the trail, I stopped to admire an incomprehensible goldenrod; I'm familiar with Showy, Grey, Canada, Zigzag, Elm-leaved and Grass-leaved Goldenrods, but this particular plant was unlike any of those species. The flower arrangement most resembled Zigzag, but the leaves were too narrow. There probably was an answer, but I didn't know it. Nearby, at the edge of the footpath, clusters of Pepto-Bismol-pink, puff-ball-like fungi sprung out of a rotting log. Only it wasn't a mushroom, rather it was a slime mold, Wolf's Milk Slime Mold.

Hetaerina americana – American Rubyspot
Petrobiinae – Bristletail
Cisseps fulvicollis – Yellow-collared Scape Moth
Solidago sp. – Goldenrods
Solanum nigrum – Black Nightshade
Lycogala epidendrum – Wolf's Milk

94°
73°

September 22

CARLETON COLLEGE STUDENT NATURALISTS

I was invited by Nancy Braker, director of the Cowling Arboretum, to join the Cole Student Naturalist group on their weekly walk at Carleton. Nancy, Matthew Elbert, the Arb manager, a dozen students and I braved the afternoon heat.

Right off, we encountered a spider wasp and a paralyzed wolf spider on the sandy trail leading into the arboretum. The wasp was stuck, struggling to drag the spider, which was much larger than the wasp, out of a depression in the sand. This allowed everyone to get a good look at the wasp. Eventually, Nancy took pity on the wasp and moved its spider to the edge of the trail. A little further on, a student found a Dekay's Brownsnake, a tiny snake with an outsized temper.

We hiked to the Turtle Pond, talking and looking at plants and insects along the way. Then to the banks of the Cannon River, where I'd recently found Pygmy Mole Crickets. And then back on the trail along the restored prairie. At a tree at the edge of the prairie, we saw a Giant Ichneumon Wasp.

A number of Wandering Gliders and Black Saddlebags flew overhead at one point on our hike back through the prairie. After pointing them out, I managed to net one of the saddlebags with a surprisingly athletic leap, certainly one of my luckiest netting performances. The Black Saddlebags was passed around and admired, hand to hand. This inspired a number of students to attempt netting some of the dragonflies as well.

Storeria dekayi – Dekay's Brownsnake *Hylephila phyleus* – Fiery Skipper
Araneus marmoreus – Marbled Orbweaver Pompilidae – Spider Wasps
Tramea lacerata – Black Saddlebags
Megarhyssa atrata – Black Giant Ichneumonid Wasp

September 23

SOMETHINGS CHANGE

90°
—
73°

Driving from Mankato to Northfield after a volleyball tournament, my wife and I stopped, briefly, at Lily Lake. I've stopped at this small lake many times in the past, finding it somewhat idyllic. Water lilies, great beds of them. Sedges along the shore. And, surprisingly, a haze of wild rice. All combined to give this small southern Minnesota lake a northern Minnesota feel. The dragonfly productivity has been extremely high, especially when compared to the larger, more compromised lakes. On one memorable occasion in 2010, I found hundreds of teneral Autumn Meadowhawks. It also supports a population of Elegant Spreadwings. Always, an enjoyable place to visit.

Today, however, when we stopped and looked out from the boat landing, there was no sedge and no wild rice, just a few stands of dead cattail and the blackened remains of water lily beds. And the water reeked like a feedlot. It was not a pleasant sight.

Hopefully this is only a temporary condition due to heat and recent rains. Even still, perched in the trees surrounding the parking lot, a few Autumn Meadowhawks were to be found. A little reminder that not everything has changed.

Meloe sp. – Oil Beetles
Sympetrum vicinum – Autumn Meadowhawk
Linaria vulgaris – Common Toadflax

90°
70°

September 24

PTEROMALIDS: REVISIONS

According to Thoreau, he often composed in his head, perfecting lines and sentences in his mind, perhaps while walking or rowing or skating, before writing them down.

Also, he recommended setting new work aside for a period of time, then returning to it with fresh eyes for revisions and fixes. This is good advice. And experience bears it out. There's a certain benefit to forgetting; it's much easier to be objective and critical when treating a piece of writing as if someone else wrote it.

Still, you can't polish a hedgehog and expect a diamond.

❖ ❖ ❖

I took a look at the vial containing the hoverfly puparium collected earlier in the month [Septermber 9]. Instead of a hoverfly, two Pteromalid wasps had emerged. Rearing pupae of moths and flies and wasps mirrors this method, when expectations are revised by what emerges.

Pteromalidae – Pteromalid Wasp

September 25

SLEEPY BEES

71°
—
58°

To St Olaf, late afternoon. Following a night of rain, it is a day both decidedly cool and thickly overcast. No flies, no wasps, no butterflies. In the habit of eyeing the asters, I found four bumble bees hanging from the flowers, dreaming, tarsal hooks fixed. All four bees were Common Eastern Bumble Bees (Bombus impatiens). The species name "impatiens" derives from the Touch-me-not flowers (Impatiens sp.) it sometimes visits. One of the bees kicked out a leg as I loomed over it, as though protesting this disturbance of its sleep.

In Norse Mythology "Niflheim was colder than cold, and the murky mist that cloaked everything hung heavily. The skies were hidden by mist and the ground was clouded by the chilly fog." Here at the base of the St Olaf hill on a day that fit these old descriptions, the bees were wise to sleep and wait it out, patient for the fine world and the warmth that would return.

Bombus impatiens – Common Eastern Bumble Bee

65° / 53°

September 26

MICROFISHING: SKUNKED

Another overcast day. With the bees still sleeping, I decided to try some microfishing. I readied a line and a snelled micro-hook and some tiny snips of bacon. Then headed for Spring Creek in the upper arboretum, above the lakes and, hopefully, above the populations of sunfish and bass.

I could see minnows in a narrow section of the stream. To reach them I tied the line and baited hook to a long stick found at the edge of the stream, not at all lithe. After each crude cast, I swept the stick downstream, assisting the drift of the line and hook as the current took hold and brought it down toward the minnows. Many times the minnows followed it, even struck, but I never managed to hook one. I left skunked.

❖ ❖ ❖

A couple moths showed up at the front door light after dark. A Fall Cankerworm Moth and a Corn Earworm Moth.

Alsophila pometaria – Fall Cankerworm Moth
Helicoverpa zea – Corn Earworm Moth

September 27   66° / 48°

MICRO-FISHING: SMALL SUCCESS

Today I returned to Spring Creek, determined to land one of the minnows that had eluded me yesterday. Fishing in rivers is something I've never done well. I'd tried fly fishing many years ago and mostly failed and those efforts fizzled out. More of lentic personality than a lotic, I guess. And yet, moving water, especially small natural streams and creeks, continue to fascinate and hold my attention, my curiosity, my wonder. "Here I am fishing for minnows for christ's sake," I thought to myself. But, having gifted much of my life's energies to poetry, I'm at home, comfortable even, in the realm of folly.

Reading Aristotle yesterday, I was reminded that fishing is like philosophy. "For it is owing to their wonder that men both now begin and at first began to philosophize" (*Metaphysics* 982b11)

The minnows, like dark-gray brush strokes, wait near the bottom, flash silver when they strike. Just like thoughts and ideas and worries, that suddenly move into conscious thought, snapping at the surface of the moment.

Today, with the upgrade to a portion of fly rod instead of a sodden tree branch, with a remembrance of wild water, I meet with a modicum of success. A roll-cast puts the hook across the stream; it drifts and sinks directly toward the pool where the minnows hold. The vibration on the line is high pitched, a Creek Chub not much longer than an inch in length, far different than the slice and tug fight of a larger fish.

Semotilus atromaculatus – Creek Chub

75° — 52°

September 28

BEGGARTICKS

> "Over the pass, all the way to the sky, a road
> for none but the birds.
> On the river and lakes, to the ends of the earth,
> one old fisherman."
> – Tu Fu, 'Autumn Meditation'

For an hour in the afternoon, I decided to microfish the lower reaches of Heath Creek between the railroad bridge and its confluence at the Cannon River at the edge of Sechler Park. Here the creek passes through a wooded area. The creek bed is very rocky. The current was fast and difficult to fish. I saw one trout-like rise at the edge of an eddy, otherwise no signs whatsoever of any other fish or minnow.

An unsuccessful outing. And instead of catching anything, I was caught. Dozens of stick-tights covered my pants and shirt, after fighting my way through the undergrowth along the creek banks. The clinging seeds were Devil's Beggarticks (*Bidens frondosa*). The achenes, small black tick-like seeds, have two barbed spines extending from one end. This design, the pair of prongs at the end of a long seed has given rise to another common name—Devil's Boot Jack.

While the seeds are a bit of a nuisance, they can be patiently removed. Unlike grief, a kind of bur that sticks and simply cannot be removed. I learned, this day, of two friends diagnosed with prostate cancer.

Poecile atricapillus – Black-capped Chickadee
Bidens frondosa – Devil's Beggarticks

September 29　66°

UP-AND-DOWN PARK　48°

Lisa, Lida, Charlie, and I went for a hike at the Cannon River Wilderness Area (west unit). Lida drove. Probably the one place we've hiked together the most. It's the "up-and-down park" because the trail we usually hike follows the bluffs above the river, plunging up-and-down as the path crosses ravine after ravine. One of the things I like about this park is the number of Ironwood trees. Stands of these trees are especially attractive in the fall.

On the upside, today is a dry, autumnish day and we have the trails to ourselves. On the downside, a stench from the nearby landfill roamed the hills. As we descended into several of the ravines the odor increased until we almost had to hold our breath. I flipped over some limestone shards along the trail and uncovered a bristletail, the second I'd ever seen (the first being just over a week before).

Lida ran ahead, something the young are destined to do. Accompanied by the dog, she reached the footbridge over the river and waited. The bridge was one of her familiar and remembered places. Perhaps not the place I would choose to lodge in someone's heart, but we don't really get to make these choices for others. I can see she loves this place—it's her place now, memories layered upon memories.

Enemion biternatum – False Rue Anemone
Archaeognatha – Bristletails
Rhaphidophoridae – Spotted Camel Cricket
Symphyotrichum oblongifolium – Aromatic Aster
Sympetrum vicinum – Autumn Meadowhawk

69°
—
46°

September 30

AN AUGURY OF WINTER

Lisa noticed the dragonfly on the window screen this morning. With the sun now warming the insect, I stepped outside to get a photograph before it flew away. A brisk fall morning with the temperature right at 50 degrees F. Curiously the dragonfly, a recently emerged Common Green Darner, gripped the window screen and perched at an odd angle, off-kilter or tipped as if fallen or lifted by a sudden updraft of wind. The double shadows intrigued me as well—one for the dragonfly one for me. Best of all was the happenstance of having a snowflake ornament as a backdrop, presaging the end of a season this last day of September.

❖ ❖ ❖

Attended a poetry reading at Content Bookstore in Northfield, where I heard several of my favorite local poets read, including Larry Gavin, Florence Dacey, and Rob Hardy. Rob Hardy's 'Old Field' reminded me of Loren Eiseley's poem 'Poison Oak,' which begins:

> This fallow acre lapsed to weed
> Unturned beneath the sun,
> Is nourishing a wilder seed
> Domestic furrows shun.

Anax junius – Common Green Darner

October

Wild Apples – October 13

October 1

WIDOW YELLOWJACKET

63°
—
55°

"Such formidable servants are always necessary, but often fatal to the throne of despotism..." – Gibbon, *The Decline and Fall*

Today, at the St Olaf Natural Lands, I visited a different kind of wasp tree. Instead of the standing trees (both dead and dying) that support populations of wood wasps and parasitic wasps, I stopped to look at a small, living Jack Pine. The pine's branches and needles had attracted a variety of vespid wasps. It wasn't clear what the wasps were doing here. They could be searching for food or masticating the tree's bark to make paper for their nests. Or, perhaps, on this cool and rainy day, the wasps were simply waiting out the weather amidst the needles. Some did seem asleep. In the dull autumnal landscape, the bright green of the Jack Pine and the bright yellow and black wasps looked slightly out of phase.

In *Walden*, Thoreau famously described a battle of ants in terms heroic, comparing them to the *Illiad*'s warriors. The wasps of the family Vespidae, in contrast to the warring ants, rule by fear. Their colors, the bold black-and-yellow markings, are recognized universally. They are the despots of the insect world. Like the Praetorian Guards of the Roman Empire, they alone carry weapons in the citadel, and are rewarded for it. Birds leave them alone. People run from them at fall picnics. In general, they give all wasps and bees a bad name.

And yet, if one is respectful of their armaments, there is little danger of being stung. I capture a large wasp, predominantly black with thin yellow markings, and bring it home to photograph. This turns out to be a male Widow Yellowjacket, a species I hadn't encountered before. Even though male wasps cannot sting, I handle it as though it could sting, so much respect do I have for the black and yellow.

Eris militaris – Bronze Jumper
Vespula vidua – Widow Yellowjacket

October 2

A BRACE OF LEAFHOPPERS

71°
—
55°

Between spates of rain, the drear drapery of clouds dragged on. Twilight and shadows all day. Dark as an armpit. And yet...

...a quick walk in a raincoat yields a few leafhoppers. A brace of Red-striped Sharpshooters, a scattering of other species, hanging out in the rain. A leafhopper's existence is leaf-to-leaf. Launched into flight by the surprising, mouse-trap-like-contraption of oddly-positioned, spring-loaded legs, they catapult themselves. Their hind legs, which are cantilevered over the front and middle legs when stationary, have rows of movable spines.

In the shadowy, autumnal afternoon, as the weather spattered and diffused, a kind of late-night exhaustion expanded like a mist. The raincoat draped over the back of a kitchen chair dripped water onto the floor.

Myrmica sp. – Spine-waisted Ant
Pachypsylla celtidismamma – Hackberry Nipple Gall
Graphocephala sp. – Red-striped Sharpshooter
Cicadellidae – Leafhoppers

72°
—
52°

October 3

UNCENTERING

To St Olaf Natural Lands for another look at the wasp tree. Facing so many vespid wasps, one gains a strange perspective, less self-centered, not entirely human. It's a useful experience to put oneself on the level with the insects. I'm put in mind of the poetry of Robinson Jeffers, which often challenges the reader to expand and change their perspective. For instance these lines from his poem 'Carmel Point':

"Their works dissolve. Meanwhile the image of the pristine beauty
Lives in the very grain of the granite,
Safe as the endless ocean that climbs our cliff.—As for us:
We must uncenter our minds from ourselves;
We must unhumanize our views a little, and become confident
As the rock and ocean that we were made from."

— Robinson Jeffers

Vespula maculifrons – Eastern Yellowjacket
Sympetrum obtrusum – White-faced Meadowhawk
Dolichovespula maculata – Bald-faced Aerial Yellowjacket
Sympetrum vicinum – Autumn Meadowhawk
Polistes fuscatus – Dark Paper Wasp
Vespula vidua – Widow Yellowjacket

October 4

RED OCTOBER

60°
—
46°

Recently I'd been lamenting how little time I've spent with the red dragonflies this year. Easily the least amount of time in the field with the Sympetrum dragonflies in ten years. In fact, it seemed likely that I'd not see three of Minnesota's nine species: no Red-veined Meadowhawks, no Black Meadowhawks, no Saffron-winged Meadowhawks. And I nearly missed Variegated Meadowhawks as well, seeing no spring migrants, but two late-season vagrants. However, today's outing to Lake Byllesby County Park near Cannon Falls helped to relieve these red dragonfly blues.

Each year this series of spring-fed ponds produces a stupendous number of Band-winged Meadowhawks. Visiting the site earlier this year and finding it completely dry, I had worried that the streak would come to an end, that this source population would disappear. By the time I'd crossed the green space between the parking lot and the trailhead, a dozen or so Band-winged Meadowhawks had flown in front of me. The purpose of the visit today was to check on these dragonflies. They made it through the dry months just fine. The population was as strong as ever.

In addition to the multitudes of Band-winged Meadowhawks, there were a few White-faced Meadowhawks (how unusual for this species to be in the minority), a number of Autumn Meadowhawks, and dozens of Saffron-winged Meadowhawks. It was my first encounter with Saffron-winged Meadowhawks this year.

Saffron-winged Meadowhawks have the most distinct "waist" (the constriction at the start of the abdomen) of all North American meadowhawks. As such they are a good example of what British entomologist Edward Newman had in mind when he coined the genus name "Sympetrum" in 1833. A footnote to the original description explains the origin of the word to be constructed of συμπιέζω and ητρον, the Greek verb for compress and the word for abdomen, referencing the narrowness of the abdomen sections 4 and 5. Unfortunately, this name gets mistranslated and misinterpreted because it would seem (if one didn't reference the original description) to be a construction of *sym* and *petra*, a friend of stones.

Hemileuca maia – Buck Moth
Syrphus sp. – Flower Fly
Promachus vertebratus – Giant Robber Fly
Lycaena phlaeas – American Copper
Oenothera gaura – Biennial Beeblossom
Hygrocybe – Waxcaps
Boisea trivittata – Eastern Boxelder Bug
Dianthus armeria – Deptford Pink
Augochlorini – Green Metallic Bees
Microlinyphia – Sheetweb Spider
Sympetrum costiferum – Saffron-winged Meadowhawk
Sympetrum obtrusum – White-faced Meadowhawk
Aeshna constricta – Lance-tipped Darner
Sympetrum vicinum – Autumn Meadowhawk
Sympetrum semicinctum – Band-winged Meadowhawk

October 5   65°/48°

COBWEB SPIDER

The day is diminished by weather, cool, cloudy, controlled by a curl in the atmosphere. The hours go by in gray minutes. The fallen leaves lay sullenly on the concrete, muted, less vivid than the day before.

Among the wide, yellowing leaves of a Common Milkweed in our garden, I noticed one leaf was bent, buckled in the middle, and slightly folded. A cobweb spider had cinched the edges with silk, creating refuge and resource, home and hunt.

Small, bright yellow, perpetually upside-down, Theridion. Theridion waiting. What's more patient than a spider? Eventually, a meal will happen along and get itself stuck. Like the spider, the patient writer waits, hoping to trap the needed words.

Theridion sp. – Cobweb Spider

60°
—
48°

October 6

SUMAC LEAF BLOTCH MINER

I overslept. Lida woke me by calling out, "Dad? Breakfast?" She had been up for some time, watching television in the living room. Lisa left an hour earlier and was already at school, preparing to teach. I get up, let the dog out, brew some coffee, and set out breakfast. 6:55 A.M. Once Lida leaves for the bus, I retreat to the reading chair in the basement. The cat leaps into my lap and I look at the piles of books. I read several pages of *The Table* by Francis Ponge, a warm-up for the mind. Then it's on to *Tiger Beetles: The Evolution, Ecology, and Diversity of the Cicindelids*.

❖ ❖ ❖

Knowing that the scrubbing October rains were to resume in the afternoon, I took an early walk to the St Olaf Natural Lands. Most of the leaf color was among the ash and walnut and cottonwood trees, a pallid yellow like the foxing of pages in old books. The sumac, however, like bright claret in a glass, startled the eye with it rich reds.

Among the sumac leaves, a number of leaflets were rolled up at the ends. Unrolling one of them, a small green caterpillar was disclosed. Inside another, a pupa. Sumac Leaf Blotch Miners. The caterpillars of these small moths mine the interior of the leaf, then exit the leaf before pupating.

Nearby, I encountered a mason wasp nectaring on goldenrod. It's just this kind of wasp that hunts leaf-roller caterpillars, pinching the folded leaf until the caterpillar leaves the safety of the curled shelter.

Caloptilia rhoifoliella – Sumac Leaf Blotch Miner Moth
Ancistrocerus adiabatus – Pathless Mason Wasp

October 7   67°

AN EARLY OCTOBER NIGHT   54°

Neighbor and philosopher David Fowler's 80th birthday. Rain in the morning, cleared by mid-afternoon.

❖ ❖ ❖

An early October night, doubly so—both early at night and early in the month. I turned on the mothing light to see what (if anything) might show up. (Oh, a homing lion perhaps—Thomas McGrath)

After two hours, I shut it off. Two moths—a Corn Earworm Moth and a Common Tan Wave—that was all. A winter crane fly. A half-dozen or so Brown Lacewings. And one stupendous bug—a Giant Water Bug—a carnivorous flying flip-flop, that had slapped itself down on the wet concrete among the fallen leaves. Large and flat, with an oiled-leather look to its wings, the Giant Water Bug is intimidating. The muscular front legs and motionless stance give the impression that this petite monster is the lost guardian of some unknown temple, misplaced somehow to the sill of our garage door. "None shall pass!"

Pleuroprucha insulsaria – Common Tan Wave
Tipulomorpha – Crane Flies
Hemerobiidae – Brown Lacewings
Lethocerus sp. – Giant Water Bug

October 8

AFTER APPLE-PICKING

Minnesota grows some of the finest apples on the continent, thanks in part to the apple breeding program at the University of Minnesota, thanks in part to the climate, thanks in part to the market. Lisa, Lida and I met her brother and his family at an apple orchard in Lakeville in the morning. With clear skies and temperatures in the 70s at 10 A.M., it was a fine autumn day. Hundreds of other people made their way to the orchard as well. Snowsweet, Honeycrisp, Fireside, Haralson, Connell Red, Sweet Sixteen, Yellow Gold were some of the varieties available for picking.

While walking through rows of apple trees, I sampled several of the crab apples we happened upon. These are interspersed throughout the orchard to aid in the pollination of the hybrid apple trees, which are not self-fertile. The smallish, egg-shaped crab apples, possibly Dolgo, reminded me of the pair of crab apple trees that grew just outside the back door of my grandparents' farmhouse. They tasted tart and had a cidery smell after they began to drop to the ground. And I remembered a crystalline quincunx of half-pint jars on the countertop, each filled with rose-colored crab apple jelly.

After apple-picking, which was neither too strenuous nor repetitive enough to worry my thoughts or sleep as it did Robert Frost (as it did in his famous poem by this title), I went for a walk. I took the standard loop, walking from our house to the St Olaf Natural Lands and back. Much of the time was spent fending off Thirteen-spotted Lady Beetles which were in flight in great numbers. Saw a number of Autumn Meadowhawks, several tandem pairs. A bright green planthopper, an inchworm, and an ichneumon wasp all on the same tree trunk.

Ichneumonidae – Ichneumonid Wasps
Acanalonia conica – Green Cone-headed Planthopper
Geometridae – Geometer Moths

57°
—
40°

October 9

AHEAD OF THE COLD FRONT

Racing the clock. The circling seasons coming to the final days and hours before the first frosts.

With frost in the forecast, with clouds visible to the west, I raced the end of the dragonfly season. I hurried to the trailhead at Lake Byllesby County Park, then hoofed it to the end pond. I arrived at 11:30 A.M. with the eastern ⅔ of the sky overhead still blue, the sun still shining.

Soon, I had found several male Saffron-winged Meadowhawks. As the weather held and the temperature warmed, a tandem pair of Saffron-winged Meadowhawks began ovipositing on the pond. Then another pair. After taking a couple of short videos with my camera, I decided I should try to net one of the pairs and collect some eggs. This, not surprisingly, was a task easier decided upon than accomplished. Saffron-winged Meadowhawks oviposit in tandem in open water usually some distance from the shoreline, the female touching the tip of her abdomen to the water intermittently. I took up a position at the shoreline and waited for one of the tandem pairs to oviposit near enough to net. It was a long wait, during which I repositioned my ambush site several times. The dragonflies seemed to be actively avoiding me. Occasionally, a mosaic darner would circumnavigate the small pond. I turned my attention to netting one of these in order to identify it, a Shadow Darner. I then returned to my previous task and vigil.

Eventually a pair ventured near enough and was netted. I'd not collected eggs from this species before. Unlike White-faced,

Ruby, and Cherry-faced Meadowhawks, this female didn't release eggs when held over the vial. Since it oviposited in water, I decided to fill the vial with water and dip her abdomen into it. The result was astonishing. The moment the tip of her abdomen was immersed eggs streamed out. A continuous double stream of tiny white eggs entered the water and settled to the bottom of the vial. In half a minute there were well over a hundred eggs in the vial. Another difference, these eggs were oval, more egg-shaped, than the perfectly round eggs of the White-faced, Ruby, and Cherry-faced eggs I'd collected previously. [see August 22 and September 18]

Enchenopa binotata – Twomarked Treehopper
Sympetrum obtrusum – White-faced Meadowhawk
Sympetrum semicinctum – Band-winged Meadowhawk
Aeshna umbrosa – Shadow Darner
Sympetrum costiferum – Saffron-winged Meadowhawk
Sympetrum vicinum – Autumn Meadowhawk
Argiope trifasciata – Banded Garden Spider
Oecanthus forbesi – Forbes' Tree Cricket
Aix sponsa – Wood Duck

October 10

SIDEWALK ART

Near freezing this morning with only the slightest of frost touching parts of the neighborhood. The first morning since last spring that I was able to see my breath when I stepped outside. A few hours later, as the temperature approached 50 degrees, I walked to the library.

Along the way, I saw a number of Autumn Meadowhawks perched flat on the cement of the sidewalks, like fossils or sidewalk art. This species persists longer than all others, surviving the first frosts and freezes, so this was not a surprise. Most years the Autumn Meadowhawks makes it through October and into the first weeks of November. Once the cold weather sets in, these hardy dragonflies will often be found perching on sunlit vertical surfaces, tree trunks, the sides of houses, even people wearing dark colored clothing will do, any place optimal to the uptake of solar energy.

Sympetrum vicinum – Autumn Meadowhawk

October 11   56°

NEXT TO NOTHING   38°

Whorled milkweed seeds.

         Rough horsetail.

                 Silted spectaclecase.

Pondweed.

         Forktail.

                 Silence.

Yellow willow leaves.

         Rainpool.

                 Frost in the morning.

Potamogeton – Pondweed
Ischnura verticalis – Eastern Forktail
Sympetrum obtrusum – White-faced Meadowhawk
Ligumia recta – Black Sandshell
Equisetum hyemale – Rough Horsetail
Asclepias verticillata – Whorled Milkweed
Diabrotica undecimpunctata – Spotted Cucumber Beetle

64°
—
53°

October 12

ROADSIDE GEOLOGY

On the road to a volleyball match in Red Wing, I stopped to look for fossils. Ahead of schedule, I had half an hour to look for remnants of ancient sea life among the scrabble of shale at a highway cut through an Ordovician outcropping. Some extra time to rummage through time.

What does it mean to find a fossil? The strangeness, the real incomprehensibility of the vast amount of time involved weathers away the mind. My thoughts fissure as cars speed by on the highway behind me—crescendos and shadows.

As I pick among fossil crinoids, early mollusks still working out the mechanics of a spiral, chips of trilobite, and the twig-like fragments of bryozoans, a living fossil slowly legs it across the crumbles of shale and shell. This living fossil is a Harvestman, its long legs traversing eons.

Asclepias syriaca – Common Milkweed
Daucus carota – Queen Anne's Lace
Tetramorium sp. – Pavement Ants
Opiliones – Harvestmen

October 13

WILD APPLES

61°
—
45°

I recalled gathering several nice apples a few years ago from a tree in the woods, a tree either long forgotten or growing wild. I set out to find them again, hoping to satisfy an uncultivated taste, a more authentic taste of autumn than the hybrid standards from the local orchard. Wild fruit, even in the smallest portions, justifies the search.

Tacitus, describing the German tribes north of Rome, wrote that "Their food is simple—wild apples, fresh game, and curdled milk" (*Germania*, section 23).

I'd come too late. I found the apple tree but no apples on the branches and none on the ground. Maybe it had been an off year, the weather or the timing off for pollination. Turning, resigned to finding no wild apples, I see a tall, wizened tree with small apples. It's on the other side of the trail, deeper in the woods. Most of the fruit had fallen; what hadn't was shriveled and soft. After a little searching, I located two additional crab apple trees and found several partially firm apples to taste.

Later, arranging five wild apples for a photograph, I recall the famous Chinese painting, 'Six Persimmons.' This Song Dynasty painting is treasured for the masterful brushstrokes and the perfected simplicity. If Muqi, the painter, had lived in North America, I suspect the work might have been 'Six Crabapples.'

Malus sp. – Apples
Micrathena mitrata – White Micrathena
Storeria occipitomaculata – Redbelly Snake
Euryuridae – Flat-backed Milliped
Gnaphosa – Ground Spider
Depressaria – Flat-bodied Moth

59° / 44°

October 14

ALL DAY RAIN

The water was speaking all day. Beginning with whispers in the morning on the window glass, then raging off and on in the afternoon across the metal roof, the rain eventually lost its voice after dark, murmuring and repeating itself as dripped from the trees, as if water might talk in its sleep.

The Welsh poet Dylan Thomas understood such weather. Here are a few lines from his poem 'Especially when the October wind':

> "Some let me make you of the vowelled beeches,
> Some of the oaken voices, from the roots
> Of many a thorny shire tell you notes,
> Some let me make you of the water's speeches."

Ulmus sp. – Elms
Gleditsia triacanthos inermis – Thornless Honey Locust

October 15

A FLOCK OF SISKINS

53° / 41°

I could hear birds in the trees, high above the house. Lots of them. Robins barreled through the lower branches, but smaller birds flitted in the top branches. I tried for a while to get a look at them, but couldn't ever find more than a speck, a dark silhouette against the bright sky. Then I noticed some splashing in the water at the bottom of the driveway where a number of the small birds were bathing.

Spinus pinus – Pine Siskin
Sciurus carolinensis – Eastern Gray Squirrel

66°
—
39°

October 16

A HUDDLE OF BOXELDER BUGS

To St Olaf. Recent cold weather put an end to most of the insects. However, rounding the end of the wooded pond, I notice on the opposite side of the trail, on the sunlit wood of a fallen tree, a small group of Boxelder Bugs. Arranged in a small clump, all with their heads pointed inward, as if conferring about the day or a team discussing a play in a huddle.

One of the great, oddball books written by a Minnesotan is *Boxelder Bug Variations* by Bill Holm. This book, beautifully illustrated by R. W. Scholes, contains poems, songs (including scores) and cautionary tales, all on the theme of Boxelder Bugs.

> "Two things in this world you ought to accept peacefully: dandelions and boxelder bugs"
> – Bill Holm

❖ ❖ ❖

Drove to Sakatah State Park, a midway-point between Northfield and Mankato, to deliver books to Lorna Rafness. Took a few photos of several Marbled Orbweavers under the eaves of one of the park buildings.

Araneus marmoreus – Marbled Orbweaver
Syrphus sp. – Flower Fly
Bombus impatiens – Common Eastern Bumble Bee
Vespula maculifrons – Eastern Yellowjacket
Glischrochilus fasciatus – Picnic Beetle
Boisea trivittata – Eastern Boxelder Bug
Sympetrum vicinum – Autumn Meadowhawk
Gracillariidae – Leaf Blotch Miner Moths
Limoniidae – Limoniid Crane Flies

October 17

BLACK-SHOULDERED DRONE FLY

73°
—
47°

To St Olaf and back through the church garden. Mid October and even in full sun the insects are lethargic, slow to take flight as if gravity were stickier, the dark, undersides of leaves gripping upwards.

Now, the stolid black fly becomes a spell, a charm, a trap I get caught in, my eye regretting its being, my curiosity as oddly inexplicable as the spurious vein in the wing. It must mean something? There must be some reason?

There are secrets here, on nearly every flower, every petal of sunlight. Yet, the air thins this time of year as if one has climbed a mountain peak very late in the day.

I captured a Black-shouldered Drone Fly (*Eristalis dimidiata*) for some close-up photographs. This black fly seems to be an inexact mimic of a Bald-faced Hornet. A fairly common fly.

Eristalis dimidiata – Black-shouldered Drone Fly
Halyomorpha halys – Brown Marmorated Stink Bug

73° / 48°

October 18

PATTERNEYE

To St Olaf. Autumn Meadowhawk at the wooded pond. Walked back through the church garden and photographed a variety of hoverflies on the chrysanthemums and late-blooming white asters.

Most interesting to me this day, was the Wavy Patterneye (Orthonevra nitida), a tiny hoverfly. As the name suggests, this species has a distinct pattern visible on its eyes. Other species in this genus have different patterns. Because of its size, it is very challenging to photograph the fly itself and even more difficult to get a clear picture of its eyes. The eye pattern consists of a thin horizon line over-layered and crossed by two vertical scribbles. This pattern, in a simplistic way, echoes the wing venation—black veins in an amber membrane. This little fly also has a bronze-metallic thorax. The eyes, the body color, the wings make the Wavy Patterneye a strikingly beautiful insect...when magnified and looked at closely.

Sympetrum vicinum – Autumn Meadowhawk
Zadontomerus – Small Carpenter Bee
Syritta pipiens – Thick-legged Hoverfly
Syrphus sp. – Flower Fly
Toxomerus geminatus – Eastern Calligrapher
Orthonevra nitida – Wavy Patterneye
Eristalis tenax – Drone Fly
Eristalis transversa – Transverse Flower Fly
Hyla sp. – Holarctic Tree Frogs
Phymata sp. – Jagged Ambush Bugs

October 19

LOWLY BEGINNINGS

74°
—
43°

"Two white asters, the common ones, not yet quite out of bloom." – Thoreau, from *Journals* (October 19, 1954)

Because of some home renovation commitments, I didn't get to take a hike, so the day's observations were limited to photographs of a fly and a bee, both captured on asters and both brought back from yesterday's walk and stored in the refrigerator overnight (which is not much different than being outside overnight this time of year).

Drone flies (genus *Eristalis*) are large and common hoverflies. Their larvae are known as rat-tailed maggots and are found in such lowly places as stagnant ponds and ditches, farm manure pits, outhouses, and carrion. Among the North American species, *Eristalis transversa* is the only black-and-yellow wasp mimic (there are several black wasp mimics and several bumble bee mimics). The common name for this species is the Transverse Flower Fly or, somewhat better, in my opinion, the Transverse-marked Drone Fly. It is sexually dimorphic, with the male abdomen pattern differing from that of the female by the addition of extra yellow spots on the 2nd tergite. The abdomen shape is more rotund than other drone flies.

A common and widespread species of eastern North America, Telford in *The Syrphidae of Minnesota*, reports it from eight counties: Clearwater, Todd, Olmstead, Ramsey, Anoka, Houston, Cass, and Le Sueur. To which I can now add Rice County. Probably occurs statewide.

Eristalis transversa – Transverse Flower Fly
Augochlora pura – Pure Green Augochlora

78°
—
53°

October 20

HOURGLASS FLY

Brought a load of construction debris to the landfill, all the heavy steel and cast-iron pipes from the bathroom remodel, a large derelict shelf that had been in the basement long before we purchased the house, off-cut lumber and shards of sheetrock. After this, I took a hike to the St Olaf Natural Lands.

A great wind blew throughout the afternoon hours, gusting multiple directions, causing house-high swirls of fallen leaves. From the natural lands, I walked to the garden at the Northfield Retirement Community. Here, on a few remnant flowers, I found a number of hoverflies.

Something like feedlot deer, this congregation of flies on cultivated flowers seems slightly unnatural. Nonetheless, I'm grateful. New this day, an Hourglass Fly (*Eristalis arbustorum*), also know as the European Drone Fly.

Hesperus sp. – Rove Beetle
Eristalis transversa – Transverse Flower Fly
Eristalis tenax – Drone Fly
Eupeodes sp. – Flower Fly
Eristalis arbustorum – European Drone Fly

October 21

69°

HELPER BUG

52°

The day woke with a thunderstorm just before dawn. I was up, moving a few items into the garage that had been left out: sawhorses, tools, a box of papers. The first raindrops were large, spattering my shoulders, each drop landing on the metal roof of the house and garage with a discernible thwack. A flash of lightning followed less than a full second later clarified the groggy mind. For as long as I can remember, I've counted the seconds...one thousand one, one thousand two...between the flash and the sound, the size of the gap providing an estimate of how far an approaching storm might be. This one was about to roll into town from the south.

A Saturday morning rain is a reader's delight. I started out with a short nature essay by the Welsh writer Llewelyn Powys, 'The First Fall of Snow' from his book *Earth Memories*. "Do we retain in our round skulls, in our square skulls, in our narrow, hatchet-shaped skulls certain subconscious recollections of the appalling struggles in far-off glacial periods?" From Wales in winter, I turned to pre-WWII England, reading the first nineteen sections of Louis MacNeice's *Autumn Journal*, an account, in the form of a long poem, of the last five months of 1938. "Now it is morning again, the 25th of October, / In a white fog the cars have yellow lights; / The chill creeps up the wrists, the sun is sallow, / The silent hours grow down like stalactites."

After the morning reading, the remainder of the day was spent on some home renovations. On one of my many trips outside to inside, a small helper bug must have hitched a ride. I noticed later as it ambled along the newly installed shower door as if inspecting the work. Recently, I'd been a little envious of some nearby observations of the remarkably colored Smooth Green Snake, a snake I've never encountered. But this Pale Green Assassin Bug assuaged that small amount of discontent.

Zelus luridus – Pale Green Assassin Bug

October 22
THE EYES OF AUTUMN

64° / 48°

I've noticed these faces before, not really faces but masks. Silver Maple leaves when they fall, curled and upside-down, have eyes. Gaping underfoot, gazing vacantly through the bare branches above them at the empty sky, they stare quietly up from the ground at passersby. And these eyes, like all the other eyes of autumn, are secretive and filled with shadows.

Acer saccharinum – Silver Maple

61°
—
43°

October 23

THE RIGHT PROFILE

Worked on book layout and a few press matters most of the day. Sid Gershgoren called. He informed me of some memory loss issues and had been on a new diet, excluding sugars. Either the new diet was working or the memory loss didn't extend to poetry because after I mentioned the poem by Louis MacNiece, 'The Sunlight in the Garden,' Sid proceeded to recite the first stanza from memory.

Photographed a Wild Indigo seed pod, inside and out. A small circular hole on one side of the pod indicated the seeds had been eaten by a weevil. And the pod was, indeed, empty.

The hoverfly season continues. Today's species is the Drone Fly (*Eristalis tenax*), a convincing Honey Bee mimic. The Drone Fly was introduced from Europe at some point mid-century in the 1800s. A synanthropic species, its larvae thrive in nutrient-rich aquatic environments. It finds a happy home in stockyards, catchment ponds, and uncleaned gutters.

One character used to identify and differentiate hoverfly species is the facial profile. And *Eristalis tenax* has a distinct face which protrudes below the eyes almost like a short trunk and which is white with a wide, vertical black stripe in the center. Two additional characters that distinguish this fly are a vertical stripe of dark hair across the eyes and a curved and inflated hind tibia.

Baptisia sp. – White Indigo
Eristalis tenax – Drone Fly

October 24

SNOW IN THE FORECAST

51°
—
40°

To St Olaf. Geese on the big pond at sundown, the prairie grass in the sunlight almost a blaze. With the leaves down, the aerial yellowjacket nests are now visible. I noticed a large one high in a tree on the south side of the pond.

I admire the paper these wasps fabricate from chewed wood and other plant material. Examined closely, the paper is layered, striped gray and white and tan. This is due to the varying sources, but also a result of the pulp being added mouthful-by-mouthful by the individual wasps.

Branta canadensis – Canada Goose
Vespinae – Hornets And Yellowjackets

58°
—
38°

October 25

SLEEP-SWIMMING

I approached the shoreline as stealthily as I could, which was not stealthy enough to avoid detection by the Canada Geese. Long before I even neared the edge of the pond, the geese were on the move, moving further out, heads up and vigilant. I stood motionless hoping the geese might forget I was there and maybe drift in closer. And while I waited I noticed a duck moving near shore slowly gliding toward me.

In the viewfinder of the camera, I could see that the duck's head was tucked back in sleeping position. Yet one eye was open and it was paddling. A sleep-swimming Hooded Merganser. If I were to make a sudden movement, it would surely wake and take flight. Surveying the hundreds of geese across the pond, a number of sleeping geese were drifting along with the more vigilant members of their flocks.

Recent research on Frigatebirds proved that birds sleep while flying. However, they don't plummet out of the sky because they rest half of the brain at a time, alternating hemispheres. They keep one eye open to avoid collisions and tend to fly in circles while sleep-flying. Just as Mallard ducks are known to keep one eye open for predators while asleep. Marine mammals also have the ability to half-sleep, the waking half allowing them to swim while sleeping.

So if birds can sleep while flying, surely it's not difficult for them to sleep while swimming. As I watched the Hooded

Merganser it swam in lazy circles, powered by a single foot (the other asleep I presume). Eventually, it woke and swam off, its crest of feathers upright, all hemispheres firing.

Melospiza georgiana – Swamp Sparrow
Lophodytes cucullatus – Hooded Merganser
Branta canadensis – Canada Goose
Spatula clypeata – Northern Shoveler

58°
—
35°

October 26

LATE BLOOMER

To Cowling Arboretum. Found an American Bellflower still in bloom. Several Mallard ducks on Hay Creek.

◆ ◆ ◆

An American Crow, with an injured wing, hopped along the ground, then up, branch-by-branch, to a perch in the Burning Bush outside the front window of our house. It stayed there for a long while. When it left, it jumped to the ground, walked across the yard then climbed the large cedar, again ascending the tree hop by hop. The crow seems to have adapted to its injury quite well. Even still, one wonders how long it will be able to survive, not being able to fly?

Corvus brachyrhynchos – American Crow
Campanulastrum americanum – Tall Bellflower
Anas platyrhynchos – Mallard

October 27

FIRST SNOW

36°
—
32°

To St Olaf. The first snowfall! Photographed Wild Cucumber, the spiky seed pod. Also collected a pair of seeds from one of the pods. Most pods were empty; their seeds had fallen to the ground. One of the pods, however, still held a couple seeds. The seeds were shaped, I thought, like the overemphasized eyes of pharaohs.

The seeds, when held in hand, were as inert as stones. They could be planted or (if I dropped them) they might sprout from the soil with all the others. Awakening next year, the seeds will become vine and flower. Through nonexistence, the seed quietly carries the future.

Echinocystis lobata – Wild Cucumber

34° / 29°

October 28

A COLD DAY IN OCTOBER

To St Olaf. Temperature at 3 p.m. was 35 degrees. I brought my benthic net but forgot it in the car. As a result, I had to resort to a cruder method, pulling a submerged branch out of the water at the pond and gleaning a few benthic invertebrates from its surface. Weed-wrapped, the black stick glistened in the air. Many damselfly nymphs clung to it. A few amphipods—like thick, flexing, gray commas—punctuated the length of the branch. A single leech balled and folded itself into a living, animal bud.

Strangely, this reminded me of an exclamation made by the Chilean poet, Gonzalo Rojas, after retrieving a large stick protruding from mud flats, hands and clothes soiled by the effort: "That's what I like. Impure poetry!"

Placobdella sp. – Jawless Leech
Coenagrionidae – Narrow-winged Damselflies
Blarina brevicauda – Northern Short-tailed Shrew

October 29
PRETTY PROBLEM

44°
—
29°

Turtlehead. Looking into the open mouth of the seed pod, I found several seeds. The seeds are unusual in shape, something like dried up fried eggs or tiny, vegetable flying saucers meant to travel (more likely wobble) a foot or two in the wind. No need to get too carried away.

Burning Bush. This ornamental shrub, a favorite choice for landscaping, has become a pretty problem. Birds consume the berries and disperse the seed (Cedar Waxwings seem to be quite fond of the berries). Next to invasive Buckthorn, Burning Bush is nearly as successful at spreading and establishing itself.

So we have decided to remove it from our property and replace it with something native, perhaps one of the sumacs which would retain the same autumn color.

Euonymus alatus – Winged Euonymus

40°
—
31°

October 30

THE GINKGO

Ginkgo trees seem to require a solid, sustained freeze before any leaves begin to fall. Prior to this happening, the leaves are green going gold and uncaring of the season. Though when the proper time arrives, the leaves drop all at once, sometimes nearly all fall in a single day, forming a circle at the base of the tree like an apron or a skirt dropped to the ground.

Because the Ginkgo's leaf-fall depends on a hard freeze, nights with temperatures well below freezing, the date of this annual occurrence corresponds closely to the end of the flight season of our hardiest dragonfly, the Autumn Meadowhawk. Most years these dragonflies survive the first frosts, wet snows, and brief overnight dips in temperature below freezing. It's not uncommon to find a few Autumn Meadowhawks on warm sunny days in November. Seeing the neighborhood Ginkgo trees losing their leaves, I'm afraid the recent cold spell has been cold enough to put an end to the dragonfly season.

Needing to visit the campus library (for a title about Tacitus), I detoured on my walk home to pass by what may be my favorite tree in Northfield, a very large Ginkgo on the south side of the philosophy building. With the recent construction and landscaping work at this building, I wanted to make sure the tree still stood. Wizened and elegant the tree still stands. With most of the green-going-gold leaves still on the tree, located alongside a castle-like stone building, the Ginkgo forms an iconic visage of autumn. As a tonic for this momentary romantic idyll, I picked up one of the fleshy fruits from the sidewalk and took a cautious sniff and the instant snap of rancid butter made me wince. Obviously, the purpose of such repellent "fruit" must be to prevent the nuts, the embryos of future trees, from being eaten before reaching the ground where they can grow.

Ginkgo biloba – Ginkgo

35°
—
27°

October 31

HALLOWEEN

A moment of whimsy. To decorate our table, I drew jack-o-lantern faces upon a handful of acorns gathered on my walk. More goofy than scary.

Quercus rubra – **Northern Red Oak**

November

The Hills – Maplewood State Park – November 24

November 1   37°
                              —
DISCOMBOBULATED   26°

Discombobulated, down-in-the-mouth, dour. Emotionally, today was one of those days when the heedless actions of neglectful people really knifed at my soul. Maybe it's just that it's the sixth day in a row with snow and clouds. Maybe it's just that I didn't get to say "Thank you" and knowingly take my leave of this year's dragonflies. Or maybe it's the steady rumors of war and other wrong-doing that insinuates dismay. No matter what, some kind of acrid smoke clouds and curtains my disposition.

I took a short walk through thick snow, hoping for a cure. At the catchment pond at the corner of St Olaf, I stepped close to the thatch of cattails. As I did a Snipe suddenly flew up, circled twice, and left. I was sorry to have scared it off.

Dia de los Muertos seems an appropriate moment to reflect on endangered species and the current, human-caused mass extinction event. I know this is a human-centric holiday, but why not extend it to the animal and plant kingdoms, why not demote ourselves that small amount more?

If (despite all reason and good sense) there is an afterlife, I hope it's not a cattail-choked retention pond, or, worse, an endless cornfield. To be a muerto entomologist stationed in a northern bog were Emerald dragonflies are flying or alongside a small stream with clubtails would be just fine. But a late summer meadow riddled with red dragonflies would be a more welcome fate; I could wait that out.

Typha angustifolia – Narrow-leaved Cattail

40° — 34°

November 2

BUTCHER-BIRD

Quiet, gray skies. The temperature a few degrees above freezing. I hiked the western edge of the natural lands, following game trails through the tall prairie grasses, enjoying the chance to wander a bit. As I passed a small oak tree, I noticed a vole dangling from the branches. Here was evidence that the Northern Shrikes have arrived. These birds nest in the far north at the border of taiga and tundra, then migrate south to over-winter. This bird is sometimes referred to as the Butcher-bird because of its habit of impaling its prey on tree spikes or wedging the prey between branches.

Approximately thirty years ago, snowshoeing at Maplewood State Park, I found a White-footed Deer Mouse wedged in the branches of a basswood tree. I remember how surprised I was to find it there. I examined the frozen body and placed it back in the tree. At the time I assumed it was an owl or some odds-defying drop from a hawk passing overhead. Now I recognize it as the work of the butcher-bird.

The Northern Shrike preys upon songbirds in addition to mice and voles. The following observation made by a resident of Hutchinson, Minnesota, is found in Thomas' *The Birds of Minnesota* (1932): "I saw a Northern Shrike eating a Horned Lark. When flushed it flew to a near-by tree-top. The Lark was stuck on a stick of willow bush and the Shrike watched me while I was examining the partly eaten bird. When I returned a half-hour later I found that the bird had returned and eaten all but the intestines."

Microtus – Meadow Voles
Asclepias incarnata – Swamp Milkweed

November 3
AGROSTOLOGY

35°
—
27°

Saw the sun today, but only at sunrise. Snow accumulating over much of the state to our north. After mailing a few packages at the post office, I stopped at the Cowling Arboretum for a walk.

In between the busy insect days and the cloistered silences of winter is the limbo of late autumn, an interregnum between the last asters and the flattening snow. This is a good time to turn one's attention to the grasses, to become for a day or two an agrostologist. With the plants still standing, it's easy to examine inflorescence and spikelets, leaves and glumes, before November gales and December storms break the stems and scatter the seeds.

I walked to the edge of the retention pond and found the water covered by a slushy film of first ice. Around me on the very edge of the pond was a stand of Prairie Cordgrass, each plant four to five feet in height, the once-flat leaves rolled up to form elegant fronds. The grass sways gently from wind on a windless day.

Sporobolus michauxianus – Prairie Cordgrass

November 4

A FEW LAST MINUTE INSECTS AND A SPIDER

Driving home from Rochester in the early evening dark, a moth flitted across the road in front of us. The first insect I've seen outside in about a week. The outside temperature as reported on the dash panel is 43 F, warmer than it's been in a while. As soon as we were home, I turned on the front house light, hoping to attract another moth or two. Eventually, a single Fall Cankerworm Moth turned up.

Alsophila pometaria – Fall Cankerworm Moth
Hemerobiidae – Brown Lacewings
Pholcidae – Cellar Spiders

November 5   43°

WAYBREAD   25°

The dreariness continues. While the forecast foretold of the sun making an appearance around noon, the clouds persisted. Anticipating a bright afternoon, I set about some early morning housework; now that the bathroom renovations were completed, I began the monstrous task of reorganizing my books. Not hundreds, but thousands. The Persian poet Fereydoun Faryad confessed in a Greek documentary film about his life and work to being "greedy for books." When he visited us in 2011, the year before he died of cancer, he brought a suitcase full of books, many of which he gave to me as a gift, beautiful Modern Greek editions of poetry. Well, as it turns out, I too must confess to this same weakness for books. It began long ago in college when I discovered used book stores, many were the days that I bought an old book instead of buying lunch.

So I set about culling titles to be given away, clearing entire shelves, discovering many neglected and misplaced favorites. One of Lida's favorite games, when she was a toddler, was to play library. This game, however, had nothing in common with the librarian's task of organization or the use of the alphabet. She revelled in the physicality of the books, sometimes using them as bricks to build elaborate forts, sometimes sorting them by color. By three in the afternoon, I had only made a larger mess. Books were scattered in heaps about the basement, looking not entirely different than the leaf piles Lisa had raked up in the backyard.

The Book of Field and Roadside: Open-Country Weeds, Trees, and Wildflowers of Eastern North America by John Eastman surfaced, came to hand, during the morning labors. A book long on lore and ecological interactions, a book of patterns and botanical traditions, a book that aims to "enlarge one's perspectives." This is an open-minded and sage account of our most common plants, many of which are (ironically) designated alien species. "In North America, changes in the plant components proceeded rapidly, mostly as accidental side effects of trade, settlement, wars, territorial acquisition, and discovery. Explorers not only mapped new territory but, in a sense, created it."

Walking the trails at the St Olaf Natural Lands, thinking of Eastman's book, I stopped to photograph some plantain. Is there a more common, more familiar weed? Perhaps Dandelion, but I can think of few others. According to Richard Mabey (in his recent book *Weeds*), plantain was one of the nine sacred herbs of the Anglo-Saxons. They called it Waybread because it was a broad-leaved plant of the waysides. "Its tough, elastic leaves, growing flush with the ground, are resilient to treading." The section of the 'Lay of the Nine Herbs' pertaining to Waybread contains the following lines: "So withstand now the venom that flies through the air, / And the loathed thing which through the land roves." This is a reference to Waybread's use as a remedy for bee stings and snake bites. Stings can be treated by applying a spit poultice, chewing up some plantain leaf and covering the affected area. Betsy Mead, a friend and herbalist, first alerted me to this remedy when she learned I was studying wasps.

Eastman informs us that "plantain leaves make a tasty cooked or salad green when collected very young." He also relates that "plantains have prominent associations with human feet. The word *plantago* derives from a Latin word meaning 'sole' or 'footlike'"... Supposedly common plantain followed the Roman legions wherever they set foot. Native Americans, observing the plant's spread, carried the analogy further by naming it Englishman's Foot and White-man's Footprint."

Setaria sp. – Foxtails And Bristlegrasses
Panicum sp. – Panicgrasses
Elymus canadensis – Canada Wild Rye
Schizachyrium scoparium – Little Bluestem
Trifolium sp. – Clovers
Plantago rugelii – American Plantain
Dolichovespula – Aerial Yellowjackets

33°
—
22°

November 6

KITCHEN SPIDER

I found a small spider in a corner of the kitchen, captured it, photographed it, and released it. The Triangulate Comb-Foot is a species of the genus *Steatoda*, known as False Widow Spiders, and part of the family of Cobweb Spiders (Family Theridiidae).

This is the second time I've photographed this species in our house, the first time some years ago, in 2011.

Steatoda triangulosa – Triangulate Comb-foot

November 7

FREEZE-DRIED FLOWER

36°
—
28°

Sunshine at last, after a lapse of ten days, I left the house and took a walk under blue skies. A bright, warm-looking day, but the looks were deceiving—the temperature peaked at 40 degrees. Still, it felt good.

The wooded pond was iced over. The edges, where duckweed had been concentrated by a long-ago wind, caught the sun energy, slight as it was, and began to melt.

Here and there, a dab of color remains at the trail's edge, bright yellow goldenrod, blue star asters, freeze-dried, preserved into winter.

Joe Paddock called. Joe, who recently turned 80, was a founding member of the Land Stewardship Project. He's also a poet, an oral historian, a Jungian, and a biographer. He's someone I've admired for many years and it was a pleasure to speak with him and tell him a little about my current project of following the earth around.

Calyptratae – Calyptrates
Fungi – Fungi Including Lichens

November 8

ΚΟΛΕΟΠΤΕΡΑ

A little Greek goes a long way in entomology. For instance, the origin of the word "insect" can be traced back to a descriptive name coined by Aristotle. Because the abdomens of insects are segmented they appear to have a number of lines cut into them, like the notches cut into the top of a loaf of bread. Aristotle combined the prefix ἐν with the verb τέμνειν to get ἔντομο. Later the Romans constructed a similar word from their word "insecare" which means "to cut into."

At some point, during my college days as an engineering student, I encountered the book *On Growth and Form* by the Scottish biologist and polymath, D'Arcy Thompson. The illustrations of water droplets and jellyfish, the comparisons of the cells in honeycombs to the cells in the wing of a dragonfly, the application of mathematics to the morphology of the natural world profoundly changed the way I looked at the world. (The only other book that unleashed a similar reformatting of my perceptions was *Chaos* by James Gleick.)

Thus I was intrigued to learn that D'Arcy Thompson had translated Aristotle's *History of Animals*, the work that contains mention of insects and beetles. Here's a short passage from Book IV: Part 7:

"...as in the case of those that have their wings in a sheath or shard (εν κολεῶ), like the cockchafer, the carabus or stag beetle, the cantharis or blister-beetle."

Amara sp. – Sun Beetles

November 9   30°/12°

КАЛИНКА, КАЛИНКА, КАЛИНКА МОЯ!

The first truly cold day. Clear skies, slant sunlight, and a biting north wind greeted me as I walked a loop of trails through the St Olaf Natural Lands. By the time I was done, my nose and cheeks were numbed and raw, my ears ached from the cold wind. An early serving of winter.

I'm still adjusting to the sudden emptiness of the woods and fields, now that the insects are gone. My eye, still drawn to movement in the air, follows the descent of falling leaves and the drift of seeds. A Red Squirrel shimmies down the trunk of an old oak where not so long ago Spine-waisted Ants traveled and worked. Further on, I noticed the frozen berries of the Geulder Rose, bright red against the drab backdrop of winter woods.

The Geulder Rose or European Cranberry is an important symbol in the Ukraine. And in Russia there's a popular folk song named for this tree, the snowball tree (Viburnum opulus) and its red berries, Kalinka, kalinka, kalinka moya! The song has a quick tempo, speeding up with each repetition. Little red berry, red berry, red berry of mine!

Viburnum opulus – Cranberry Viburnum

November 10

WIND CHILL

Nothing like single digit wind chills to clarify the mind... and make the nose run... and numb the fingers. Even still, I essay into the cold weather, knowing there are darker days and colder days to come. It's an attempt at honesty, not bravado. The way seedlings can be hardened off to wind and outdoor conditions in the spring, I sometimes think one can prepare for winter. Maybe, by simple exposure to the elements, an alchemy of attitude and metabolism might occur? I realize it's a little simplistic, for each winter the long nights eventually close in on the mind, a grip I've never fully been able to escape.

With the wind gusting out of the south, the woods on the north slope of the St Olaf hill were quiet and calm. I stopped at one of the wasp trees along the trail. Hub of so much insect and bird activity during the warmer months, the remains of this tree now stand derelict, like a ruined mansion. I reached out and took hold of one of a number of small shelf fungi protruding from the trunk of the tree. Covered in wood dust and bark chips, I was uncertain if it was a frozen mushroom or bracket fungus. It was solid. I snapped off a chunk like breaking a piece off a cookie. A small beetle larva dangled out of the broken edge. I placed the larva and the piece of bracket in my pocket to examine later at home.

Cautiously, I stepped onto the ice of the wooded pond. Near shore, where yesterday's fierce north wind hadn't kept the water open, the ice was several inches thick. It held my weight easily. Bubbles trapped under the ice moved with each step; the ice must be bending ever so slightly. During the calm overnight hours the open water had frozen as well, but that ice was quite thin.

| Coleoptera – Beetles | Cicadellidae – Leafhoppers |

November 11 — 44° / 24°

ODES TO SILENCE

Woke to a dusting of snow. Warmer than yesterday, the snow was almost all gone by late morning when I went for a walk. Still, portions of the trail were snow-covered, fallen trees in the woods kept a coruscation of snow, and the iced-over pond remained blindingly white, the powdery crystals protecting the pond's ice from the sun's heat.

I walked onto the ice. The covering snow concealed the ice so that I couldn't guess at its thickness. I'd hoped to get closer to a set of tracks where a fox had walked across the pond, but when the ice began to give under my weight I retreated.

There's something elemental about fox paw prints in new snow. The pond had only recently frozen, yet a fox was quick to take advantage of this extension of its territory. The prints left behind are an elegant trace of life, like a fragment, a single line, from a lost ode to silence.

Peromyscus leucopus – White-footed Mouse
Canidae – Canines

36°
—
24°

November 12

BIRCHES

Northern Lights volleyball tryouts. 7 A.M. to 1 P.M. What an ordeal for the kids trying out. Lida did so well in this anxiety-ridden atmosphere—I'm simply amazed and, of course, very proud. Thirteen setters tried out for nine positions and she made the cut. We certainly felt some of the sadness of tryouts as well, as three of her previous teammates were cut and didn't make a team.

Dinner at Lisa's mom's house in the afternoon. This was Bruce's birthday. With his recent death, this was a second emotional event.

A dead robin amidst some fallen leaves, orange breast feathers lifted gently by fingers of invisible breezes, lay to the side of the cement steps. Tried to photograph the Gold Finches in the Paper Birch using my phone. The quality of the image was not good enough to use as an observation of the birds, so I settled for an observation of the tree.

Betula papyrifera – Paper Birch

November 13   46°/25°

FLIES

To St Olaf. 3:15 to 4 PM. A brisk lap through the natural lands. Sunny. Temperature in the low 40s. Opposite the trail from the wooded pond, on the bare, sunlit branches of a toppled elm, several small flies dallied in the slant light.

At the very fringe of winter, a few flies

Cardinalis cardinalis – Northern Cardinal
Sciomyzidae – Marsh Flies

45°
—
39°

November 14

EASTERN PONDHAWK

Cowling Arb. Overcast. Misting rain. I searched the edge of the Cannon River in a few places, hoping to find a clubtail nymph, risking the rather vile mud at the margins. The late-season rains, runoff from decaying farm fields and rotting lawns, plus the usual (more natural) litter of leaves created a smell like manure, a cloudiness to the water, and a layer of soot-black silt in the shallows. The Cannon River: devious and shallow, carrying cowslop and carpwater, the main vein of our disregard for the natural landscape.

Not finding any dragonflies, only a few large crane fly maggots, I moved inland, and took a couple scoops at the edge of the retention pond. Amid dozens of damselfly nymphs, a couple of diving beetles, and a water scorpion, I found several dragonflies, all skimmers, including a blocky, green-colored Eastern Pondhawk nymph.

Skimmer nymphs are ambush hunters. As such, they spend most of their time motionless. This allows a flora of algae and a fauna of tiny animals to use the dragonfly's exoskeleton as a foundation. Certainly, it aids the dragonfly in its hiding, but it makes for messy photographic portraits, often obscuring characters necessary for identification. Not surprisingly, these nymphs look something like a couch would look if it sat under the same water for a month or two—a mess!

Erythemis simplicicollis – Eastern Pondhawk
Leucorrhinia intacta – Dot-tailed Whiteface

November 15   46°

WATER DRAGON   27°

The life of a catchment pond, and certainly this one, is dramatic. To begin with, an impressive variety of plants and animals find their way to the newly constructed pond, immediately it is filled with water. This particular catchment pond, located along one of my most frequented walking routes, has been easy to observe and follow through its maturation. I've stopped here many times and have followed its progression from barren beginnings to its current cattail-choked and willow-ringed fulfillment.

The catchment pond receives water from two cement culverts—the origins of both are unclear. One enters level with the pond on its south side. The other culvert empties its water above the pond, spilling storm-water runoff into a ravine filled with large, limestone rocks. During heavy rainfall, the pond level rises dramatically. I've observed rings of debris nearly six feet above its drained level. Local miscreants rolled a tractor tire into the pond. For a while, a street sign (sheared off by a car?) reclined in the shallows. Other flotsam include the occasional soccerball (once a rugby ball), plastic bottles, and dining hall trays.

Out of a muddy pool, I scoop up a few small dragonfly nymphs, probably Common Whitetails which have been so prevalent at this catchment pond this summer. And also several damselfly nymphs, Eastern Forktails, a species that can live in nearly any habitat.

Libellulidae – Skimmers

November 16

HAWK VS. WOODPECKER

Upper Arb for a short hike.

In the woods near the footbridge spanning Spring Creek, a Pileated Woodpecker railed and rattled like a barnyard chicken. Not being its usually monkey-like call, I watched it for a while. The woodpecker made its way up the trunk of a tree. Then it was attacked. A hawk, probably a Sharp-shinned, swooped down from a nearby branch, aiming at the woodpecker. This happened repeatedly over a long period, the Pileated Woodpecker continuing to make a kind of cackling sound moving tree to tree, the smaller hawk repeating its half-hearted assaults. I left not understanding what was actually going on. Perhaps some territorial dispute? Perhaps the woodpecker was leading the hawk away from a nest?

I photographed several kinds of grass along the trail. Hairy Cupgrass (Eriochloa villosa) and Virginia Wild Rye (Elymus virginicus). Hairy Cupgrass is a non-native species, imported from eastern Asia, currently limited to a small area in the midwest. Virginia Wild Rye is a native, closely related to Canada Wild Rye and Bottlebrush Grass.

Eriochloa villosa – Hairy Cupgrass
Asparagus officinalis – Wild Asparagus
Elymus virginicus – Virginia Wildrye

November 17 — 39° / 34°

MOODMOLDED

Though a spate of days, weeks even, the weather pools and spits. My outlook is moodmolded by rain, cold rain and close clouds.

An exfoliate landscape.

The leaves underfoot, the yellow-brick road
    being trampled to mud.

The emptiness, the absent
    cling to the bare branches
        ramified

The frost remembers
    as it grows crystal leaves
        on the window glass

Accretive. The land, a month too long in the bathtub, its limbs and toes and fingers all wrinkled and sodden. Followed, all of a sudden, by the slow down, the cold season files in.

❖ ❖ ❖

It occurs to me, that if I were to write a memoir it might properly be titled *What Not To Do*. It might also be extremely uninteresting. Like the word "solecism," which originates in the Greek soloikos—speaking incorrectly, perhaps there is a parallel word that means living incorrectly?

Andropogon gerardii – Big Bluestem

37°
—
21°

November 18

FLAT RED BARK BEETLE

Ah, sunshine. A day without many clouds, a rarity during this last month. I looked forward to having a peek at the landscape, the late autumn meadows and wetlands illuminated instead of shadowed.

Previously, I'd noticed some peculiar lichen on several trees at the northeast corner of the Natural Lands. I went there to get a photo. Bushy lichens of the genus Ramalina have flat, strap-like branches that extend away from the bark on which it grows, in contrast to the flat, lobed, surface-hugging lichens.

In the woods, nearby, a trio of brown mushrooms provided dots of color to a blackened stump. At the base of a dead tree, its snapped and fallen trunk was dotted, warted with the bronze aethalia of Wolf's Milk slime mold. At the base of another dead tree, wedged behind a piece of loose bark, I discovered two Flat Red Bark Beetles.

These bright red beetles, the color of Chinese red lacquer and flat as fingernails, are predators of wood beetles and, most impressively, have been studied for their cold tolerance. By dehydrating themselves and producing a protein antifreeze, these beetles have survived temperatures of -150 C in the laboratory!

Cucujus clavipes – Red Flat Bark Beetle
Lycogala epidendrum – Wolf's Milk
Ramalina sp. – Bushy Lichens
Agaricales – Gilled Mushrooms
Didelphis virginiana – Virginia Opossum

November 19
WILD LETTUCE GALL

41°
—
18°

We visited the big pond at St Olaf Natural Lands, Lisa, Charlie, and I. In addition to stretching our legs, the reason for the visit was to return several dragonfly nymphs to the pond. I had to break through the ice.

Collected a gall from the stem of Canada Wild Lettuce. A Lettuce Tumor Gall caused by a gall wasp of genus Aulacidea. Slightly elongated, spindle-shaped, leafy. Each gall that is new to me, gives the opportunity to dip into a favorite book, Margaret Redfern's prodigious monograph *Plant Galls* (New Naturalist Series).

The gall itself is a simplified representation of a complex process, like a book cover, a hand closed around a secret object, or a mask. Especially in winter—the host plant dead, flowerheads and leaves broken off and worn away by weather—the stem galls with their hidden occupants towering like vegetative space needles above a landscape of snow a hidden minimalism...and yet, when one sets in motion the biological interactions and phenological sequences that cycle and encircle (incircle?) these simple, lignified spindles one fronts confusion, complexity and awe. Rings of interlinked cycles, a madness of biochemical reactions, living clockwork, both happenstance and fixity ticking gallwise, the genome of wasp instructing the genome of the plant.

Aulacidea sp. – Gall Wasp

48°
—
28°

November 20

COFFEETREE

I had never seen a Kentucky Coffeetree until I moved to Northfield. The city has planted many along the city streets. For six years one of these trees greeted my daughter and me at the far side of the crosswalk at her elementary school. Though I no longer see it almost daily, it remains one of my favorite trees. The two of us, Lida and I, collected seed pods and extracted the stone-like seeds. And for many years kept a container of these black, vegetable marbles. Slightly irregular in shape, somewhat silky surface, the seeds are satisfying to hold.

Biogeographically, Northfield is at the northern limit of the species' range. It has been speculated that the now extinct Giant Sloth played an important role in the dispersal of this tree. Something large and powerful was needed to scarify the tree's stone-like seeds to enable germination. A metal file will do if you don't have a giant sloth.

Here's an excerpt from the beginning of a children's story I began but never finished:

> "Look. Witch's fingers," Anna's father pointed.
> "Are there really witch's fingers?" Anna asked, slightly concerned.
> "Do you see that tree with the big seed pods? Some people think the branches look like witch fingers. And some call it the dead tree because it doesn't have any little branches in the winter."
> "They should call it the big old ugly ear tree," Anna said with a laugh. Her father smiled and laughed as well.

Gymnocladus dioicus – Kentucky Coffeetree

November 21

FRESHWATER SEA MONSTER

42° / 18°

Up close photos of the tentacle-like gills of an aquatic crane fly larvae. This larva was collected at the shore of the big pond at the St Olaf Natural Lands in the mucky sediment beneath the ice.

The genus *Tipula* includes many and varied species. They are often referred to as Giant Crane Flies. The adults, of which, look like super-sized mosquitoes.

The gills retract and expand. At full extension, and under magnification, the larva makes a convincing sea monster.

Tipula sp. – Crane Fly

33°
—
15°

November 22

HOME FOR THE HOLIDAYS

To Pelican Rapids. To my father's house on Lake Lida. This is the first visit since Christmas the previous year—usually, we make it up to the lake several times each summer. We left Northfield, Lisa, Lida, Charlie, and I, just after 9:30 A.M., and picked up my aunt Ginger in Crystal at about 10:30. We had lunch when we reached Alexandria. Lida drove the remainder of the way. We arrived at the south shore of North Lake Lida just after 2 P.M. Here and there were traces of snow on the ground. The lake was frozen.

It's good to have an anchor point (or several) on earth. Though I haven't lived here since 1983, this place, an irregular circumference centered at the farm where I had lived (no longer owned by the family), still feels like home. The shape of the hills, the scattering of lakes, even the placement of trees remains familiar.

Lisa and I took the dog for a long walk. Following a circuit of back roads, we passed the north edge of the old family farm where I'd spent my teenage years. Strong memories are associated with this place and surface each time I've had the opportunity to pass by in recent years. Lunchtime, a happy moment, isolated out of a day of labor, weeks and years of oblivion to either side. I remember it was a hot, sunny summer day, that I wasn't alone, no doubt joined by my father, possibly my brother and sister. We sat in the shade beneath a large cottonwood at the edge of a small field, where we had been baling hay or harvesting oats. That tree still stands. I crossed the ditch and placed my hand on the tree as though I might feel my own heartbeat.

Just down the hill from the now fallow field and ancient tree, at the edge of a small wetland, Lisa and I admired several bright orange sprays of Bittersweet, the vines clambering through the bare branches of sumac.

❖ ❖ ❖

At some point, after I'd begun reading poetry, I discovered the work of Cesare Pavese, his book of poems *Hard Labor* and his evocative novels like *The Hills*. I appropriated "The Hills" to refer to this place, my one-time home between Lake Lida and the hills of Maplewood State Park, the glacial hills of the Alexandria terminal moraine. Here's a short elegy, written about that time:

The Hill

Often he returns to the hill that houses his mother's ashes.
He stands there, empty as a hand. He tries to understand the grass
    sawing at his knees, understand the smell of distant hay and lake
    water.
He is conscious of his hair above the land, the leaves sprung from
    the tortured oak.
Always, the moment contains both day and night.
With his tongue, he tries to feel the inside of his skull.
He knows the tangle of stars is in his head, and any meaning is
    simple, like the moment's unexplained sumac trees and the one,
    mysterious cattail stem to which the red-winged blackbird keeps
    returning.
A land of humid shadows and greenish winds in summer, a hill of
    frigid noise and star-lit bones in winter.
He stands, one foot in spring, the other in autumn, trying to under-
    stand.

Celastrus scandens – American Bittersweet   *Asclepias incarnata* – Swamp Milkweed
Schoenoplectus – Club-rushes & Bulrushes   *Phragmites sp.* – Common Reed Grass
Celastrus scandens – American Bittersweet   *Planorbidae* – Ramshorn Snails
Schoenoplectus acutus – Hardstem Bulrush

49°
—
23°

November 23

MAPLEWOOD STATE PARK

Lisa and I hiked the Bass Lake trail after the traditional holiday meal. Being Thanksgiving, the park was abandoned, the quiet woods welcoming. We were in the woods and on the trail during the half-hour before the sunset and an hour of dusk.

Established in 1963, Maplewood State Park is a mosaic of re-wilding homestead farms. The Ag-industrial economics undermined the viability of these small holdings. The landscape is too steep and too pock-marked by wetlands and lakes to ever be consolidated into large farms, unlike the huge, square-mile tracts not far to the west in the Red River Valley.

In the late 70s, after my family had settled on the 180-acre farm (the old Drayton farm) across the highway from the state park, my father bid on a granary from one of the decommissioned farms in the park. I remember the excitement of going to look at it, the dreams of filling it with oats and corn harvested from our own fields. A curious optimism, in retrospect, given the vaults of deep shadow.

Ostrya sp. – Hop-hornbeams
Cladonia chlorophaea – Mealy Pixie Cup
Gracillariidae – Leaf Blotch Miner Moths
Populus grandidentata – Bigtooth Aspen
Hepatica americana – Round-lobed Hepatica
Flavoparmelia caperata – Common Greenshield Lichen
Ramalina sp. – Bushy Lichens
Teloschistes chrysophthalmus – Golden-eye Lichen
Elymus hystrix – Bottlebrush Grass

November 24   60°

THE HILLS   40°

Rain. Read Michael Gregory's poem 'This Far' — a very appropriate poem to read on Black Friday.

> "tired of the political can't whenever
> we mention economic democracy,
> everyone sucked into the cash nexus
> struggling to keep their heads above water
> in pursuit of money as happiness,
> who can't imagine what a self might be
> except in terms of private property"

Private property is a problem here. What once belonged to my family and people we knew now belong to strangers. "No Trespassing" signs fence the perimeter of the farm I grew up on.

During these years, I often roamed the adjacent land of the park. It became a steadfast retreat and a vast reserve of unexplored and (most importantly) unowned land. It remains the only land, the only portion of those childhood hills that I can revisit and return to.

The skies cleared around noon. The sun came out and the temperature rose to 50 degrees. We ate lunch at the Muddy Moose in Pelican Rapids, then hiked at Maplewood State Park. Unexpectedly, on our way up Hallaway Hill, we met Debra Hovland and Doug Dorow and their daughter Olivia and one of her friends on their way down. Friends we hadn't seen or spoken to in years. What a good place to meet!

Later after summiting the hill, we visited the swimming beach. It was warm in the sunshine and out of the wind. Found a few insects to photograph along the sunlit beach, at the edge of the lucent ice.

❖ ❖ ❖

"Portes ouvertes sur les sables, portes ouvertes sur l'exil" – Saint-John Perse

---

Tilia americana – Basswood
Calyptratae – Calyptrates
Gastropoda – Gastropods
Bivalvia – Bivalves
Planorbidae – Ramshorn Snails
Araneae – Spiders
Limoniidae – Limoniid Crane Flies
Bembidiini – Ground Beetle
Dreissena polymorpha – Zebra Mussel
Lycosidae – Wolf Spiders
Hydrophorinae – Longlegged Flies
Cirsium sp. – Thistles

November 25

SWAMP THISTLE

41°
—
29°

Travel day. Pelican Rapids to Northfield. The Interstate 94 transect from Fergus Falls to Maple Grove. With a stop at a grocery store in Monticello for cheese and crackers and hummus and apples to eat in the car, a high-speed picnic. We dropped my aunt at her house in Crystal. We made it home around 4 PM.

Photographed a thistle head and its seeds which I'd collected at the park. What I thought was a Swamp Thistle because of its height and because it had few flowerheads was an atypical Canadian Thistle.

I also photographed what I presumed to be a moth cocoon, though I'm not entirely sure. It could be that I'm wrong and this is the pupae of a wasp. Found under bark and embedded in what appeared to be a spider web? The only sure way to know what made it and what's currently inside would be to keep it and see what emerges. So it sits in a vial in a cool spot on the windowsill in my office.

Cirsium sp. – Thistles
Lepidoptera – Butterflies And Moths

51°
—
24°

November 26

WHITE POPLAR

Sunday. A warm November day with temperatures in the upper 40s.

Returning to Northfield after a family gathering in Apple Valley, I noticed, out the car window, a Shrike perched on a power line just north of town. When I went for a walk a little while later, I decided to hike back to that spot and see if the bird was still there. It wasn't.

On the way out to look for the Shrike, I took my usual route through part of the St Olaf Natural Lands. For years, I'd only half-noticed the white leaves fallen among the brown-tinted oak and maple leaves along a certain stretch of the trail. Today, for the first time, I paused to photograph one of the fallen leaves.

I assumed the leaves belonged to a poplar. Checking *Trees of Minnesota* it was quickly apparent the tree to which this leaf belonged wasn't included in the *Populus* section. The discussion for this genus mentioned a number of hybrids and imports. After a little further research it appeared the leaf belonged to a White Poplar tree (*Populus alba*) or a Grey Poplar, a hybrid between a White Poplar and a Common Aspen.

Populus alba – White Poplar

November 27

60°

NORTHERN PAPER WASP

32°

A November heat wave. Temperature was near 60 degrees F today. Warm enough that a couple of paper wasps came out of hibernation. I saw two of these wasps on the sidewalk when I visited the library. These early winter wasps reminded me of this poem written after finding a wasp near the end of December some years ago.

NORTHERN PAPER WASPS (*Polistes fuscatus*)

These last weeks of December,
the snowiest on record,
I've been indoors reading
Howard Ensign Evans,
his *Wasp Farm*, chapters
on the spider wasps,
the great golden diggers,
the beewolves.

Now comes the release
of a mid-winter thaw,
then, more surprising still,
a Co-op of wasps found
scattered on a sidewalk
like a handful of small caliber
rifle bullets. They are hibernating
Northern Paper Wasps
knocked down from the roofline
by birds or a collapse of roof-ice,
the pale sun on red brick

not nearly enough to wake them.
I pick one up gently,
carefully hold it in my fingertips.
This warm-blooded grip stirs
the sleeping queen
to stretch out a yellow leg
as though it were spring.

Back home, I take up the book,
flip forward through unread pages—
sure enough—the wasp
is waiting there as well,
the name and pronunciation:
po-LIST-eez fus-CATE-us.
I say it over and over—
the Greek meaning
founder of a city, the Latin
black, for its smoke-colored wings.

Polistes fuscatus – Dark Paper Wasp

November 28   53°/33°

SHORT-BELTED ICHNEUMON

"I wrote a line where I spoke about an adjective that was sprouting grass. A number of years later, in Paris, I saw on a headstone in the Montparnasse Cemetery an adjective covered with grass. Prophesy of Poetry." – César Vallejo

At 9:09 A.M., I learn that Joseph Conrad was a literary hero to César Vallejo. Two wires of my own get twisted together, such disparate circuits that I can almost feel the confluence and snap of the breaker. How can *Heart of Darkness* and *Trilce* fit together? Vallejo encountered Conrad in French translation in Paris in 1925, translating and appropriating favorite phrases into his own writings afterward.

"A work that aspires, however humbly, to the condition of art should carry its justification in every line." – Joseph Conrad

There can be little doubt the weather is about to change. Under the clear pivot of a high-pressure system comes a confusion of winds. Crows in a cottonwood.

When I stepped off the narrow footpath in the direction of the sunlit base of a limestone outcropping, it was with a certain expectation, not grand, not overly hopeful; having watched a bottle fly disappear, I simply hoped to see where it had landed. Instead of the fly, the warm microclimate at the base of the rocks yielded a surprise. A small parasitoid wasp, mimicking one of the spider-hunting wasps, flicked its wings and searched among a few crispy brown leaves. Huh! I'm amazed to see something other than a fly active this late in the year.

Btw, there's a ridiculous video (on YouTube) of this wasp riding a cutworm caterpillar the way a clingy drunk might ride an electric, barroom bull.

Vulgichneumon brevicinctor – Short-belted Ichneumon

November 29   46°

HACKBERRY NIPPLE GALLS   25°

The path to winter differs year to year. The route is never a smooth, continuous descent into darkness and cold weather. The changes often arrive precipitously—from sixty degree days to weeks of snow and freezing temperatures. But then come days of reversal when winter loosens its grip. This sudden string of bright, clear days at the end of November lifts the spirit, makes one feel fortunate, like being dealt a hand full of aces.

❖ ❖ ❖

Photographed a Hackberry Nipple Gall. The occupant of this small gall is neither fly nor wasp but a tiny Psyllid, *Pachypsylla celtidismamma*. Small enough to fly through our porch screen, we've had hundreds of these little bugs in our house during peak times of emergence. In shape and structure, they look a lot like minuscule cicadas.

Pachypsylla celtidismamma – Hackberry Nipple Gall

48°
—
31°

November 30

BIG WOODS

Taking advantage of the continued mild weather, I drove to Nerstrand-Big Woods State Park in the afternoon. There seemed to be more cars in the parking lot than when I'd visited at the height of the spring wildflower season [see May 4th]. And yet, contrary to the number of cars, the trail to Hidden Falls was empty.

Below the falls, Prairie Creek flows through a debris field of flat limestone shards. Where the sun struck the bank, bright through the bare-limbed trees, it was warm. Several Liminid Crane Flies (also known as Winter Crane Flies) took flight as I passed near. Flipping over a few stones in the shallows of the creek revealed a few snails and a few amphipods.

This park is one of the few places with a population of Red-headed Woodpeckers. After hiking back from the waterfall, I sat on a bench outside the visitor center and watched the feeders for a while, hoping one might show up. No such luck. Instead, I enjoyed the company of a number of Nuthatches, Downy Woodpeckers, and Chickadees.

Poecile atricapillus – Black-capped Chickadee
Quercus alba – White Oak
Limoniidae – Limoniid Crane Flies
Physidae – Bladder Snails
Gammarus pseudolimnaeus – Northern Spring Amphipod

December

Aphid Wasp Larvae – December 21

December 1

GRIMACE

51°
28°

Sometimes at the onset of winter, days, landscapes, and the souls of resident humans seem to adopt a resigned grimace, knowing that four months or more of snow and cold weather must pass before the return of green leaves and spring flowers. Today, however, was not one of those days. It was fifty degrees in the afternoon when I visited the St Olaf Natural Lands where I stood at the edge of the wooded pond and watched several muskrats enjoying the open water.

Araneae – Spiders
Solidago speciosa – Showy Goldenrod
Dalea – Prairie Clovers

47°
—
30°

December 2

BOG PANEL

To Groveland Gallery in Minneapolis for the opening of Meg Ojala's exhibition 'From the Bog.' As part of the opening, John Latimer, a phenologist from Grand Rapids, Katrina Vandenberg, a poet from Saint Paul, and I presented our varied knowledge and connections to bogs. My contribution consisted of two parts: my biography done in bogs and an annotated list of common bog dragonflies. I've included the first part of my presentation below.

BOG DRAGONFLIES: PART ONE

I've held a lifelong affinity for wetlands: lakes, sloughs, swamps, bogs. A half mile from my childhood farm was a beaver dam, behind which stretched a mysterious tamarack swamp. I was always fascinated by that place, probably its inaccessibility, but also its beauty.

In college, beginning in the late 1980s, I spent some time at the Trout Lake Station in northern Wisconsin. And began, at that time, to learn more of the names of the bog plants and boreal wildflowers, having purchased my first guidebook, Fassett's *Spring Flora of Wisconsin*.

Then came a long interim of poetry, a digression of many years, a path I'm still wandering upon.

In 2001, I became a stay-at-home dad. Over the subsequent years, my daughter and I began spending a lot of time outside. In trying to teach her the names of plants and animals, I suddenly realized how little I knew. So I threw myself into the study of natural history. At this time the opportunity arrived to join the Minnesota Odonata Survey Project under the direction of Kurt Mead. That was ten years ago, in 2007.

As part of the MOSP, I participated in a number of week-long surveys. Conducted by a crew of at least four dragonfly experts, these field weeks took place in the under-surveyed parts of the state. Several very memorable trips centered around the bogs and fens and peatlands of northern Minnesota. One particular day on one of these trips I visited a bog with Ken Tennessen, a dragonfly expert from Wisconsin, and this led to a correspondence in haiku and eventually the publication of our co-authored book, *Dragonfly Haiku*.

In 2011, I received a research grant to study *Sympetrum madidum* in northwest Minnesota. Where I fell in love with the Tallgrass Aspen Parkland and the boggy road ditches.

In 2013, I gave a talk at the Dragonfly Society of the Americas annual meeting in Prince Albert, Saskatchewan. Field trips after the meeting included visits to the lakes and bogs north of the Churchill River.

More recently, I had the opportunity to help with surveys of the vast Red Lake Peatland SNA. The Red Lake Peatland contains the largest and most diversely patterned peatland in the United States. The home base for these outings was Norris Camp, a MNDNR outpost that was originally a CCC camp, located south of Warroad.

In 2016, driving Highway 1 to Finland, Minnesota, returning from a trip to North Dakota (a state that has no bogs), I stopped at a near perfect roadside bog and finally had my chance to observe North America's smallest dragonfly, the Elfin Skimmer.

This year I feel slightly bog-deprived, my bog-time being limited to three days of dragonfly work at the beginning of June near Hoyt Lakes assisting some Minnesota Dragonfly Society researchers. But those three days and the bogs visited were certainly a highlight of this year's circle.

◆ ◆ ◆

Walking across a bog has been wonderfully described as similar to walking on a wet sponge. John Latimer's description that walking on a bog is like walking on a waterbed may even be more accurate. I also like the effect of walking out of the forest past shorter and more stunted trees and into the open, which is a miniature version of trekking north of the tree line. And making it to the very edge of the open eye and peering over into the deep water always brings to mind Dostoyevsky. He wrote about the sensation of being at the edge of a precipice and that little part of your mind that wonders what it would be like to go over. The sudden depth of the bog's eye, the light penetrating just a few feet into the peat stained waters, is an apt metaphor for anything the human mind can't answer or comprehend.

BOG DRAGONFLIES: PART TWO

(Many species can be found at bogs and fens, this is only a partial list of some notable species.)

Subarctic Bluet: The range of this darkest of bluets only dips into the northernmost parts of Minnesota. Found at the edge of bog and fen pools.

Emerald Spreadwing: A large metallic green damselfly.

Ebony Boghaunter: These small, black dragonflies emerge right away in spring, probably the first dragonflies to do so. Because of their preferred habitat and early emergence, this dragonfly is surely under-observed. It has been spotted less than ten times in Minnesota.

Belted, Crimson-ringed, Frosted, and Hudsonian Whitefaces: These small dragonflies are probably the most frequently encountered skimmers at bog and fen habitats.

Four-spotted Skimmer: A circumpolar species. Common at many habitats in boreal and alpine regions in North America, Europe, and Asia.

Elfin Skimmer: A true bog-o-phile. Also the smallest North American dragonfly. A wasp mimic about the size of a horse fly.

Black Meadowhawk: A circumpolar species. A late season species. A personal favorite.

Lake Darner: A very large dragonfly. Despite its name, this dragonfly is often found at bog ponds and fens.

Sedge Darner: Not yet known from Minnesota, but has been observed just to the north in both Manitoba and Ontario. A circumpolar species.

Subarctic Darner: Known from only a few sites in Minnesota. One site is a bog near Isabella, Minnesota. A circumpolar species.

Zig-zag Darner: A fen-o-phile that seems to have an affinity for pools and runnels where Buckbean grows.

Delicate Emerald: For many dragonfly enthusiasts, Emeralds are highly sought after dragonflies because of their metallic green body coloration and neon-electric green eyes and because of their remote habitats. Some of the Emeralds, like the Delicate Emerald, might live their entire lives in the water held in a Moose hoofprint, only moving if it runs out of food.

Lake Emerald: A large dragonfly of boreal, bog-fringed lakes. Quite a challenge to net. It took two attempts over two years and the efforts of close to ten people to net one of these elusive dragonflies at Winter Lake near Warroad.

Quebec Emerald: Disjunct populations of this rare dragonfly have been found in Minnesota at the Sand Lake Peatlands and the Red Lake Peatlands.

Didymops transversa – Stream Cruiser   *Quercus rubra* – Northern Red Oak

50° / 29°

December 3

RESPECTABLE WEEDS

Visited Hampton Hills Tree Farm to get a Christmas tree. Walking the rows of farmed pines and firs and spruce, I couldn't help but admire some of the respectable weeds growing in between the trees. Being adjacent to McKnight Prairie, Big Bluestem and other prairie grasses had seeded the tree farm, carried south from the prairie remnant. Common Evening Primrose, Great Mullein, Whorled Milkweed were a few of the wildflowers growing among the trees.

Asclepias verticillata – Whorled Milkweed
Verbascum thapsus – Great Mullein
Oenothera biennis – Common Evening-primrose

December 4

63°

HALL'S CREEK

27°

To Black River Falls, Wisconsin, with Ami Thompson. I met Ami in Hudson and she drove from there. Our destination was her property along Hall's Creek, a late-season expedition in search of dragonfly nymphs, knowing a drastic change in the weather was on the way. We arrived around 10 A.M., and were in the river for nearly four hours.

I'd talked to Ami at the Bog Panel at the Groveland Museum and, afterward, at dinner with her and several other dragonfly aficionados. Prior to that, I hadn't seen Ami since early June during a survey in northern Minnesota [see May 31 - June 3]. She had been working throughout the summer on her graduate research, a study of the migration patterns and overwintering populations of the Common Green Darner. Ami and her assistants had collected, identified, and counted several thousand dragonfly exuviae gathered from the edges of a series of small ponds. This work had kept her from visiting her property, until now.

Hall's Creek, where we were searching for nymphs, was clear and swift. Flat shelves of bedrock formed much of the creek's bottom, but there were many areas of sand and accumulated silt. All very good habitat for clubtail dragonfly nymphs. The main reason I wished to visit Ami's creek was to find a Common Sanddragon (*Progomphus obscurus*), which had been found in the creek previously. A sub-reason, was to collect a variety of clubtail nymphs to rear overwinter. A sub-sub-reason was simply to enjoy this unseasonably warm day, before the predicted cold weather arrived with a vengeance overnight.

We had pretty good luck finding dragonflies, but we also had some luck catching minnows, photographing a fair number in a mini-aquarium then releasing them.

◆ ◆ ◆

While we were enjoying the river, its diversity of insects and the beauty of the surrounding landscape, Donald Trump, no friend to natural habitat, announced reduction to Bears Ears National Monument (85%) and Grand Staircase Escalante National Monument (46%), opening the previously protected lands to gas and oil speculation.

Plecoptera s – Stoneflies
Isonychia – Slate Drakes
Ichthyomyzon sp. – Freshwater Lamprey
Cladonia – Pixie Cup Lichens
Polytrichum juniperinum – Juniper Polytrichum Moss
Dendrolycopodium dendroideum – Prickly Tree-clubmoss
Etheostoma nigrum – Johnny Darter
Rhinichthys obtusus – Western Blacknose Dace
Etheostoma flabellare – Fantail Darter
Cyprinidae – Minnows And Carps
Catostomus commersonii – White Sucker
Dolomedes tenebrosus – Dark Fishing Spider
Cambaridae – Crayfish
Trichoptera – Caddisflies
Belostomatidae – Giant Water Bugs
Lampsilis cardium – Plain Pocketbook
Macromiidae – Cruisers
Calopterygidae – Broad-winged Damselflies
Scleroderma – Earthballs

December 5

HALL'S CREEK: PART TWO

27°
—
16°

A strong cold front swept through as predicted. From the seventies of the afternoon at the creek on the day before to freezing temperatures and snow. I photographed three of the dragonfly nymphs collected from Hall's Creek. And one stonefly.

The labial palps of the Spiketail nymph are amazing and rather wicked looking. They fit together not symmetrically but along a jagged, asymmetrical fissure, the outline of which forms a kind of black lightning bolt, a natural symbol, perhaps, for the speed at which the labial palps rocket out and capture unsuspecting prey.

If you've never seen one of the adults, I recommend that you go look for one. They are among the most impressive dragonflies. The huge, boldly-striped males cruise up and down tiny streams and creeks. They are unmistakable and seeing one is both spell-binding and adrenaline-pumping. The only thing that can top it is seeing one of the females ovipositing, using the spike at the end of her abdomen to plant eggs in the soft silt at the water's edge.

Cordulegaster sp. – Spiketails
Stylurus sp. – Hanging Clubtails
Plecoptera sp. – Stoneflies
Ophiogomphus sp. – Snaketails

20° / 16°

December 6

HALL'S CREEK: PART THREE

I photographed three more of the dragonfly nymphs collected from Hall's Creek: a Snaketail (Ophiogomphus sp.), a Mustached Clubtail (Hylogomphus adelphus), and a Cruiser (Macromidae).

Hylogomphus adelphus – Mustached Clubtail
Macromiidae – Cruisers
Ophiogomphus sp. – Snaketails

December 7

SPARE TIME

21°

7°

"I am not exactly retired, because I never had a job to retire from. I still work, though not as hard as I did. I have always been and am proud to consider myself a working woman. But to the Questioners of Harvard my lifework has been a "creative activity," a hobby, something you do to fill up spare time." Ursula Le Guin, from *No Time to Spare*

Parent, writer, entomologist, printer, publisher—vocations or hobbies? There are no clear divisions when there's no "job" to speak of, when the day to day work is purposeful but not for profit, where the rewards are great but not monetary—books, sentences, photos and time well spent with family and friends and our natural surroundings.

❖ ❖ ❖

Today, I used a portion of my spare time to photograph a freshwater isopod. The order Isopoda includes sowbugs as well as their aquatic equivalents, known simply as freshwater isopods. The latter are not as commonly encountered as their basement dwelling relatives. Under magnification and adequate lighting, the exoskeleton reveals an interesting and, certainly, attractive pattern. This tiny armored being, collected from Hall's Creek in Wisconsin and reminiscent of some of the fossil creatures found in the Burgess Shale, has a small head and two small black eyes that are nearly inconspicuous compared to the massive plate at the other end of its body, making it possible to mistake which end is which. Aquatic isopods of the genus *Caecidotea* feed on detritus, dead organic matter such as fallen leaves and senescent algae.

Caecidotea sp. – Freshwater Isopod

32°
—
19°

December 8

ASTERISMS

"*Aster drummondii* [now Drummond's Aster, *Symphyotrichum drummondii* (Lindl.) G.L. Nesom var. drummondii] is said to be hardly distinct from *Aster sagittifolius* [now *Symphyotrichum urophyllum*]; but, with me, *A. drummondii* has larger, thicker leaves, larger and darker blue flowers, a less brittle stem, and a more gregarious habit." – Eloise Butler, from 'Asters in the Wild Garden' (1915)

We had a number of these asters this year in our backyard garden, all volunteers. One of the reasons I collected the seed was that the flowers had been pollinated by an endangered species, the Rusty-patched Bumble Bee, that visited the flower while it was in bloom in September. Similar to a collection of seed, is a collection of quotes, in this case, asterisms.

> Chide me not, laborious band,
> For the idle flowers I brought;
> Every aster in my hand
> Goes home loaded with a thought.
> – Ralph Waldo Emerson, from 'The Apology'

> The aster greets us as we pass
> With her faint smile.
– Sarah Helen Whitman, from 'A Day of the Indian Summer'

> A drowned drayman was hoisted on to the slab.
> Someone had jammed a lavender aster
> between his teeth.
> – Gottfried Benn, from "Little Aster"

I end not far from my going forth
 By picking the faded blue
Of the last remaining aster flower
 To carry again to you.
– Robert Frost, from 'A Late Walk'

Symphyotrichum drummondii – Drummond's Aster

29°
—
19°

December 9

LEAF MINES

I photographed a Black Raspberry leaf, crispy and freeze-dried, dark on top, downy-white underside. What was most interesting about this particular leaf was the presence of a leaf mine, a large area excavated by the larva of a fly or moth feeding inside the leaf, leaving the eaten part of the leaf hollow and somewhat transparent.

Backlighting the leaf revealed the mine to be empty. Not surprising since it was late in the year. A few pellets of frass remained. The leaf mine is blotch shaped rather than a long linear trail, which might be an indication that the inhabitant was a moth rather than a fly. Either way, it's probable that the mature larva exited the leaf, dropped to the ground, and pupated in the soil where it will overwinter before emerging next year.

Many leaf miners are host specific. Knowing the plant and the mine shape often allows you to identify the insect to species. The naturalist Charley Eiseman has been collecting and rearing leaf miners for years and is preparing a multivolume publication on his findings, including the descriptions of a staggering number of new species.

Insecta – Leaf Miner

December 10

SKELETONIZER

29°
—
20°

To St Olaf. Temperature near freezing. The wooded pond, frozen now for the third or fourth time, seems frozen for good now. The surface is smooth and free of snow, perfect for skating. Fox tracks and mink tracks left on a shoreline margin of snow enticed me to try out the ice as well. Just six days ago the pond was open, but the subsequent cold weather closed it, providing a clear four inches or more of ice.

I crossed the pond and entered the woods on the north side. Just over the hill is a stand of Red Oak trees. The fierce winds of the last week had removed most of the leaves across the tops of the trees, though many red-orange leaves remained on the lower branches. I examined a couple of trees, looking for leaf mines, noticing the wind had broken many of the stems, finding only a few incomplete mines or blotches left by leaf skeletonizers.

Later, photographing one of the leaves, back-lighting the mined blotches, I couldn't help but see the network of veins as similar to those in the wings of dragonflies.

Quercus rubra – Northern Red Oak
Insecta – Insects
Herpyllus ecclesiasticus – Eastern Parson Spider

35° / 15°

December 11

TWO TWILIGHTS

"The day is short; it seems to be composed of two twilights merely; the morning and the evening twilight make the whole day. You must make haste to do the work of the day before it is dark." – Henry David Thoreau, from *Journals* (December 11, 1854)

> Winter for a moment takes the mind; the snow
> Falls past the arclight; icicles guard a wall;
> The wind moans through a crack in the window;
> A keen sparkle of frost is on the sill.
> Only for a moment; as spring too might engage it,
> With a single crocus in the loam, or a pair of birds;
> Or summer with hot grass; or autumn with a yellow leaf.
> Winter is there, outside, is here in me:
> Drapes the planets with snow, deepens the ice on the moon,
> Darkens the darkness that was already darkness.
> – Conrad Aiken, from *Preludes for Memnon*

Agromyza sp. – Leaf Miner Fly

December 12

27°

6°

PHOTOGRAPHING SNOWFLAKES

A cold day. Snowflakes drift into view intermittently with a slower frequency, almost, than falling stars in the night sky. The picturesque crystals hit the ground here and there, landing on the concrete of the driveway, on the waxy surface of a fallen leaf, glittering specks of dust. While walking the dog I began to notice the isolated flakes, placed in different places as if on display. So after we (the dog and I) returned to the house, I took the camera and flash and extension tube and went back outside in search of snowflakes to photograph.

Strangely, I found myself holding my breath as I leaned close to the ground, positioning the camera lens and flash, concerned that the warmth of exhaled air might melt the subject flake. Photographing a flake on one small green leaf, a second flake appeared in the image though it remained nearly invisible to the eye, like a watermark on a fine sheet of paper. This ghost snowflake must have fallen sometime earlier, its dendritic arms losing molecules of water through sublimation until an incredibly thin structure of ice remained.

These microscopic, six-legged creatures of ice are perhaps even more ephemeral than any insect, subject to transformations beyond the material metamorphoses of butterflies and dragonflies—think for instance of the social migrations that form rivers and clouds, the deep collective power of the oceans. And yet, despite their molecular fragility, their potential to shape-shift and vanish, snowflake populations will arrive and persist these next months, adding their reflective whiteness to winter.

Sciurus carolinensis – Eastern Gray Squirrel
Corvus brachyrhynchos – American Crow

30°
―
17°

December 13

BLACK WALNUT

Nuez: sabiduría comprimida,
diminuta tortuga vegetal,
cerebro de duende
paralizado por la eternidad.

Nut: compressed wisdom
small vegetable tortoise
midget goblin brain
paralyzed for eternity.

- Jorge Carrera Andrade, from *Microgramas* (1929)
[translation by Scott King]

Juglans nigra – Eastern Black Walnut

December 14

TOWNSEND'S SOLITAIRE

24°/15°

Yesterday, Dan Tallman alerted me to the presence of an unusual visitor at Lake Byllesby County Park, a Townsend's Solitaire. So today I went looking for it.

What I found were other bird watchers. The few people I spoke with near the parking lot hadn't found the bird. Nonetheless, I set out on the snowy trails hoping for the best. I'd been instructed to look for the bird perched at the top of trees so that is where I focused my attention. Several times an elusive bird perched at the top of a distant tree. It always flew before I could get close enough to see what it was. There were other birds: Cedar Waxwings, House Finches, Blue Jays, and Canada Geese flying overhead.

If you're left wondering about the presence of the Townsend's Solitaire, so am I.

Branta canadensis – Canada Goose
Cyanocitta cristata – Blue Jay
Haemorhous mexicanus – House Finch
Bombycilla cedrorum – Cedar Waxwing
Juniperus virginiana – Eastern Redcedar

30°
—
22°

December 15

HORSE WITH A TIE

Today, I began preparations for a talk I will be giving in January on the biogeography of dragonflies, comparing studies of Siberian dragonflies with what is know of the dragonflies of Minnesota. In latitude, the state of Minnesota, if slid around to the Russian side of the globe, lies south of Siberia, say somewhere in northern Mongolia. However, there exist some climatic ties, as well as similarities, in ecosystems. For instance, consider the boreal forest biome that surrounds the two largest lakes in the world, Lake Superior and Lake Baikal. And there's the literal connection between the two places on the occasions when the Bering Land Bridge is extant.

Boris Belyshev, like his predecessor Bartenov, published an immense number of papers and books about dragonflies relating to four decades of work.

In 1936, Belyshev was sentenced to ten years in Vorkuta Gulag, working the coal mines. During this time he attempted to retain his dignity by staying shaved and always wearing a white shirt with a tie. This habit combined with his strength earned him the nickname of "a horse in a tie".

Стрекозы Сибири (Three Volumes). The first two volumes include species accounts. The third volume addresses distribution patterns and biogeography. The first of these volumes was published in 1973.

Gymnosporangium juniperi-virginianae – Juniper-apple Rust

December 16

HIBERNACULUM

31°
—
23°

To the Anderson Center in Red Wing. Visited the Red Dragonfly Press print shop, needing to cut some paper and grab a few lines of metal printing type, preparations for some printing at a friend's print shop the upcoming week.

The print shop building, converted from a farm granary, retains its allure to various insects as a hibernaculum, especially Boxelder Bugs and Lady Beetles, but also the occasional Northern Paper Wasp queen. Once, I found a Milbert's Tortoiseshell butterfly overwintering, which was a small mystery since I'd never observed others on the surrounding grounds. There are other observations from nearby, but my nearest observation was several hundred miles to the north.

Boisea trivittata – Eastern Boxelder Bug
Harmonia axyridis – Asian Lady Beetle

28° / 18°

December 17

MID-WINTER SLEEP

Lisa, Lida, and I, and Lisa's brother's family gathered at Lisa's mom's house for a day of holiday baking. Playing games, eating meals, sampling cookies.

In the evening, after we returned home, I brought in a trap nest from the front garden and opened it. The diameter of the nest and the architecture of the cells indicated the occupants were most likely an Aphid Wasp (genus *Passaloecus*). Under the microscope, the remains of aphid prey were visible, mostly legs and antennae. Of the seven cells in the nest, three had full-grown larvae, one had a pupa, and three were empty. The missing cells contained the remains of whole aphids (dehydrated and moldy), so it seems that the wasp eggs never hatched or the larvae died in these cells, three conjoined cells in the middle of the nest. These wasps are known to overwinter as larvae and pupate in the spring, therefore, the pupa in this nest is most likely a different species. Either the nest was superseded, briefly, by another hunting wasp or it was parasitized by a cuckoo wasp.

Passaloecus sp. – Aphid Wasp

December 18 — 42° / 24°

KNOTWEED

Craig Kotasek agreed to let me use his Chandler & Price printing press to print some letterhead. What would have been a long and arduous task on the Vandercook proofing press at the Anderson Center, turned out to be a quick and enjoyable job on Craig's powered platen press.

A folk musician, an artist, and a printer, Craig built his own house on his grandfather's land overlooking the wide valley of the Minnesota River near Le Sueur. In addition to the house, he also built a print shop. Carrying on a Minnesota tradition of repurposing farm outbuildings, he converted the farm's soy-bean granary into his print shop, giving it the name Tin Can Valley Printing Company. Rough-sawn oak posts and beams provided the sturdy framework for the building. The reclaimed wood on the walls and the rows of old type cabinets and vintage printing equipment gave the shop its warmth and lent an out-of-time aura to the space. It was a wonderful place to work and spend an afternoon. There was even a friendly, tortoiseshell mouser to keep us company.

Later, after I had finished my printing and while Craig was picking up his son, Charles, from school, I had a chance to walk around the property. Narrow and long, matchstick-like on the map, the property extends upland from the river bottoms. Highway 169, a divided highway, bisects the property at the edge of the river bottoms just before the rise of the bluffs. The farm was situated on a level area above that first sharp rise in the land. And the fields rose even further. Craig had shown me a few plants he had collected with the intention of drawing the seeds and writing a small guide to the clinging and sticking seeds found around his farm, plants such as Beggar's Tick (*Bidens sp.*) and Burdock. In honor of that commendable project, I photographed some of the interesting seeds encountered on my walk: Virginia Stickseed, American Germander, American Bittersweet, and Japanese Knotweed.

Celastrus scandens – American Bittersweet
Fallopia japonica – Japanese Knotweed
Teucrium canadense – American Germander
Hackelia virginiana – Stickseed

December 19

SHORE FLY

38°
—
24°

A premature mid-winter warm-up. A January thaw in December. Sunshine, with the temperature well above freezing, but windy. I visited the upper Arboretum because it's well protected from the wind and because the trail and creek bank would be in the sun. It was a good choice.

Winter insects are few and far between; from the cast of thousands present in the summer months, only a handful of flies tough it out in December. Like an explorer lost in the Arctic, a Snow Midge crawls across the sunlit surface of the snow. Near the river's edge, a Shore Fly made short flights, not more than several inches at a time.

I was crouched down looking at this fly when a man approached along the trail. He seemed a kind of revenant, holding a coffee cup importantly in front of him, a large, wispy beard leading him forward. He made several questioning glances in my direction, his pace hesitating ever so slightly, then he proceeded onward. I stood up, hoping the camera in my hand would be visible and deflect any awkward assumptions made on the part of the stranger walking by. I'd become a kind of shore fly myself.

Ephydridae – Shore Flies
Diamesa sp. – Snow Midge

24°
—
18°

December 20

WINTER SUNSET

The year's end is the year's beginning. This is nothing new. And yet with the year about to loop around, this daily accounting about to come full circle, some reflection, some summing up seems necessary. Let me begin with a recap of the year in flies.

I was able to photograph flies every month of the year, all year round. The barren mid-winter months were spanned by the Snow Midges that venture out of the fast-moving water on sunlit days, that crawl and take flight above the snow. The warmer months of the other three seasons abounded in flies. In all, many different kinds of flies were observed in 2017. From the spectacular wasp-mimic Mydas Fly at the summit of Angel's Landing in Zion National Park to the tiny, hummingbird-like Eastern Band-winged Hoverfly found nectaring at a wood sorrel flower in our backyard, it was a good year for flies.

Solidago flexicaulis – Broad-leaved Goldenrod
Branta canadensis – Canada Goose

December 21   24°/19°

ACARINARIUM

Solstice occurred at 10:27 A.M.

I opened another wasp nest from the front yard, this time in a cut length of Joe-Pye weed stem. Observing a number of mites in the antechamber, I became curious about their role.

It turns out, certain wasps and bees have evolved a metasomal acarinarium, a hollow space on the outside of the abdomen in order to transport mites to their nests. This complicit dispersal suggests some mutual benefit to the wasps and bees. It has been suggested that the mites scavenge and clean the larval chambers, keeping them free of fungi and pests.

Passaloecus sp. – Aphid Wasp
Acari – Mites And Ticks

27° / 18°

December 22

SCUTTLE FLY

I noticed a tiny fly on the window pane in the kitchen. Not the most handsome of flies, but at this time of year, one can't be too picky about their insects. Phorid flies (family Phoridae) are small, hunch-backed flies which tend to "scuttle" away instead of taking flight. Perhaps this was the kind of fly William Blake drew with a human face and animates in his poem 'The Fly'?

> Little fly,
> Thy summer's play
> My thoughtless hand
> Has brushed away.
>
> Am not I
> A fly like thee?
> Or art not thou
> A man like me?
>
> For I dance
> And drink and sing,
> Till some blind hand
> Shall brush my wing.
>
> If thought is life
> And strength and breath,
> And the want
> Of thought is death,

Then am I
A happy fly,
If I live,
Or if I die.

Phoridae – Humpbacked Flies

23° / 11°

December 23

RIVERBANK

Lisa's birthday. And as part of her day, we took a family walk in the afternoon at the Cowling Arboretum, Lisa, Lida, myself and the dog.

The recent cold weather has constricted the open water to the main river channel and the swift tributary streams; most of the ponds and lakes are now frozen. This concentrates the Canada Geese and Mallard Ducks overwintering in the area and during our walk the sound of thousands of honking geese flying overhead and congregating at the last open water on Lower Lyman Lake embrightened the quiet winter woods.

Along the river, beavers have been harvesting small trees. Here and there the stumps of trees stick out of the ground like thick pencils freshly sharpened with a knife. Their work trails and riverbank slides are as wide as cattle paths and evidence of their persistent labor.

Neovison vison – American Mink
Castor canadensis – American Beaver
Vitis riparia – Riverbank Grape
Branta canadensis – Canada Goose

December 24

TWO BEETLES

22°
—
4°

Christmas Eve's Day. Read Gillian Clark's poem, 'The Christmas Wren.' Cleaned the house, wrapped gifts. Lisa's mom arrived in the afternoon. The four of us cooked and watched a movie and played some dice.

I photographed, early in the day, two small beetles, both dead, that had been found in my office, hoping they could be identified, hoping (also) that they were not feeding on specimens kept there.

The wing sheaths, the elytra are not completely shut like obstructed cupboard doors. One of the beetle's final acts must have been flight, or the attempt at flight. The membranous wing tips protrude, a bit of an embarrassment for the living onlooker.

It's pebbled back, hunched shoulders and tucked head, its six legs clutching its antennae beneath its body, take on a recognizable aspect, a penitent perhaps, or simply the wish for better circumstances, say an open window and a way to escape. Smaller than a grain of rice, its suffering was on the same scale with any other of the living.

Anthocomus equestris – Malachite Beetle
Ptinidae – Deathwatch, Spider, And Wood-borer Beetles

4°
—
-6°

December 25

SUBZERO

Christmas Day. We opened gifts in the morning. Lisa and Lida gave me a mechanical pencil, a Russian cookbook, and new wool mittens for my winter choppers. Karen gave me a subscription to *Orion* magazine and a fleece-lined flannel shirt—'The Flapjack.' All good winter gifts.

The first truly arctic air moved in, the temperature descending from single digits to subzero during the day. In addition to memories of winter cold stored like castles of crystal in my very being, a ghost-ache in each of my fingers as I thought back to breaking bales in the dark, removing the tight loops of twine from the tightly packed hay, on some deep winter morning bright cities of stars overhead, cattle lumbering in on frozen hooves, exhaling entire clouds of frost in the twenty-below shadows, so close I could pat the curly hair on their anvil heads. Darkness without any terror or sadness, without any knowledge of Gogol's Russian sorcery.

"A wintry, clear night came. The stars peeked out... Here smoke curled from the chimney of one cottage and went in a cloud across the sky, and along with the smoke rose a witch riding a broom." – Nikolai Gogol, from 'The Night Before Christmas'

I was outside for minutes only, long enough to walk the dog, load Karen's car, and take a few quick photos, but it was long enough to have raw cheeks and a burning cold nose. The photos were of the empty seed pods of Blue-eyed Grass pointing through the snow in the front yard, the spare stalks of which looked like empty candle holders.

Sisyrinchium – Blue-eyed Grasses

December 26

DIARY OF A MADMAN

Having near at hand Gogol's collection of stories, I read (for the first time in many years) the short and bizarre 'Diary of a Madman.' This story contains some very amusing passages, plus there are striking similarities to Dostoyevsky's 'Notes from the Underground' and Melville's 'Bartleby, the Scrivener." Perhaps, at some point, I'll return to these journal pages and recognize that same mad grin invented by these mad protagonists... 'Journal of a Mad Naturalist.'

Another frigid day. The forecast is for the final week of 2017 to be below zero, consistently cold. I set out some dried mealworms for the Chickadees and Nuthatches but they have yet to find them. My order of live blackworms arrived at the pet shop, food supply for the overwintering dragonfly nymphs in my office. I dropped a small, writhing glob and as they dispersed slowly through the cold water I watched, with satisfaction, the large Macromia nymph capture one with ease.

Photographed another trap nest. Again it looks to be an Aphid Wasp nest, but the innermost cells held empty pupae, so the nest had held another species earlier in the year, the Aphid Wasp that provisioned this nest was the second wasps to make use of this particular segment of Joe-Pye Weed stem.

I also photographed a batch of exuviae that Janet Nelson had collected in June from Lake Kabetogama. This batch consisted of four *Ladona julia* and twelve Epitheca nymphs (probably *E. spinigera*).

Passaloecus sp. – Aphid Wasp

5°
—
-8°

December 27

DRAGONFLY FOOD

The day was spent running errands with Lisa and Lida—first to Burnsville, next to Eagan, then back to Burnsville for a three-hour marathon session at the device store (no kidding, three full hours to trade in and update our phones). The payoff, we left almost cutting edge and no poorer than we arrived. Also stopped at the pet store to pick up an order of live Black Worms, food for the dragonflies.

I've been enjoying the behavior of the large cruiser dragonfly nymph collected earlier this month. This large nymph has commandeered a large rock in the makeshift, rearing aquarium. It shares the space with a number of gomphid nymphs and a spiketail nymph, but those all burrow down and hide in the sand. With its long legs spread out across the surface of the rock, it resembles, quite remarkably, a large fishing spider.

◆ ◆ ◆

Dinner at a local Mexican restaurant with the Arnold family, our neighbors who live across Oxford Street from us.

Lumbriculidae – Black Worms
Macromia illinoiensis – Swift River Cruiser

December 28

VADE MECUM

Vade mecum is defined as a handbook or guide that is kept constantly at hand for consultation. The Latin literally says "goes with me."

Across a lifetime, there has been a series of important items I've kept close at hand, mostly books, but there have been years when a certain pen or a new insect net has been consulted as often as a familiar book. And, in recent years, I'm tempted to name places, lakes and trails, landscapes that have become treasured texts.

No book altered the course of my life more than the immense, book-length poem *Letter to an Imaginary Friend* by the North Dakota poet Thomas McGrath. The writing and its arguments taught me to think, refined my Great Plains dialectic, combining threads of agrarian common sense, folk surrealism, and an innate wonder for the natural world. There's little doubt that this book was, has been, and will remain my *vade mecum*.

Eutrochium – Joe-pye Weeds

4°
—
-6°

December 29

MOON IN THE TREES

This is how the year should end, I thought to myself while walking the dog in the dark just after 8 pm. The temperature was dropping, headed somewhere well into the negative digits. The sky clear. The moon caught overhead in the bare branches of a mid-winter tree. The night calm and quiet but for the squeak of newly fallen snow beneath my boots.

On similar nights during other winters I've heard owls (though usually later in the night and later in the season). I suddenly felt the longing to hear them again, to stand and listen as long as it might take. Like the poet Thomas McGrath awaiting the "blue star Kachina," I would keep vigil for the voice of a Great Horned Owl. I stopped when the dog stopped, and I listened. The sound of a car passing many blocks away. Then the night was quiet once again.

Carex sp. – True Sedges

December 30   -6°

WINTER OAK   -15°

Drove to Eagan to have lunch with my sister, Sheri, and Marlo, and Marlo's sister's family.

A cold winter day, low near -20 degrees F, and the high topping out at -7 degrees. Driving to and from Eagan, we traveled on the familiar stretch of Highway 19 between town and the freeway. In the middle of one of the expansive farm fields to the south of the highway, a large Bur Oak stands, bare winter branches silhouetted against hundreds of acres of snow-covered ground. Glanced at over and over, noticed countless times, this oak becomes a landmark. Here's a small poem of mine:

> Oaks
>
> Of all trees, the oaks
> carry the most sky.
>
> In winter, they remind us
> of water,
> the great branches reaching
> the horizon in ripples.

◆ ◆ ◆

The day's mail included a copy of *Consolations of the Forest* by French travel writer, Sylvain Tesson, an account of six months in a cabin on the northwest shore of Lake Baikal. Cold as it is here in Minnesota, winter is longer and far colder in Siberia.

Quercus macrocarpa – Bur Oak

-5°
—
-16°

December 31

"SONG I REMEMBER"

One complete year, making observations, writing descriptions, wandering in and out of familiar territory, following the earth around. Now part of the "song I remember."

Ready to do it all over again.

♦ ♦ ♦

"One crow
Slowly goes over me — a hoarse coarse curse — a shrill
Jeer: last of the past year or the first of the new,
He stones me in appalling tongues and tones, in his tried
And two black lingoes. A dirty word in the shine,
A flying tombstone and fleering smudge on the winter-white page
Of the sky, my heart lightens and leaps high: to hear
Him. And the silence. That sings now: out of the hills
And cold trees. Song I remember."

- Thomas McGrath, from *Letter to an Imaginary Friend*

Cornus sericea – Red Osier Dogwood

Indices

GENERAL INDEX

Acorn, John Mar-20 Apr-3
Aiken, Conrad Dec-11
Alekhine, Alexander May-24
Alexander, Floyce Mar-9
Allen David Broussard Catfish Creek
 Preserve State Park Jun-18
Anderson Center May-5 Jun-3 Dec-16
Archilochus Sep-15
Aristotle Nov-9
Aurora-Hoyt Lakes May-30, Jun-1, 2
Beadle, David Apr-26
Belyshev, B. F. May-24, 25 Dec-15
Benn, Gottfried Dec-8
Blake, William Dec-22
Bly, Robert Feb-13 Jun-3
Bolt, Clay Sep-12
Boyd Sartell WMA May-7, 25
Bradley, Richard Feb-23, 27
Braker, Nancy Sep-22
Bronstein, David May-24
Brown, George Mackay Jan-5
Browne, Janet Jan-26
Bryce Canyon National Park Jul-16
Burns, Robert Jan-25
Buson, Yosa Feb-4
Butler, Eloise Dec-8
Byrd, William Apr-1
Caddy, John Jun-26
Cameron, Robert Jan-11
Cannings, Robert Jun-13
Cannon River Wilderness Area Feb-12
 Mar-28 Apr-14 Sep-28
Cappelen, Bodil Jun-3
Carrera Andrade, Jorge Dec-13
Carroll, David M. Mar-4
Cash, Johnny May-20, 25
Chadde, Steve Jan-13
Chigorin, Mikhail May-24
Clare, John Aug-11
Clarke, Gillian Jan-6 Dec-24
Clash, The Sep-13
Conrad, Joseph Aug-7 Nov-28
Corbet, Philip Apr-25
Crane, Hart Jul-29

Dacey, Florence Sep-30
Darwin, Charles Jan-26 Mar-11 May-11
 Aug-7
Dia de los Muertos Nov-1
Dickens, Charles Feb-25
Dijkstra, K. D. Feb-26
Dorow, Doug Nov-24
Dostoevsky, Fyodor May-24 Dec-2, 26
DuBois, Bob May-31 Apr-24
Eastman, John Nov-5
Edwards, Robert Jul-16
Eiseley, Loren May-7, Sep-30
Eiseman, Charley Jan-27 Dec-9
Eliot, T. S. Apr-30
Elliott, Lee Jan-21
Elytis, Odysseus Sep-15
Emerson, Ralph Waldo Dec-8
Empson, William Jul-28
Etter, Dave Apr-21
Evans, Howard Ensign Jan-21 May-3,
 12
Fabre, J. Henri May-30
Faryad, Fereydoun Mar-10 Nov-5
Fassett Dec-2
Fletcher, Jim Aug-27
Florida Jun-18, 19, 20, 21, 22
Fossils Oct-12
Franklin, Benjamin Jan-23
Frost, Robert Oct-8 Dec-8
Gavin, Larry Sep-30
Geology May-23, Oct-12
Gershgoren, Sid Jul-24, Oct-23
Gibbon, Edward Oct-1
Gleick, Jame Nov-9
Goethe, J. W. von Feb-2 May-19 Jun-
 24
Gogol, Nikolai Dec-25, 26
Goo Goo Dolls Feb-26
Gooding, Brian Aug-9
Gray, Asa May-4
Gregory, Michael Nov-24
Grote, Augustus Radcliffe Aug-13
Haag, Mitch May 31 June-1, 2
Haag, Jason May 31 June-1, 2

Hall's Creek Dec-4, 5, 6, 7
Hardy, Rob Sep-30
Hauge, Olav Jun-3
Hedges, Mary May-4
Hedin, Robert Jun-3
Herbert, George Mar-14
Hewitt, Susan Jan-11 Mar-27
Hippocrates of Chios Jan-16
Holden, Molly Jan-13
Holm, Bill Oct-15
Hooke, Robert Mar-17
Hovland, Debra Nov-23
Hume, David Feb-13
Icelandic Sagas Mar-6
Illiad Oct-1
Jacobson, Dale Jan-2
Janovy, John Jr. Feb-17
Jeffers, Robinson Oct-3
Karpov, Anatoly May-24
Kasparov, Garry May-24
Kieschnick, Sam Feb-26
Kjar, Arlene Sep-13
Koester Prairie Jul-8
Kotasek, Craig Dec-18
Krombein, Karl Apr-6
Lake Byllesby County Park Oct-4, 9
Latimer, John Dec-2
Le Guin, Ursula Dec-7
Lopez, Barry Jul-17
Lyell, Charles May-23
Mabey, Richard Nov-5
MacCaig, Norman Apr-2
Macfarlane, Robert Jan-18
MacNeice, Louis May-11 Oct-21, 23
Maplewood State Park Nov-2, 23, 24
Marshall, Stephen Apr-22
Martinson, Harry Apr-23
Matthiessen, Peter Mar-23
McAlister, Erica Apr-17
McGrath, Eugenia Jul-24
McGrath, Thomas Jan-19 Mar-11 Oct-7 Dec-28, 29, 31
McKnight Prairie Jun-15
McLaughlin, Walt Mar-4
Mead, Betsy Nov-5
Mead, Kurt Dec-2
Melville, Herman Aug-9 Dec-26
Merwin, W. S. Jan-15
Michener, Charles Apr-27
Milosz, Czeslaw Jun-20
Milton, John Mar-28
Minnesota Dragonfly Society May-25, 31 June 1, 2 Dec-2
Mitchell, Erika Jan-13
Montale, Eugenio Mar-8
Muqi Oct-13
Myaskovsky, Nikolai May-24
Nabokov, Vladimir May-23 Jul-9
Needham, James Apr-25
Nelson, Janet Aug-20
Nerstrand-Big Woods State Park May-4
Neruda, Pablo May-2
Niflheim Sep-25
Oien, Curt May-31 June 1, 2
Ojala, Meg Apr-5 Dec-2
Old English Riddle Mar-17
Olsen, Don Aug-9
Paddock, Joe Nov-7
Parish, Roger Aug-26
Pavese, Cesare Nov-22
Peale, Titian May-3
Peckham, George and Elizabeth Jul-2
Pelican Rapids Nov-22, 23, 24
Perse, St.-John Jan-15 Aug-21
Phenology Apr-8, May-22, Dec-2
Pielou, E. C. Jan-5
Ponge, Francis Jan-16 Feb-13 Oct-6
Powys, Llewelyn Oct-21
Primack, Richard Apr-8, 20
Rafness, Lorna Oct-15
Rau, Philip Jul-20
Redfern, Margaret Jan-27 Nov-19
Rezmerski, John Aug-19
Ritsos, Yannis Jan-20 Feb-16 Apr-11 May-1
River Bend Nature Center May-26
Roberts, Thomas S. Jan-6, 18 Feb-21

Nov-2
Rojas, Gonzalo Oct-29
Rustad, Orwin Apr-8
Rylander, Edith Mar-10
Saramago, José Jul-6
Say, Thomas May-3
Scholes, R. W. Oct-15
Schultz, Jennifer Mar-7
Shakespeare, William Feb-28 Apr-29
Shepard, Sam Jul-27
Shirao, Kaya Jul-23
Shostakovich, Dmitri May-24
Sjoberg, Fredrik Aug-29
Skevington, Jeff May-22, Apr-17
Smith, Thomas R. Feb-13
Soul Asylum Feb-15
Sterling, Phil Apr-26
Stork, Harvey Feb-1
Tacitus Oct-13, 30
Tallman, Dan May-11, 24, 26 Dec-14
Taneyev, Sergei May-24
Teale, Edwin Way Jan-4 Mar-28 Apr-14, 28 Aug-4
Telford, Horace S. May-6, 7 Oct-19
Tennessen, Ken May-31 Apr-24 Jul-6 Dec-2
Tesson, Sylvain Dec-30
Thomas, Dylan Oct-14
Thomas, Edward Sep-17
Thompson, Ami May 31 June-1, 2 Dec 2, 4
Thompson, D'Arcy Nov-8
Thomson, Robin Mar-7
Thoreau, H. D. Jan-1, 3, 8, 31 Feb-14 Mar-5, 11, 25 Apr-20, 21, 29 May-4, 7, 23, 27, 28 Jun-8 Jul-12 Aug-12 Oct-1, 19 Dec-11
Tinbergen, Niko May-30
Tolstoy, Leo May-24
Trump, Donald Jan-20 Dec-4
Tu Fu Feb-21 Sep-17, 28
Ustvolskaya, Galina May-24
Vallejo, César Sep-5 Nov-28
Vandenberg, Katrina Dec-2

Velvet Underground, The Jun-9
Wagner, David Aug-13, 23
Walker, Edmund Apr-24, 25
Wallace, Alfred Russel May-11
Wasp Watchers Program Mar-7 Jun-30 Jul-1, 2, 3, 4, 5
Weinberg, Mieczysław May-24
Whedon, Arthur D. May-26
Whitman, Sarah Helen Dec-8
Whitman, Walt Intro
Whittier, John Greenleaf Mar-1
Wilson, E. O. Apr-12
Woolf, Virginia Jan-11
Xerxes Society Sep-12
Yeats, W. B. May-10, 12 Aug-3
Yevtushenko, Yevgeny Jul-17
Zion National Park Jul-12, 13, 14, 15 Dec-20

SPECIES INDEX

A

Aberrant Cellophane Bee   Aug-8
Abia sp.   May-5, 12
Abies concolor   Jul-14, 15
Acanalonia conica   Oct-8
Acari sp.   Dec-21
Accipiter cooperii   Sep-6
Acer sp.   Mar-2
Acer grandidentatum   Jul-14
Acer negundo   Feb-5, Apr-19
Acer saccharinum   Oct-22, Mar-3
Acmaeodera   Jul-16
Acorn Ants   Apr-23
Acorus americanus   Jun-2
Acraspis macrocarpae   Sep-19
Acris blanchardi   Jun-7
Acrobat Ants   May-16, Jun-14, Jul-22
Acroneuria   Jan-2
Acronicta americana   Aug-28
Actaea rubra   Jun-1
Acute Bladder Snail   Feb-7
Aerial Yellowjackets   Sep-7, Nov-5
Aeshna sp.   Feb-22, May-8, Jun-30
Aeshna constricta   Sep-7, Oct-4
Aeshna tuberculifera   Sep-1
Aeshna umbrosa   Sep-1, Sep-8, Oct-9, Oct-9
Aeshna verticalis   Aug-22, Aug-29, Aug-30
Aeshnidae   Jul-13
Agapostemon sp.   Aug-7, Aug-18, Sep-17
Agapostemon splendens   Jun-22
Agapostemon virescens   Jun-17
Agaricales   Nov-18
Agave sp.   Jul-12
Agelaius phoeniceus   Feb-19, Mar-24
Agnetina flavescens   Jan-2
Agrilus cyanescens   Jun-3
Agrilus quadriguttatus   Jul-1, Jul-4
Agrilus vittaticollis   Jul-3
Agromyza   Dec-11
Agromyzidae   Apr-29
Aix sponsa   Oct-9
Alisma triviale   Feb-5

Allegheny monkeyflower   Jul-29
Allium tricoccum   May-4
Allograpta obliqua   Jun-14, Jul-30
Alobates pensylvanica   Jan-19
Alsophila pometaria   Sep-26, Nov-4
Altica   Apr-5
Alticini   Feb-21
Amanda's Pennant   Jun-18
Amara sp.   Apr-9, Nov-8
Amber Snails   Mar-29, May-5
Ambiguous Moth   May-29
Amblycorypha   Aug-22
Amblyscirtes hegon   Jun-2
Ambrosia artemisiifolia   Aug-11
Ambrosia trifida   Aug-10
Ambrosia trifida   Aug-14
Ameiurus natalis   May-13
American Angle Shades   May-27
American Beaver   Dec-23
American bittersweet   Mar-14, Nov-22, Nov-22, Dec-18
American Carrion Beetle   Sep-7
American Copper   Oct-4
American Crow   Jan-30, Oct-26, Dec-12
American Dagger   Aug-28
American Dog Tick   May-25
American Elm   Mar-13
American Germander   Jul-19, Dec-18
American Giant Water Bug   May-14
American Goldfinch   Aug-15
American Hazelnut   Mar-18
American Heineken Fly   May-7, May-15
American Idia Moth   Aug-5
American Lady   May-3, Jun-15
American lopseed   Jun-27
American Mink   Dec-23
American Pelecinid Wasp   Aug-16
American Plantain   Nov-5
American Plum   Jan-24
American Robin   Jan-6, Mar-18
American Rubyspot   Jun-15, Sep-9, Sep-21
American Sand Wasp   Jun-18, Jul-8
American Toad   May-13, May-25,

Jun-28
Ammodramus leconteii   Apr-19
Ammophila sp.   May-30, Sep-20
Ammophila procera   Jun-18
Amphiagrion abbreviatum   Jul-8
Amphibolips cookii   Sep-3
Anas platyrhynchos   Jan-5, Mar-19, Oct-26
Anatis labiculata   Mar-6
Anax junius   Apr-8, Apr-16, Apr-21, May-24, May-25, Aug-4, Aug-25, Sep-7, Sep-17, Sep-30
Anaxyrus americanus   May-13, May-25, Jun-28
Anaxyrus terrestris   Jun-22
Ancistrocerus sp.   May-11, May-22, Aug-27
Ancistrocerus adiabatus   May-28, Sep-10, Oct-6
Ancyloxypha numitor   Aug-4
Andrena sp.   Apr-17, Apr-21, May-2, May-3, May-12, May-15, May-22, May-25, Jun-12
Andrena carlini   Apr-27
Andrena dunningi   Apr-14, Apr-15
Andrena erigeniae   May-4
Andrena helianthi   Aug-29, Sep-1
Andrena miserabilis   Apr-23
Andrena vicina   May-3, May-25
Andropogon gerardii   Aug-11, Nov-17
Anemone acutiloba   Mar-28, Apr-5, Apr-14, May-4
Anemone cylindrica   Jun-15
Anemone patens   Apr-7
Anemone quinquefolia   May-4, May-31
Anguispira alternata   Mar-28
Anhinga anhinga   Jun-22
aniseroot May-28
Anolis sagrei   Jun-20
Anoplius depressipes   Sep-11
Antennaria neglecta   May-2
Anthidium oblongatum   Sep-14
Anthocomus equestris   Dec-24
Anticlea elegans   Jun-15
Antigone canadensis   Feb-21
Antigone canadensis   Mar-24
Antigone canadensis   Jun-18
Antigone canadensis   Jun-19
Antigone canadensis   Jun-20
Antlions   Jun-18
Antlions   Jun-20
Antlions   Jun-22
Antmimic and Ground Sac Spiders   Mar-4
Antmimic and Ground Sac Spiders   Mar-4
Ants   Apr-7
Ants   May-3
Apantesis virguncula   Jun-16
Aphaenogaster   May-28
Aphaenogaster   Sep-19
Aphilanthops frigidus   Jul-1
Aphilanthops frigidus   Jul-4
Aphilanthops frigidus   Jul-6
Apiocera Jul-15
Apiosporina morbosa   Mar-5
Apis mellifera   May-5
apples   May-9
apples   May-10
apples   Oct-13
Aquilegia canadensis   May-11
Aquilegia chrysantha   Jul-15
Aralia nudicaulis   Jun-1
Araneae   Mar-27
Araneae   Mar-28
Araneae   May-30
Araneae   Jun-2
Araneae   Nov-24
Araneae   Dec-1
Araneoidea   Feb-12
Araneoids   Feb-12
Araneus marmoreus   Sep-22
Araneus marmoreus   Oct-16
Araneus pratensis   May-3
Araneus thaddeus   Sep-4
Araneus trifolium   Sep-3
Arboridia Mar-28
Archaeognatha   Sep-29
Arctia caja   Jul-16
Arctium lappa   Jan-6
Arctium lappa   Jan-14
Arctium minus   Jan-13
Arctium minus   Jan-14
Ardea herodias   Aug-25

Argia	Feb-6	
Argia	Jul-13	
Argia	Jul-16	
Argia apicalis	Jun-26	
Argia fumipennis	Jun-18	
Argia fumipennis atra		Jun-22
Argia lugens	Jul-13	
Argia lugens	Jul-15	
Argia plana	Jul-15	
Argia tibialis	Jun-5	
Argiope aurantia	Aug-4	
Argiope trifasciata	Oct-9	
Argus Tortoise Beetle		May-23
Argyrotaenia velutinana		Apr-26
Arigomphus cornutus		May-26
Arigomphus cornutus		Jun-3
Arigomphus cornutus		Jun-5
Arion	Mar-28	
Arisaema triphyllum		May-4
Arizona thistle	Jul-13	
Armoracia rusticana		Apr-18
Armoracia rusticana		May-26
Arnoglossum reniforme		Aug-14
Aromatic Aster	Sep-29	
Arrowheads	Jun-18	
Arroyo Bluet	Jul-16	
Asarum canadense	Apr-18	
Asarum canadense	May-4	
Asclepias incarnata	Jul-29	
Asclepias incarnata	Nov-2	
Asclepias incarnata	Nov-22	
Asclepias syriaca	Jul-7	
Asclepias syriaca	Oct-12	
Asclepias verticillata		Aug-25
Asclepias verticillata		Oct-11
Asclepias verticillata		Dec-3
Asclera ruficollis	Apr-14	
Ashy Clubtail	Jun-2	
Asian Lady Beetle	Dec-16	
Asilinae	Jul-16	
Asilinae	Jul-19	
Asilinae	Jul-21	
Asilinae	Jul-28	
Asparagus officinalis		Nov-16
Aspen Bracket	Mar-5	
Aspidoscelis tigris	Jul-13	
Astata	Jul-11	

Astata	Jul-27	
Astata	Aug-25	
Astata unicolor	Jul-19	
Astata unicolor	Sep-1	
Astragalus crassicarpus		Jun-15
Atalopedes campestris		Jun-17
Atalopedes campestris		Aug-22
Atalopedes campestris		Sep-7
Atherix	Jan-2	
Atherix	May-28	
Athyrium filix-femina		May-4
Athyrium filix-femina		May-5
Atlantic Bluet	Jun-18	
Atlantic Bluet	Jun-20	
Augochlora pura	Jul-9	
Augochlora pura	Sep-14	
Augochlora pura	Oct-19	
Augochlorella aurata		May-16
Augochlorini	May-25	
Augochlorini	Aug-10	
Augochlorini	Oct-4	
Augochloropsis	Aug-29	
Aulacidea	Nov-19	
Aunt Lucy	May-26	
Aunt Lucy	May-28	
Australian Cockroach		Jun-22
Autumn Meadowhawk Jul-23, Jul-27, Aug-7, Aug-18, Aug-22, Aug-25, Sep-1, Sep-7, Sep-8, Sep-9, Sep-17, Sep-19, Sep-23, Sep-29, Oct-3, Oct-4, Oct-9, Oct-10, Oct-16, Oct-18		
Aythya collaris	Mar-19	

B

Baetisca sp.	Apr-25	
Bald Eagle	Jan-23	
Bald-faced Hornet see Aerial Yellowjacket		
Bald-faced Aerial Yellowjacket		Oct-3
Baldcypress	Jun-19	
Balsam Poplar	Jun-2	
Balsam Ragwort	May-28	
Baltimore Oriole	May-9	
Band-winged Grasshoppers		Mar-4
Band-winged Meadowhawk Jul-19, Jul-27, Aug-8, Aug-22, Aug-25, Oct-4, Oct-9		
Banded Garden Spider	Oct-9	

Banded Pennant	Jun-18	Bicyrtes quadrifasciatus	Jul-5
Banded-wing Flies	May-24	Bicyrtes quadrifasciatus	Jul-19
Baptisia	Oct-23	Bicyrtes quadrifasciatus	Jul-27
Barbarea vulgaris	May-4	Bicyrtes quadrifasciatus	Jul-28
Bare-eyed Mimic	May-28	Bicyrtes quadrifasciatus	Aug-18
Barn Swallow	May-19	Bicyrtes quadrifasciatus	Sep-1
Barypeithes pellucidus	May-15	Bidens cernua	Sep-5
Basiaeschna janata	Jun-9	Bidens frondosa	Sep-7
Baskettails	Jun-9	Bidens frondosa	Sep-28
Bassaniana	Mar-22	biennial beeblossom	Oct-4
basswood	Feb-12	big bluestem	Aug-11
basswood	Nov-24	big bluestem	Nov-17
Beautiful Patterneye	Jun-12	big bur-reed	Feb-5
Beaverpond Baskettail	Jun-2	Big-headed Flies	Jun-14
Bee Flies	Jun-24	bigtooth aspen	Nov-23
Bee Flies	Jul-14	bigtooth maple	Jul-14
Bee Flies	Jul-15	bird's foot violet	May-3
Bee Flies	Jul-24	Bittacomorpha clavipes	May-8
Bee-mimic Robber Flies	May-28	Bittacomorpha clavipes	May-22
Bee-mimic Robber Flies	Jun-8	Bittacomorpha clavipes	Sep-15
Bee-mimic Robber Flies	Jun-24	bitternut hickory	Jul-19
Bee-mimic Robber Flies	Jul-7	Bivalves	Nov-24
Beetles	Feb-11	Bivalvia	Nov-24
Beetles	Nov-10	Black Blister Beetle	Aug-25
Beewolf	Jun-29	Black Blister Beetle	Sep-20
Beewolf	Jun-30	Black Blow Fly	Feb-22
Beewolf	Jul-2	Black Blow Fly	Apr-21
Beewolf	Aug-25	black cherry	Mar-2
Beewolf	Sep-2	Black Dancer	Jun-22
Belostoma	May-26	Black Flies	Feb-7
Belostomatidae	Dec-4	Black Giant Ichneumonid Wasp	Sep-22
Bembidiini	Nov-24	black knot	Mar-5
Bembidion	Feb-21	black locust	Aug-2
Bembix americana	Jun-18	black nightshade	Sep-21
Bembix americana	Jul-8	black raspberry	Jan-16
Bembix americana spinolae	Aug-8	black raspberry	May-22
Bembix texana	Aug-8	Black Saddlebags	Sep-22
Berberis repens	Jul-15	Black Sandshell	Oct-11
Betrothed Underwing	Jul-26	black spruce	Jun-1
Betrothed Underwing	Aug-2	Black Swallowtail	Jun-6
Betula papyrifera	Mar-4	Black-and-gold Bumble Bee	Jun-27
Betula papyrifera	Nov-12	Black-and-gold Bumble Bee	Sep-1
Bibio femoratus	May-11	Black-and-yellow Mud Dauber Wasp	Jun-16
Bicolored Striped Sweat Bee	Jun-17		
Bicyrtes	Jun-18	Black-and-yellow Mud Dauber Wasp	Jul-21
Bicyrtes capnoptera	Jul-13		
Bicyrtes capnoptera	Jul-13	Black-bindweed	Aug-18

Black-capped Chickadee   Jan-1
Black-capped Chickadee   Sep-28
Black-capped Chickadee   Nov-30
Black-dotted Spragueia Moth   Jun-18
black-eyed Susan   Aug-8
Black-footed Globetail   May-25
Black-footed Globetail   Sep-7
Black-headed Grosbeak   Jul-15
Black-legged Meadow Katydid   Aug-27
Black-legged Meadow Katydid   Aug-30
Black-shouldered Drone Fly   May-2
Black-shouldered Drone Fly   Sep-14
Black-shouldered Drone Fly   Sep-20
Black-shouldered Drone Fly   Oct-17
Black-tipped Darner   Sep-1
Blackside darter   Jan-2
bladder campion   Jul-23
Bladder Snails   Nov-30
Blanchard's Cricket Frog   Jun-7
Blarina brevicauda   Oct-28
Blood Bees May-3
Blood Bees May-22
Blood Bees Jun-11
Blood Bees Jul-9
Blood Bees Sep-20
bloodroot   Apr-10
bloodroot   Apr-14
bloodroot   Apr-30
bloodroot   May-4
blue cohosh   Apr-28
Blue Dasher   Jun-18
Blue Dasher   Jun-24
Blue Dasher   Jul-7
Blue Dasher   Jul-19
Blue Jay   Jun-16
Blue Jay   Aug-9
Blue Jay   Dec-14
blue phlox   Apr-25
blue phlox   May-4
blue vervain   Jan-20
Blue-eyed Darner   Jun-9
Blue-eyed Darner   Jun-30
Blue-eyed Darner   Jul-11
blue-eyed grasses   Dec-25
Blue-fronted Dancer   Jun-26
Blue-green Sharpshooter   Jul-13
Blue-tipped Dancer   Jun-5

blueberries, cranberries, and allies   May-31
Bluegill   Aug-3
Bluets   May-11
Bluets   May-31
Bluets   Jul-16
bluntleaf sandwort   May-11
bog myrtle Jun-2
Bogbean   Jun-1
Boisea trivittata   Oct-4
Boisea trivittata   Oct-16
Boisea trivittata   Dec-16
Boloria selene   Jun-1
Boloria selene   Jun-2
Bombus affinis   Sep-12, 13, 14, 15, 16, 17
Bombus auricomus   Jun-27, Sep-1
Bombus bimaculatus   Apr-16, 28, May-21, Jul-20
Bombus borealis   Jun-6
Bombus citrinus   Jun-16, Aug-15, Sep-17
Bombus fervidus   Aug-11
Bombus griseocollis   Jul-9, 28
Bombus griseocollis   Aug-8
Bombus griseocollis   Aug-22
Bombus griseocollis   Aug-23
Bombus huntii   Jul-16
Bombus impatiens   May-17
Bombus impatiens   Jun-6
Bombus impatiens   Jun-18
Bombus impatiens   Sep-14
Bombus impatiens   Sep-14
Bombus impatiens   Sep-25
Bombus impatiens   Oct-16
Bombus ternarius   May-31
Bombus ternarius   Jun-1
Bombus vagans   Jul-24
Bombycilla cedrorum   Dec-14
Bombyliidae   Jun-24
Bombyliidae   Jul-14
Bombyliidae   Jul-15
Bombyliidae   Jul-24
Bombylius major   May-3
Boreal Bluet   May-25
Boreal Bluet   May-25
Boreal Chorus Frog   Mar-31
bottlebrush grass   Aug-24
bottlebrush grass   Nov-23

boxelder maple	Feb-5	
boxelder maple	Apr-19	
Brachycera	Apr-14	
Brachyceran Flies	Apr-14	
Brachymesia gravida	Jun-19	
Brachymesia gravida	Jun-20	
Brachymesia gravida	Jun-21	
Brachypalpus oarus	Apr-17	
bracted sedge	May-1	
Bradynotes obesa	Jul-17	
brambles	Jun-1	
Branta canadensis	Mar-21	
Branta canadensis	Oct-24	
Branta canadensis	Oct-25	
Branta canadensis	Dec-14	
Branta canadensis	Dec-20	
Branta canadensis	Dec-23	
Brickellia eupatorioides		Aug-4
Bridge penstemon	Jul-15	
Bristletails	Sep-29	
bristly buttercup	Apr-25	
bristly buttercup	May-4	
Bristly Cutworm Moth		May-29
broad-leaved goldenrod		Jan-17
broad-leaved goldenrod		Sep-4
broad-leaved goldenrod		Dec-20
Broad-winged Damselflies		Feb-10
Broad-winged Damselflies		Dec-4
Broad-winged Thistle	Aug-11	
broadleaf arrowhead	Aug-27	
Bronze Jumper	Sep-1	
Bronze Jumper	Oct-1	
Bronzed Tiger Beetle	Jun-6	
Bronzed Tiger Beetle	Sep-7	
Bronzed Tiger Beetle	Sep-7	
Bronzed Tiger Beetle	Sep-11	
Brown Anole	Jun-20	
Brown Creeper	Mar-6	
Brown Elfin	May-31	
Brown Elfin	Jun-1	
Brown Lacewings	Oct-7	
Brown Lacewings	Nov-4	
Brown Marmorated Stink Bug		Oct-17
Brown-belted Bumble Bee		Jul-9
Brown-belted Bumble Bee		Jul-28
Brown-belted Bumble Bee		Aug-8
Brown-belted Bumble Bee		Aug-22
Brown-belted Bumble Bee		Aug-23
Brown-winged Striped Sweat Bee		
	Jun-22	
Bryopsida	Jan-17	
Bryopsida	Mar-25	
Bubulcus ibis	Jun-22	
Bucephala albeola	Mar-21	
Buck Moth	Oct-4	
Bufflehead	Mar-21	
bull thistle	Jan-25	
bull thistle	Aug-11	
Bumble Bee-mimic Robber Fly		Jun-8
Bumble Bee-mimic Robber Fly		Jun-13
Buprestis consularis	Jul-4	
bur oak	Jan-24	
bur oak	Dec-30	
bur-reeds	Aug-27	
bushy lichens	Nov-18	
bushy lichens	Nov-23	
Buteo jamaicensis	Jan-7	
Butorides virescens	Jul-27	

C

Cabbage White	Jul-23	
Caddisflies	Dec-4	
Caddisflies	Dec-4	
Caecidotea	May-8	
Caecidotea	Dec-7	
Calidris minutilla	May-11	
Calla palustris	Jun-1	
Calligrapha alni	Jun-1	
Calliopsis andreniformis		Jul-1
Callophrys augustinus		May-31
Callophrys augustinus		Jun-1
Callopistromyia strigula		May-28
Callopistromyia strigula		Sep-2
Calopterygidae	Feb-10	
Calopterygidae	Dec-4	
Calopteryx maculata		Jun-5
Caloptilia rhoifoliella		Oct-6
Caltha palustris	May-4	
Caltha palustris	May-5	
Calyptratae	May-2	
Calyptratae	May-15	
Calyptratae	May-15	
Calyptratae	Nov-7	

Calyptratae	Nov-24	Cardinalis cardinalis	Mar-21
Calyptrates	May-2	Cardinalis cardinalis	Nov-13
Calyptrates	May-15	Carduus acanthoides	Aug-11
Calyptrates	May-15	Carduus nutans Jun-7	
Calyptrates	Nov-7	Carex Mar-2	
Calyptrates	Nov-24	Carex Dec-29	
Cambaridae	Dec-4	Carex albursina	May-1
Camel Crickets, Cave Crickets, and Cave Weta	May-3	Carex blanda	May-4
		Carex gracillima	Jun-1
Camel Crickets, Cave Crickets, and Cave Weta	Sep-29	Carex limosa	Jun-1
		Carex pensylvanica	Apr-14
Campaea perlata	Jun-6	Carex radiata	May-1
Campanulastrum americanum Jul-21		Carex sprengelii	May-1
		Carlin's Mining Bee	Apr-27
Campanulastrum americanum Oct-26		carnations	Dec-25
		Carolina draba	May-3
Campsomeris	Jun-18	Carolina Grasshopper	Sep-1
Campsomeris dorsata	Jun-18	Carolina Ground Cricket	Aug-26
Canada goldenrod	Aug-4	Carolina Saddlebags	Jun-18
Canada Goose	Mar-21	Carterocephalus palaemon	Jun-1
Canada Goose	Oct-24	Carterocephalus palaemon	Jun-1
Canada Goose	Oct-25	Carya cordiformis	Jul-19
Canada Goose	Dec-14	Cassius Blue	Jun-18
Canada Goose	Dec-20	Castor canadensis	Dec-23
Canada Goose	Dec-23	Cathartes aura	Mar-24
Canada wild lettuce	Jul-19	Catharus guttatus	May-18
Canada wild rye	Nov-5	Catocala innubens	Jul-26
Canadian bunchberry	Jun-1	Catocala innubens	Aug-2
Canadian wild ginger	Apr-18	Catogenus rufus	Sep-16
Canadian wild ginger	May-4	Catostomus commersonii	Dec-4
Canadian wood betony	May-4	cattails Jan-9	
Canadian yew	Mar-16	Cattle Egret	Jun-22
Canarsia ulmiarrosorella	May-29	Caulophyllum thalictroides	Apr-28
Canidae Jan-4		Cecidomyiinae	Feb-21
Canidae Nov-11		Cedar Waxwing	Dec-14
Canines Jan-4		Celastrina ladon	Jun-1
Canines Nov-11		Celastrina ladon	Jun-11
Canis latrans	Jan-22	Celastrus scandens	Mar-14
Canyon Rubyspot	Jul-13	Celastrus scandens	Nov-22
Canyon Tree Frog	Jul-15	Celastrus scandens	Nov-22
Capsella bursa-pastoris	May-9	Celastrus scandens	Dec-18
Carabidae	Jun-22	Celithemis amanda	Jun-18
Carabus maeander	Jun-2	Celithemis bertha	Jun-18
Carabus nemoralis	May-31	Celithemis bertha	Jun-20
Cardamine concatenata	Apr-25	Celithemis eponina	Jun-18
Cardamine concatenata	May-4	Celithemis eponina	Jun-21
Cardamine parviflora	Jun-1	Celithemis eponina	Aug-20

Celithemis fasciata	Jun-18		Chalcosyrphus nemorum		May-4
Cellar Spiders	Jun-22		Chalcosyrphus nemorum		May-7
Cellar Spiders	Nov-4		Chalcosyrphus nemorum		May-11
Cellophane Bees	May-3		Chalcosyrphus nemorum		Sep-1
Celticecis celtiphyllia		Jun-25	Chalcosyrphus nemorum		Sep-8
Celtis occidentalis	Feb-10		Chalepini	Jul-9	
Celtis reticulata	Jul-13		Chalk-fronted Corporal		Jun-9
Centaurea stoebe	Aug-11		Chalybion californicum		Jun-28
century plants	Jul-12		Chamaecrista fasciculata		Aug-8
Cerastis tenebrifera	Apr-9		Chaoborus		Mar-30
Cerastium fontanum		Apr-17	Charadrius vociferus		Mar-23
Ceratina calcarata	May-22		Checkered White	Jun-18	
Ceraunus Blue	Jun-20		Cheilosia	May-9	
Ceraunus Blue	Jun-21		Cheiracanthium	Feb-23	
Cerceris clypeata	Jul-1		Chelone glabra	Mar-4	
Cerceris fumipennis	Jun-30		Chelymorpha cassidea		May-23
Cerceris fumipennis		Jun-30	Chequered Skipper	Jun-1	
Cerceris fumipennis		Jul-1	Chequered Skipper	Jun-1	
Cerceris fumipennis		Jul-2	Chilocorus stigma	Apr-3	
Cerceris fumipennis		Jul-3	Chinavia	Jul-31	
Cerceris fumipennis		Jul-5	Chipping Sparrow	Apr-26	
Cerceris fumipennis		Jul-5	Chloropidae	Apr-18	
Cerceris fumipennis		Jul-6	chokecherry	May-12	
Cerceris fumipennis		Jul-7	Chordeiles minor	May-25	
Cerceris fumipennis		Jul-11	Chortophaga viridifasciata		Mar-4
Cerceris fumipennis		Jul-19	Chrysemys picta	Apr-4	
Cerceris fumipennis		Jul-23	Chrysemys picta	May-26	
Cerceris halone	Jul-9		Chrysididae	Jul-7	
Cerceris tolteca	Jun-20		Chrysididae	Jul-9	
Cercopoidea	May-28		Chrysididae	Jul-9	
Cercyonis oetus	Jul-16		Chrysididae	Jul-27	
Cercyonis oetus	Jul-16		Chrysis	Sep-10	
Ceropales maculata		Jul-27	Chrysobothris	Jul-6	
Cerotainia albipilosa		Jul-28	Chrysopilus ornatus		Jun-14
Cerotainia albipilosa		Jul-28	Chrysopsinae	Jul-21	
Cerotainia macrocera		Jul-21	Chymomyza amoena		May-3
Cerotainia macrocera		Jul-24	Cicadellidae	Oct-2	
Cerotainia macrocera		Jul-27	Cicadellidae	Oct-2	
Certhia americana	Mar-6		Cicadellidae	Oct-2	
Ceuthophilinae	Apr-7		Cicadellidae	Nov-10	
Chaetopsis	May-24		Cicindela	Jun-18	
chain speedwell	Jul-29		Cicindela limbalis	May-31	
Chalcidoid Wasps	Jul-7		Cicindela limbalis	Jun-2	
Chalcidoid Wasps	Sep-2		Cicindela punctulata		Jul-27
Chalcidoidea	Jul-7		Cicindela punctulata		Sep-2
Chalcidoidea	Sep-2		Cicindela purpurea	Sep-2	
Chalcosyrphus nemorum		May-4	Cicindela repanda	Jun-6	

Cicindela repanda Sep-7	Coelioxys Sep-16
Cicindela repanda Sep-7	Coenagrion resolutum Jun-1
Cicindela repanda Sep-11	Coenagrionidae Jan-2
Cicindela scutellaris lecontei May-3	Coenagrionidae Mar-30
Cicindela sexguttata May-22	Coenagrionidae May-8
Cimbex americana Jun-2	Coenagrionidae May-8
cinquefoils Jun-26	Coenagrionidae Oct-28
Cirsium Jul-16	Colaptes auratus Mar-26
Cirsium Nov-24	Coleomegilla maculata Mar-20
Cirsium Nov-25	Coleomegilla maculata May-5
Cirsium arizonicum Jul-13	Coleomegilla maculata May-26
Cirsium discolor Jan-25	Coleoptera Feb-11
Cirsium discolor Aug-7	Coleoptera Nov-10
Cirsium flodmanii Aug-8	Colias Jul-16
Cirsium vulgare Jan-25	Colias eurytheme Sep-20
Cirsium vulgare Aug-11	Colias philodice Sep-20
Cisseps fulvicollis Sep-21	Collared Ants May-28
Citronella Ants Feb-22	Collared Ants Sep-19
Citronella Ants Apr-12	Colletes May-3
Citrus Cicada Jul-18	Colletes aberrans Aug-8
Citrus Flatid Planthopper Aug-24	Colletes inaequalis Apr-14
Cladonia Mar-4	Colliuris pensylvanica Aug-21
Cladonia Mar-4	Coluber constrictor Jun-19
Cladonia Mar-26	Columba livia domestica Jan-26
Cladonia Dec-4	Columba livia domestica Feb-19
Cladonia chlorophaea Nov-23	Common Aerial Yellowjacket Aug-11
Claytonia virginica Apr-14	Common Aerial Yellowjacket Aug-15
Claytonia virginica May-4	Common Baskettail May-28
Clemensia albata May-27	Common Blue Mud Dauber Wasp Jun-28
Climaciella brunnea Jun-16	common blue violet Apr-21
Climaciella brunnea Jul-4	common blue violet May-4
Clouded Sulphur Sep-20	common boneset Sep-7
Clouded Yellows Jul-16	Common Buckeye Jun-16
clovers Nov-5	common buckthorn Feb-28
Club-rushes & bulrushes Nov-22	Common candy-striped spider Jul-30
Cluster Flies Feb-22	Common Claybank Tiger Beetle May-31
clustered black snakeroot May-25	Common Claybank Tiger Beetle Jun-2
Cobra Clubtail Jun-5	common dandelion Apr-11
Cobra Clubtail Jun-6	common dandelion Apr-21
Cobra Clubtail Jun-6	Common Eastern Bumble Bee May-17
Cobra Clubtail Jun-6	Common Eastern Bumble Bee Jun-6
Cobra Clubtail Jun-6	
Cobra Clubtail Jun-7	
Cobra Clubtail Jun-8	
Cobra Clubtail Jun-15	
Coccidae Feb-10	
Coelioxys Jul-23	

Common Eastern Bumble Bee
  Jun-18
Common Eastern Bumble Bee
  Sep-14
Common Eastern Bumble Bee
  Sep-14
Common Eastern Bumble Bee
  Sep-25
Common Eastern Bumble Bee
  Oct-16
common eastern Physocephala
  Aug-15
common evening-primrose   Jan-20
common evening-primrose   Dec-3
Common Gluphisia Moth   May-27
Common Green Darner   Apr-8
Common Green Darner   Apr-16
Common Green Darner   Apr-21
Common Green Darner   May-24
Common Green Darner   May-25
Common Green Darner   Aug-4
Common Green Darner   Aug-25
Common Green Darner   Sep-7
Common Green Darner   Sep-17
Common Green Darner   Sep-30
common greenshield lichen   Nov-23
common hackberry Feb-10
Common hemp-nettle   Aug-11
common milkweed Jul-7
common milkweed Oct-12
common motherwort   Mar-24
Common Mouse-ear Chickweed
  Apr-17
Common Nighthawk   May-25
Common purslane Jun-18
Common Raccoon Feb-14
Common Raccoon Apr-15
common ragweed Aug-11
Common Sawflies May-9
Common Sawflies May-15
Common Sawflies Jun-9
Common Side-blotched Lizard
  Jul-14
Common Silverfish Mar-17
Common Tan Wave Oct-7
Common Thread-waisted Wasp
  Jun-18
Common Tiger Beetles   Jun-18
common toadflax Sep-23
common water-crowfoot   May-26
Common Whitetail Apr-24
Common Whitetail May-16
Common Whitetail May-20
Common Whitetail May-23
Common Whitetail Jun-6
Common Whitetail Jun-9
Common Whitetail Jun-10
compass plant   Mar-25
compass plant   Jul-9
complex thalloid liverworts   Feb-12
Compost Fly   Jul-20
Condylostylus   Jun-14
Condylostylus   Jun-14
Confusing Furrow Bee   Apr-16
Confusing Furrow Bee   May-3
Conocephalum salebrosum   May-4
Conoderus auritus May-2
Convolvulus arvensis   Jan-27
Cooper's Hawk   Sep-6
Copaeodes aurantiaca   Jul-15
Copivaleria grotei   Apr-9
Coptis trifolia   Jun-1
Coral Hairstreak   Jul-8
Coras   Mar-4
Cordulegaster   Dec-5
Cordyceps clavulata   Feb-10
Cordyceps clavulata   Mar-4
Cordyceps clavulata   Mar-4
Coreopsis tinctoria Aug-22
Corinnidae   Mar-4
Corinnidae   Mar-4
Corn   Feb-20
Corn Earworm Moth   Sep-26
corn speedwell   Apr-4
corn speedwell   May-5
corn speedwell   May-22
Cornus canadensis Jun-1
Cornus sericea   Jan-3
Cornus sericea   Jan-31
Cornus sericea   Mar-4
Cornus sericea   Dec-31
Corvus brachyrhynchos   Jan-30
Corvus brachyrhynchos   Oct-26
Corvus brachyrhynchos   Dec-12

Corydalus texanus	Jul-13		Cuterebra	Jul-13	
Corylus americana	Mar-18		Cuterebra fontinella		Aug-18
Coyote	Jan-22		Cutworm Moths and Allies		Aug-23
Crab Spiders	Jun-1		Cutworm Wasps	Jul-16	
Crab Spiders	Aug-8		Cutworm Wasps	Jul-16	
Crabro	Sep-7		Cyanocitta cristata	Jun-16	
Crabronina	May-28		Cyanocitta cristata	Aug-9	
Crabronina	Aug-7		Cyanocitta cristata	Dec-14	
crack willow	May-25		Cyathus striatus	Aug-17	
Crambid Snout Moths		Jun-22	Cycloneda munda	Jun-24	
Crambidae s	Jun-22		Cylindera terricola	Jul-16	
cranberry viburnum		Nov-9	Cylindrical Thimbleweed		Jun-15
Crane Flies	Feb-11		Cynipidae	Sep-3	
Crane Flies	Oct-7		Cyprinidae	Dec-4	
Crataegus	Jan-3		Cyprinidae	Dec-4	
Creek Chub	Sep-27		Cyprinidae	Dec-4	
creeping Jenny	Jun-26		Cyprinus	Aug-27	
creeping mahonia	Jul-15		Cyrtopogon	Jul-16	
Crematogaster	May-16				
Crematogaster	Jun-14		# D		
Crematogaster	Jul-22				
Crepis tectorum	Jun-4		Dahlica triquetrella	Apr-6	
Cricetidae	Aug-12		Dainty Sulphur	Jun-21	
Crocus Geometer Moths		Aug-25	Dalea	Dec-1	
Crossidius coralinus		Jul-16	Dalea purpurea	Jun-27	
Crotalus oreganus lutosus		Jul-15	Damsel Bugs	Apr-18	
Cruisers	May-31		Danaus plexippus	May-12	
Cruisers	Dec-4		Danaus plexippus	Jul-19	
Cruisers	Dec-6		Danaus plexippus	Aug-7	
Cryptogramma stelleri		May-4	Danaus plexippus	Aug-22	
Cuban Treefrog	Jun-22		Danaus plexippus	Sep-2	
Cuckoo Leaf-cutter Bees		Jul-23	Danaus plexippus	Sep-16	
Cuckoo Leaf-cutter Bees		Sep-16	Dancers	Feb-6	
Cuckoo Wasps	Jul-7		Dancers	Jul-13	
Cuckoo Wasps	Jul-9		Dancers	Jul-16	
Cuckoo Wasps	Jul-9		Dark Fishing Spider	May-7	
Cuckoo Wasps	Jul-27		Dark Fishing Spider	Sep-11	
Cucujus clavipes	Nov-18		Dark Fishing Spider	Dec-4	
Culicidae	Aug-10		Dark Paper Wasp	Apr-1	
cup plant	Jul-21		Dark Paper Wasp	Apr-14	
Cupido comyntas	May-22		Dark Paper Wasp	Apr-21	
Cupido comyntas	Aug-25		Dark Paper Wasp	Jun-22	
Curculio	Jun-13		Dark Paper Wasp	Oct-3	
Curculio	Jul-9		Dark Paper Wasp	Nov-27	
Cuscuta gronovii	Aug-25		Dark-banded Owlet	Aug-5	
cut-leaved toothwort		Apr-25	Dark-eyed Junco	Mar-18	
cut-leaved toothwort		May-4	Dark-spotted Palthis Moth		May-29

dark-throated shooting star   Jul-15
Dark-winged Fungus Gnats   Apr-19
Darners   Jul-13
Dasymutilla quadriguttata   Jun-20
date palms   Jun-20
Daucus carota   Oct-12
Deathwatch, Spider, and Wood-borer Beetles   Dec-24
Deer Flies   Jul-21
DeKay's Brownsnake   Sep-22
Delicate Emerald   May-31
Delphacid Planthoppers   Feb-21
Delphacidae   Feb-21
Dendrolycopodium dendroideum   Dec-4
deptford pink   Jun-15
deptford pink   Oct-4
Dermacentor variabilis   May-25
Desert Stink Beetles   Jul-12
Desmodium canadense   Jul-9
devil's beggarticks   Sep-7
devil's beggarticks   Sep-28
Diabrotica undecimpunctata   Oct-11
Dialictus   Apr-14
Dialictus   Apr-15
Dialictus   May-22
Dialictus   Jun-4
Dialictus   Jul-16
Dialictus   Sep-1
Dialictus   Sep-14
Diamesa   Jan-14
Diamesa   Jan-14
Diamesa   Feb-10
Diamesa   Dec-19
Diamondback Moth   Apr-28
Diamondback Spittlebug   Jun-16
Dianthus   Dec-25
Dianthus armeria   Jun-15
Dianthus armeria   Oct-4
Dicentra cucullaria   Apr-13
Dicentra cucullaria   May-4
Dicerca   Jun-4
Dicerca   Jul-4
Dicerca   Jul-5
Dicerca   Jul-5
Dicerca   Jul-31
Diceroprocta apache   Jul-18
Dickcissel   Jun-15
Dictynidae   May-7
Dictynidae   May-12
Didelphis virginiana   Nov-18
Didymops transversa   Dec-2
Dieteria canescens   Jul-15
Differential Grasshopper   Aug-18
Differential Grasshopper   Sep-6
Dimorphic Jumper   May-16
Diptera   Feb-11
Diptera   Feb-11
Diptera   Feb-11
Diptera   Feb-11
Dissosteira carolina   Sep-1
Dolichovespula   Nov-5
Dolichovespula arenaria   Aug-11
Dolichovespula arenaria   Aug-15
Dolichovespula maculata   Sep-7
Dolichovespula maculata   Oct-3
Dolomedes scriptus   Sep-5
Dolomedes tenebrosus   May-7
Dolomedes tenebrosus   Sep-11
Dolomedes tenebrosus   Dec-4
Dolomedes triton   May-26
Doryctinae   May-29
Dot-tailed Whiteface   Feb-2
Dot-tailed Whiteface   May-6
Dot-tailed Whiteface   May-15
Dot-tailed Whiteface   May-25
Dot-tailed Whiteface   Jun-6
Dot-tailed Whiteface   Nov-14
Double-banded Scoliid Wasp   Sep-20
downy rattlesnake plantain   Mar-28
Downy Woodpecker   Feb-26
Downy Woodpecker   Mar-18
downy yellow violet   Apr-25
downy yellow violet   May-2
downy yellow violet   May-4
Draba reptans   May-3
Dragonhunter   May-31
Dreissena polymorpha   Nov-24
Drone Fly   Aug-14
Drone Fly   Sep-20
Drone Fly   Oct-18
Drone Fly   Oct-20
Drone Fly   Oct-23
Drosera rotundifolia   May-31
Drummond's aster   Sep-13

Drummond's aster	Dec-8	Eastern Comma	Apr-21
Dryinidae	Jul-10	Eastern Comma	Jul-4
Dryocopus pileatus	Jun-9	Eastern Comma	Jul-21
Dun Skipper	Jul-11	Eastern Comma	Sep-7
Dunning's Mining Bee	Apr-14	Eastern Cottontail	May-18
Dunning's Mining Bee	Apr-15	eastern cottonwood	Jan-8
Dusky-banded Forest Fly	May-4	eastern cottonwood	Feb-13
Dusky-banded Forest Fly	May-4	eastern cottonwood	Sep-19
Dusky-banded Forest Fly	May-7	Eastern Forktail	Apr-23
Dusky-banded Forest Fly	May-11	Eastern Forktail	Apr-24
Dusky-banded Forest Fly	Sep-1	Eastern Forktail	May-6
Dusky-banded Forest Fly	Sep-8	Eastern Forktail	May-13
Duskywings	May-31	Eastern Forktail	May-13
Dutchman's breeches	Apr-13	Eastern Forktail	May-15
Dutchman's breeches	May-4	Eastern Forktail	May-16
Dytiscidae s	Mar-28	Eastern Forktail	May-22
		Eastern Forktail	May-22
		Eastern Forktail	May-25

E

		Eastern Forktail	Jun-1
Early meadow-rue	May-4	Eastern Forktail	Jul-27
earthballs	Dec-4	Eastern Forktail	Aug-2
Earthworms	Apr-4	Eastern Forktail	Aug-14
Eastern Amberwing	Jun-18	Eastern Forktail	Oct-11
Eastern Amberwing	Jun-22	Eastern Giant Swallowtail	Aug-19
Eastern Amberwing	Jul-29	Eastern Giant Swallowtail	Aug-22
Eastern Amberwing	Aug-15	Eastern Gray Squirrel	Oct-15
Eastern Band-winged Hover Fly	Aug-31	Eastern Gray Squirrel	Dec-12
eastern black walnut	Dec-13	Eastern Hornet Fly	Sep-14
Eastern Bluebird	Apr-4	Eastern Least Clubtail	Jun-1
Eastern Boxelder Bug	Oct-4	Eastern Parson Spider	Mar-12
Eastern Boxelder Bug	Oct-16	Eastern Parson Spider	Dec-10
Eastern Boxelder Bug	Dec-16	eastern pasqueflower	Apr-7
Eastern Calligrapher	Apr-25	Eastern Phoebe	Apr-4
Eastern Calligrapher	May-5	Eastern Pondhawk	May-25
Eastern Calligrapher	May-7	Eastern Pondhawk	Jun-5
Eastern Calligrapher	May-17	Eastern Pondhawk	Jun-19
Eastern Calligrapher	Jul-18	Eastern Pondhawk	Jun-21
Eastern Calligrapher	Sep-1	Eastern Pondhawk	Jun-26
Eastern Calligrapher	Sep-6	Eastern Pondhawk	Jul-19
Eastern Calligrapher	Oct-18	Eastern Pondhawk	Aug-22
Eastern Catkin Fly	Apr-17	Eastern Pondhawk	Aug-22
Eastern Chipmunk	Jun-23	Eastern Pondhawk	Nov-14
Eastern Cicada Killer	Jul-19	eastern redcedar	Dec-14
Eastern Cicada Killer	Jul-23	Eastern Tailed-Blue	May-22
Eastern Cicada Killer	Jul-27	Eastern Tailed-Blue	Aug-25
Eastern Cicada Killer	Jul-31	Eastern Tiger Swallowtail	Aug-15
Eastern Comma	Apr-1	Eastern Tiger Swallowtail	Aug-22

eastern white pine	Mar-16	Enallagma civile	Sep-19
eastern woodland sedge	May-4	Enallagma doubledayi	Jun-18
Eastern Yellowjacket	Oct-3	Enallagma doubledayi	Jun-20
Eastern Yellowjacket	Oct-16	Enallagma exsulans	Jun-15
Ebony Boghaunter	Mar-31, Jun-1	Enallagma exsulans	Jun-26
Ebony Jewelwing	Jun-5	Enallagma hageni	Jun-6
Echinocystis lobata	Mar-9	Enallagma hageni	Jun-24
Echinocystis lobata	Aug-2	Enallagma praevarum	Jul-16
Echinocystis lobata	Oct-27	Enallagma s	May-11
Ectemnius	Jun-27	Enallagma s	May-31
Edwards' Hairstreak	Jun-27	Enallagma s	Jul-16
Efferia	Jun-20	Enchenopa binotata	Oct-9
Efferia	Jul-14	Endothenia hebesana	Jun-10
Efferia albibarbis	Jun-15	Enemion biternatum	Feb-14
Efferia albibarbis	Jun-15	Enemion biternatum	Apr-14
Efferia albibarbis	Jun-15	Enemion biternatum	Apr-25
Egretta caerulea	Jun-21	Enemion biternatum	May-4
Elaphrus	Jul-29	Enemion biternatum	Sep-29
Elaphrus	Sep-7	Enoclerus rosmarus	Apr-9
Elegant Grass-veneer	Jul-11	Enoplognatha	Apr-9
Elegant Spreadwing	Jun-5	Enoplognatha	Apr-29
Eleodes	Jul-12	Enoplognatha ovata	Jul-30
Ellisia nyctelea	May-26	Ensign Ground Hunter	Apr-20
Ellisia nyctelea	May-28	Epeolus	Jun-15
Ellychnia corrusca	Mar-28	Ephedra viridis	Jul-15
Elm Leaftier Moth	May-29	Ephemeroptera	Apr-25
Elm Sawfly	Jun-2	Ephemeroptera	Jun-2
elms	Feb-1	Ephydridae	Dec-19
elms	Oct-14	Epicauta pennsylvanica	Aug-25
Elophila obliteralis	May-27	Epicauta pennsylvanica	Sep-20
Elusive Clubtail	Apr-25	Epinotia vertumnana	Feb-21
Elusive Clubtail	Aug-25	Epitheca	Jun-9
Elymus canadensis	Nov-5	Epitheca canis	Jun-2
Elymus hystrix	Aug-24	Epitheca cynosura	May-28
Elymus hystrix	Nov-23	Equisetum arvense	Apr-21
Elymus virginicus	Nov-16	Equisetum hyemale	Mar-26
Emerald Euphoria	Jun-24	Equisetum hyemale	Oct-11
Emerald Moths	Jun-1	Equisetum pratense	May-4
Emerald Spreadwing	Jun-17	Eremnophila aureonotata	Jul-7
Emesinae	Mar-27	Eremnophila aureonotata	Sep-1
Emilia sonchifolia	Jun-20	Eremnophila aureonotata	Sep-17
Enallagma annexum	May-31	Erigeron philadelphicus	May-25
Enallagma boreale	May-25	Erigoninae	Mar-28
Enallagma boreale	May-25	Eriocaulon decangulare	Jun-18
Enallagma carunculatum	May-25	Eriochloa villosa	Nov-16
Enallagma civile	Sep-2	Eriophorum vaginatum	Jun-1
Enallagma civile	Sep-16	Eriosomatinae	Apr-12

Eris militaris	Sep-1	Erythrodiplax minuscula	Jun-18
Eris militaris	Oct-1	Erythrodiplax minuscula	Jun-20
Eristalis	May-31	Erythrodiplax minuscula	Jun-20
Eristalis	May-31	Erythronium	May-4
Eristalis	Aug-29	Erythronium albidum	Apr-25
Eristalis anthophorina	May-2	Erythronium propullans	May-4
Eristalis arbustorum	Oct-20	Etheostoma flabellare	Dec-4
Eristalis dimidiata	May-2	Etheostoma flabellare	Dec-4
Eristalis dimidiata	Sep-14	Etheostoma nigrum	Jan-2
Eristalis dimidiata	Sep-20	Etheostoma nigrum	Dec-4
Eristalis dimidiata	Oct-17	Etheostoma nigrum	Dec-4
Eristalis flavipes	Jul-16	Euaresta festiva	Aug-10
Eristalis flavipes	Aug-12	Euchaetes egle	Sep-3
Eristalis flavipes	Sep-14	Eufidonia	Jun-1
Eristalis stipator	May-2	Eumelissodes	Jun-17
Eristalis stipator	May-2	Eumelissodes	Aug-4
Eristalis stipator	May-22	Eumelissodes	Aug-7
Eristalis tenax	Aug-14	Eumelissodes	Aug-18
Eristalis tenax	Sep-20	Eumenes fraternus	Aug-29
Eristalis tenax	Oct-18	Eumeninae	Jun-18
Eristalis tenax	Oct-20	Eumeninae	Jun-18
Eristalis tenax	Oct-23	Eumerus funeralis	Aug-20
Eristalis transversa	May-17	Eumorpha pandorus	Sep-2
Eristalis transversa	May-25	Eunemobius carolinus	Aug-26
Eristalis transversa	Jun-24	Euodynerus foraminatus	Jul-10
Eristalis transversa	Aug-19	Euonymus alatus	Mar-15
Eristalis transversa	Sep-1	Euonymus alatus	Oct-29
Eristalis transversa	Oct-18	Eupatorium perfoliatum	Sep-7
Eristalis transversa	Oct-19	Euphoria fulgida	Jun-24
Eristalis transversa	Oct-20	Euphyes vestris	Jul-11
Erpetogomphus compositus	Jul-14	Euplexia benesimilis	May-27
Eryngium yuccifolium	Feb-17	European Drone Fly	Oct-20
Eryngium yuccifolium	Jul-25	European Earwig	Jul-9
Erynnis	May-31	European Paper Wasp	Jun-30
Erynnis brizo	May-31	European Starling	Mar-18
Erynnis brizo	Jun-1	Eurosta solidaginis	Jan-10
Erysimum cheiranthoides	May-28	Eurosta solidaginis	May-25
Erythemis simplicicollis	May-25	Euryuridae	Oct-13
Erythemis simplicicollis	Jun-5	Eustala	Feb-27
Erythemis simplicicollis	Jun-19	Eustala	Apr-22
Erythemis simplicicollis	Jun-21	Eutrochium	Dec-28
Erythemis simplicicollis	Jun-26	Evarcha hoyi	May-17
Erythemis simplicicollis	Jul-19	Exoprosopa fascipennis	Jun-20
Erythemis simplicicollis	Aug-22		
Erythemis simplicicollis	Aug-22	**F**	
Erythemis simplicicollis	Nov-14		
Erythrodiplax minuscula	Jun-18	Fall Cankerworm Moth	Sep-26

Fall Cankerworm Moth	Nov-4	Flies Feb-11	
Fallopia convolvulus	Aug-18	Flies Feb-11	
Fallopia japonica	Dec-18	Flodman's thistle	Aug-8
false boneset	Aug-4	Florida Harvester Ant	Jun-20
false golden asters	Jul-15	Florida Scrub Lizard	Jun-20
false mealworm beetle	Jan-19	Flower-loving Flies	Jul-15
False Morel	Jun-1	fluted bird's nest fungus	Aug-17
false rue anemone	Feb-14	Forbes' Tree Cricket	Oct-9
false rue anemone	Apr-14	Forficula auricularia	Jul-9
false rue anemone	Apr-25	Formica	May-9
false rue anemone	May-4	Formicidae	Apr-7
false rue anemone	Sep-29	Formicidae	May-3
Familiar Bluet	Sep-2	Forsythia	Jan-31
Familiar Bluet	Sep-16	Four-lined Hornet Fly	Sep-1
Familiar Bluet	Sep-19	Four-spotted Pennant	Jun-19
Fantail Darter	Dec-4	Four-spotted Pennant	Jun-20
Fantail Darter	Dec-4	Four-spotted Pennant	Jun-21
fawn lilies	May-4	Four-spotted Skimmer	May-15
Faxonius virilis	Jan-2	Four-spotted Skimmer	May-25
Faxonius virilis	Jan-16	Four-spotted Skimmer	Jun-1
Faxonius virilis	Feb-14	Four-spotted Skimmer	Jul-6
Faxonius virilis	May-31	foxtails and bristlegrasses	Jan-17
Feral Pigeon	Jan-26	foxtails and bristlegrasses	Nov-5
Feral Pigeon	Feb-19	fragrant bedstraw	May-4
ferns	Jun-21	fragrant evening primrose	Jul-15
field bindweed	Jan-27	Frasera speciosa	Jul-16
field horsetail	Apr-21	Fraternal Potter Wasp	Aug-29
field penny-cress	Apr-21	Fraxinus pennsylvanica	Feb-3
Field Pussytoes	May-2	fringed loosestrife	Jul-21
field thistle	Jan-25	Froelichia floridana	Aug-8
field thistle	Aug-7	Fungi	Feb-1
Fiery Skipper	Jun-18	Fungi	Nov-7
Fiery Skipper	Sep-13	Fungi Including Lichens	Feb-1
Fiery Skipper	Sep-22	Fungi Including Lichens	Nov-7
Fifteen-spotted Lady Beetle	Mar-6	Funnel Web Spiders	Mar-4
Fireflies	Feb-12		
Fireflies	Jul-18	**G**	
Five-banded Thynnid Wasp	Jul-22		
Five-banded Thynnid Wasp	Aug-8	Galeopsis tetrahit	Aug-11
Five-banded Thynnid Wasp	Aug-25	Galium mollugo	May-4
Flamed Tigersnail	Mar-28	Galium triflorum	May-4
Flavoparmelia caperata	Nov-23	Gall Wasps	Sep-3
Flea Beetles	Feb-21	Gallinago delicata	Mar-23
Flea Jumper	Apr-15	Gambel oak	Jul-14
Flea Jumper	May-7	Gambel oak	Jul-15
Flies	Feb-11	Gammarus pseudolimnaeus	Jan-2
Flies	Feb-11	Gammarus pseudolimnaeus	Nov-30

garden yellowrocket	May-4	goldenrods	Jul-14
Garita Skipperling	Jul-16	goldenrods	Sep-21
Gasteracantha cancriformis	Jun-18	Gomphurus fraternus	Jun-5
Gasteruption	Jun-14	Gomphurus fraternus	Jun-8
Gasteruption	Jul-27	Gomphurus fraternus	Jun-17
Gasteruption	Aug-12	Gomphurus fraternus	Jun-26
Gastrocopta	Mar-27	Gomphurus fraternus	Jul-9
Gastropoda	Mar-28	Gomphurus vastus	Jun-5
Gastropoda	Nov-24	Gomphurus vastus	Jun-6
Gastropods	Mar-28	Gomphurus vastus	Jun-6
Gastropods	Nov-24	Gomphurus vastus	Jun-6
Geina periscelidactylus	Jul-5	Gomphurus vastus	Jun-6
Gem Moth	Apr-23	Gomphurus vastus	Jun-7
Geometer Moths	Aug-19	Gomphurus vastus	Jun-8
Geometer Moths	Sep-2	Gomphurus vastus	Jun-15
Geometer Moths	Oct-8	Gonia	Apr-7
Geometridae	Aug-19	Goodyera pubescens	Mar-28
Geometridae	Sep-2	graceful sedge	Jun-1
Geometridae	Oct-8	Graceful Twig Ant	Jun-22
Geometrinae	Jun-1	Gracillariidae	Oct-16
Geranium bicknellii	Jun-2	Gracillariidae	Nov-23
Geranium maculatum	May-4	Grape Plume Moth	Jul-5
Geranium maculatum	May-25	Graphocephala	Aug-28
Geron	Jun-20	Graphocephala	Oct-2
Geum canadense	Jul-7	Grass Flies	Apr-18
Geum triflorum	Apr-7	Grasshoppers, Crickets, and Katydids Jul-14	
Giant Floater Mussel	Feb-15		
giant ragweed	Aug-10	Gray Fox	Mar-4
giant ragweed	Aug-14	Gray Hairstreak	Jun-20
Giant Water Bugs	Dec-4	Gray Tree Frog	Aug-27
gilled mushrooms	Nov-18	Great Basin Rattlesnake	Jul-15
ginkgo	Oct-30	Great Black Wasp	Aug-11
Ginkgo biloba	Oct-30	Great Black Wasp	Aug-14
Gleditsia triacanthos	Mar-25	Great Blue Heron	Aug-25
Gleditsia triacanthos inermis	Oct-14	Great Golden Digger Wasp	Jul-3
Glischrochilus fasciatus	Oct-16	Great Golden Digger Wasp	Aug-11
Gluphisia septentrionis	May-27	Great Golden Digger Wasp	Aug-25
Gnaphosa	Oct-13	Great Golden Digger Wasp	Aug-27
golden columbine	Jul-15	Great Golden Digger Wasp	Sep-13
Golden Northern Bumble Bee	Aug-11	Great Indian Plantain	Aug-14
Golden-eye Lichen	Apr-19	great mullein	Feb-4
Golden-eye Lichen	Nov-23	great mullein	Feb-13
Golden-eye Lichen	Nov-23	great mullein	Dec-3
Golden-winged Skimmer	Jun-18	Great Tiger Moth	Jul-16
Goldenrod Gall Fly	Jan-10	Great-tailed Grackle	Jul-18
Goldenrod Gall Fly	May-25	Greater Bee Fly	May-3
goldenrods	Jan-16	greater burdock	Jan-6

greater burdock   Jan-14
green ash   Feb-3
Green Cone-headed Planthopper   Oct-8
green ephedra   Jul-15
Green Heron   Jul-27
Green Immigrant Leaf Weevil   May-24
Green Stink Bugs   Jul-31
Green Sunfish   Aug-3
Green-striped Darner   Aug-22
Green-striped Darner   Aug-29
Green-striped Darner   Aug-30
Green-striped Grasshopper   Mar-4
Greenbottle Flies   May-22
grey-headed coneflower   Jul-8
Grote's Sallow   Apr-9
Ground Beetles   Jun-22
Ground Crab Spiders   May-13
Ground Crab Spiders   Aug-22
Ground wolf spider   Mar-19
Ground wolf spider   Apr-9
Ground-plum   Jun-15
Gymnocladus dioicus   Nov-20
Gymnosporangium juniperi-virginianae   Dec-15
Gyromitra esculenta   Jun-1

H
Habrosyne scripta   Aug-13
Hackberry Acorn Gall Midge   Jun-25
Hackberry Nipple Gall   Oct-2
Hackberry Nipple Gall   Nov-29
Hackelia virginiana   Dec-18
Haemorhous mexicanus   Dec-14
Hagen's Bluet   Jun-6
Hagen's Bluet   Jun-24
Hagenius brevistylus   May-31
Hairy Cupgrass   Nov-16
Hairy Spider Weevil   May-15
hairy sweet cicely   May-22
Half-black Bumble Bee   Jul-24
Haliaeetus leucocephalus   Jan-23
Halictus confusus   Apr-16
Halictus confusus   May-3
Halictus ligatus   Sep-14
Halloween Pennant   Jun-18
Halloween Pennant   Jun-21
Halloween Pennant   Aug-20

Halyomorpha halys   Oct-17
Hamsters, Voles, Lemmings, and Allies   Aug-12
Hanging Clubtails   Dec-5
Haplodrassus signifer   Apr-20
hardstem bulrush   Nov-22
Harmonia axyridis   Dec-16
Harpalinae   Apr-23
Harvestmen   Oct-12
Hawthorn Root Borer   Jul-3
hawthorns   Jan-3
Hedge Bedstraw   May-4
Hedgerow Hairstreak   Jul-16
Hedgerow Hairstreak   Jul-17
Helicoverpa zea   Sep-26
Helophilus fasciatus   May-2
Helophilus fasciatus   May-2
Helophilus fasciatus   May-6
Helophilus fasciatus   May-9
Helophilus fasciatus   May-22
Helophilus fasciatus   Jun-2
Helophilus fasciatus   Sep-1
Helophilus fasciatus   Sep-7
Helophilus fasciatus   Sep-20
Hemaris diffinis   Jun-1
Hemerobiidae   Oct-7
Hemerobiidae   Nov-4
Hemiargus ceraunus   Jun-20
Hemiargus ceraunus   Jun-21
Hemileuca maia   Oct-4
henbit deadnettle   May-9
Hepatica americana   Nov-23
Heriades   Jun-29
Heriades   Jul-10
Hermit Thrush   May-18
Herpyllus ecclesiasticus   Mar-12
Herpyllus ecclesiasticus   Dec-10
Hesperia colorado   Jul-17
Hesperiidae   Jul-16
Hesperus   Oct-20
Hetaerina americana   Jun-15
Hetaerina americana   Sep-9
Hetaerina americana   Sep-21
Hetaerina vulnerata   Jul-13
Heteroptera   Mar-27
Heterotheca   Jul-15
Heuchera richardsonii   Jun-15

Hirudinea	Oct-28	Hyla versicolor	Aug-27
Hirundo rustica	May-19	Hylaeus	Jun-11
hoary aster	Jul-15	Hylephila phyleus	Jun-18
hoary puccoon	May-11	Hylephila phyleus	Sep-13
Hobomok Skipper	May-22	Hylephila phyleus	Sep-22
Hobomok Skipper	May-24	Hyles lineata	Jun-16
Hobomok Skipper	Jun-6	Hyles lineata	Jul-8
Holarctic Tree Frogs	Apr-23	Hylogomphus adelphus	Dec-6
Holarctic Tree Frogs	May-25	Hypogastrura	Mar-4
Holarctic Tree Frogs	Oct-18	Hypoprepia	Mar-27
honey locust	Mar-25	Hypselistes florens	May-22
honeysuckles	Feb-25	Hypselistes florens	May-25
Hooded Merganser	Mar-19		
Hooded Merganser	May-19		
Hooded Merganser	Oct-25	**I**	
hop-hornbeams	Nov-23		
Hoplisoides	Jun-9	Ichneumon annulatorius	Mar-28
Hordnia atropunctata	Jul-13	Ichneumonid and Braconid Wasps	
Horned Clubtail	May-26	May-23	
Horned Clubtail	Jun-3	Ichneumonid Wasps	Apr-30
Horned Clubtail	Jun-5	Ichneumonid Wasps	May-15
Hornets and Yellowjackets	Oct-24	Ichneumonid Wasps	May-22
Horseradish	Apr-18	Ichneumonid Wasps	May-24
Horseradish	May-26	Ichneumonid Wasps	Jun-9
House Finch	Dec-14	Ichneumonid Wasps	Oct-8
House Sparrow	Jan-8	Ichneumonidae	Apr-30
House Sparrow	Jan-29	Ichneumonidae	May-15
House Sparrow	Feb-17	Ichneumonidae	May-22
House Sparrow	Feb-20	Ichneumonidae	May-24
House Wren	May-18	Ichneumonidae	Jun-9
Hover Flies	Sep-9	Ichneumonidae	Oct-8
Hoy's Jumper	May-17	Ichneumonoidea	May-23
Hudsonian Whiteface	May-31	Ichthyomyzon	Dec-4
Hudsonian Whiteface	Jun-1	Icterus galbula	May-9
Hudsonian Whiteface	Jun-1	Idia americalis	Aug-5
Humpbacked Flies	Dec-22	Indiangrass	Jan-5
Hunt's Bumble Bee	Jul-16	Indiangrass	Feb-17
Hydrangea	Feb-25	Insecta Dec-9	
hydrangeas	Feb-25	Insecta Dec-10	
Hydrophorinae	Nov-24	Insecta Dec-10	
Hydropsychidae	Jan-2	Insects Dec-9	
Hygrocybe	Oct-4	Insects Dec-10	
Hyla Apr-23		Insects Dec-10	
Hyla May-25		interior sandbar willow	Mar-3
Hyla Oct-18		Irpex lacteus	Jan-8
Hyla arenicolor	Jul-15	Isabella Tiger Moth	May-27
Hyla squirella	Jun-22	Ischnura ramburii	Jun-18
		Ischnura ramburii	Jun-18

Ischnura ramburii	Jun-20
Ischnura verticalis	Apr-23
Ischnura verticalis	Apr-24
Ischnura verticalis	May-6
Ischnura verticalis	May-13
Ischnura verticalis	May-13
Ischnura verticalis	May-15
Ischnura verticalis	May-16
Ischnura verticalis	May-22
Ischnura verticalis	May-22
Ischnura verticalis	May-25
Ischnura verticalis	Jun-1
Ischnura verticalis	Jul-27
Ischnura verticalis	Aug-2
Ischnura verticalis	Aug-14
Ischnura verticalis	Oct-11
Isodontia mexicana	Jun-29
Isodontia mexicana	Aug-15
Isonychia	Dec-4

J

Jack-in-the-pulpit	May-4
Jagged Ambush Bugs	Jul-28
Jagged Ambush Bugs	Aug-8
Jagged Ambush Bugs	Sep-20
Jagged Ambush Bugs	Oct-18
Japanese knotweed	Dec-18
Joe-Pye weeds	Dec-28
Johnny Darter	Jan-2
Johnny Darter	Dec-4
Johnny Darter	Dec-4
Juglans nigra	Dec-13
Jumping Spiders	May-12
Jumping Spiders	May-30
Junco hyemalis	Mar-18
juniper mistletoe	Jul-14
juniper polytrichum moss	Mar-4
juniper polytrichum moss	Dec-4
juniper-apple rust	Dec-15
Juniperus virginiana	Dec-14
Junonia coenia	Jun-16
Kalmia polifolia	May-31
Kateretidae	Apr-14
Kentucky coffeetree	Nov-20
Killdeer	Mar-23
King Skimmers	May-8

kittentails	May-3
knotweeds, smartweeds, and water-peppers	Aug-10

L

Labidomera clivicollis	May-26
Labidomera clivicollis	Jun-28
Labidomera clivicollis	Sep-15
Lacinipolia renigera	May-29
Lactuca canadensis	Jul-19
Ladona julia	Jun-9
lady fern	May-4
lady fern	May-5
Lamium amplexicaule	May-9
Lamium maculatum	Apr-29
Lampsilis cardium	Dec-4
Lampyridae	Feb-12
Lampyridae	Jul-18
Lance-tipped Darner	Sep-7
Lance-tipped Darner	Oct-4
Lanceleaf fogfruit	Jul-29
Laphria	May-28
Laphria	Jun-8
Laphria	Jun-8
Laphria	Jun-8
Laphria	Jun-24
Laphria	Jul-7
Laphria flavicollis	Jun-27
Laphria thoracica	Jun-8
Laphria thoracica	Jun-13
Large Crane Flies	Feb-11
Large Milkweed Bug	May-22
Large Yellow Underwing	Aug-15
largeflower bellwort	Apr-23
largeflower bellwort	May-4
Largeflower Rose Gentian	Jun-18
Largemouth Bass	Aug-3
Larger Pygmy Mole Grasshopper	Sep-7
Larger Pygmy Mole Grasshopper	Sep-7
Larix laricina	Jan-20
Larix laricina	May-31
Larra bicolor	Jun-20
Larrini	Jun-22
Lascoria ambigualis	May-29

Lasioglossum	May-9	Lestes inaequalis	Jun-5	
Lasioglossum	Jul-3	Lestes rectangularis	Jun-16	
Lasioglossum	Aug-23	Lestes rectangularis	Jun-24	
Lasioglossum pilosum	May-3	Lestes rectangularis	Aug-6	
Lasius	Feb-22	Lestes rectangularis	Sep-16	
Lasius	Apr-12	Lestes unguiculatus	Jul-25	
Latrodectus mactans	Jun-18	Lestes unguiculatus	Jul-27	
Latrodectus mactans	Jun-19	Lethe anthedon	Jul-23	
Lattice Orbweaver	Sep-4	Lethocerus	Oct-7	
Laurel Sphinx	Jul-4	Lethocerus americanus	May-14	
Leaf Blotch Miner Moths	Oct-16	Lettered Habrosyne Moth	Aug-13	
Leaf Blotch Miner Moths	Nov-23	Leucauge venusta	Aug-19	
Leaf-miner Flies	Apr-29	Leucorrhinia hudsonica	May-31	
Leafcutter, Mortar, and Resin Bees Jun-17		Leucorrhinia hudsonica	Jun-1	
		Leucorrhinia hudsonica	Jun-1	
Leafhoppers	Oct-2	Leucorrhinia intacta	Feb-2	
Leafhoppers	Oct-2	Leucorrhinia intacta	May-6	
Leafhoppers	Oct-2	Leucorrhinia intacta	May-15	
Leafhoppers	Nov-10	Leucorrhinia intacta	May-25	
Least Sandpiper	May-11	Leucorrhinia intacta	Jun-6	
Least Skipper	Aug-4	Leucorrhinia intacta	Nov-14	
LeConte's Sparrow	Apr-19	Leucospis affinis	Jun-16	
LeConte's Tiger Beetle	May-3	Liatris aspera	Aug-8	
Leeches	Oct-28	Liatris pycnostachya	Jul-27	
Lejops	May-6	Libellula	May-8	
Lejops bilinearis	May-8	Libellula auripennis	Jun-18	
Lejops lineatus	May-25	Libellula jesseana	Jun-18	
Lemon Cuckoo Bumble Bee	Jun-16	Libellula jesseana	Jun-18	
Lemon Cuckoo Bumble Bee	Aug-15	Libellula luctuosa	Jun-8	
Lemon Cuckoo Bumble Bee	Sep-17	Libellula luctuosa	Jun-26	
Leonurus cardiaca	Mar-24	Libellula luctuosa	Jul-9	
Lepidoptera	Nov-25	Libellula luctuosa	Aug-4	
Lepisma saccharina	Mar-17	Libellula pulchella	May-25	
Lepomis cyanellus	Aug-3	Libellula pulchella	Jun-6	
Lepomis macrochirus	Aug-3	Libellula pulchella	Jun-29	
Leptotes cassius	Jun-18	Libellula quadrimaculata	May-15	
Lepyronia quadrangularis	Jun-16	Libellula quadrimaculata	May-25	
Lespedeza capitata	Jan-7	Libellula quadrimaculata	Jun-1	
Lesser Bulb Fly	Aug-20	Libellula quadrimaculata	Jul-6	
lesser burdock	Jan-13	Libellulidae	Nov-15	
lesser burdock	Jan-14	Lichenophanes bicornis	Jun-8	
lesser periwinkle	Jan-31	Ligated Furrow Bee	Sep-14	
Lestes australis	May-22	Ligumia recta	Oct-11	
Lestes congener	Jul-16	lilac tasselflower	Jun-20	
Lestes congener	Aug-18	Limenitis archippus	Aug-22	
Lestes congener	Aug-29	Limenitis arthemis astyanax	Aug-1	
Lestes dryas	Jun-17	Limoniid Crane Flies	Apr-23	

Limoniid Crane Flies	Oct-16	Lysimachia nummularia		Jun-26
Limoniid Crane Flies	Nov-24	Lythrum salicaria	Jul-29	
Limoniid Crane Flies	Nov-30			
Limoniidae	Apr-23			

M

Limoniidae	Oct-16		
Limoniidae	Nov-24	Machimus Jun-27	
Limoniidae	Nov-30	Macromia illinoiensis Dec-27	
Linaria vulgaris	Sep-23	Macromiidae	May-31
Lithobates pipiens	May-13	Macromiidae	Dec-4
Lithobates pipiens	Jul-25	Macromiidae	Dec-6
Lithobates sylvaticus	Apr-1	Macrophya	Jun-14
Lithobates sylvaticus	Apr-4	Maevia inclemens	May-16
Lithospermum canescens	May-11	Maianthemum stellatum	May-25
Little Blue Dragonlet	Jun-18	Maize Calligrapher	Sep-1
Little Blue Dragonlet	Jun-18	Mallard	Jan-5
Little Blue Dragonlet	Jun-20	Mallard	Mar-19
Little Blue Dragonlet	Jun-20	Mallard	Oct-26
Little Blue Heron	Jun-21	Mallota bautias	May-28
little bluestem	Feb-17	Malus	May-9
little bluestem	Nov-5	Malus	May-10
Little Brown Bat	May-3	Malus	Oct-13
Little Virgin Tiger Moth	Jun-16	Mangora placida	Apr-9
Little White Lichen Moth	May-27	Mangora placida	Apr-14
liverworts Jul-15		Maple Twig Borer Moth	May-29
Long-bodied Cellar Spider	Feb-8	maples Mar-2	
Long-jawed Orbweavers	May-30	Marbled Orbweaver	Sep-22
Long-nosed Swamp Fly	May-25	Marbled Orbweaver	Oct-16
longbeak sedge	May-1	Marchantiales	Sep-11
Longlegged Sac Spiders	Feb-23	Marchantiophyta	Jul-15
Lonicera Feb-25		Marchantiopsida	Feb-12
Lophodytes cucullatus	Mar-19	Margined Calligrapher	May-6
Lophodytes cucullatus	May-19	Margined Calligrapher	May-6
Lophodytes cucullatus	Oct-25	Margined Calligrapher	Jun-11
Lucilia May-22		Margined Calligrapher	Jun-14
Lumbriculidae	Dec-27	Marsh Flies	Feb-21
Lycaena phlaeas	Oct-4	Marsh Flies	Nov-13
Lycogala epidendrum	Sep-21	Marsh Ground Beetles	Jul-29
Lycogala epidendrum	Nov-18	Marsh Ground Beetles	Sep-7
Lycosidae Mar-4		marsh marigold	May-4
Lycosidae Mar-27		marsh marigold	May-5
Lycosidae Mar-28		Masked Bees	Jun-11
Lycosidae May-3		Mason Bees	Jun-1
Lycosidae Nov-24		Mason Bees	Jun-10
Lyre-tipped Spreadwing	Jul-25	Matricaria discoidea Jun-4	
Lyre-tipped Spreadwing	Jul-27	Matteuccia struthiopteris	Mar-10
Lysimachia borealis Jun-1		mayapple May-4	
Lysimachia ciliata	Jul-21	Mayflies Apr-25	

Mayflies	Jun-2		
Meadow Sedgesitter	May-6		
Meadow Sedgesitter	May-6		
Meadow Sedgesitter	May-7		
Meadow Sedgesitter	May-12		
Meadow Sedgesitter	May-13		
Meadow Voles	Nov-2		
Meadowhawks	Aug-4		
Meal moth	Aug-26		
Mealy Pixie Cup	Nov-23		
Mecaphesa	Jul-8		
Mecaphesa	Jul-16		
Megachile	Jun-17		
Megachile latimanus	Jul-23		
Megachile latimanus	Aug-4		
Megachile pugnata pugnata		Jul-24	
Megaphorus	Jun-20		
Megarhyssa atrata	Sep-22		
Melandrena	Apr-14		
Melandrena	Apr-15		
Melandrena	Apr-17		
Melandrena	Apr-21		
Melandrena	Apr-22		
Melandrena	Apr-23		
Melandrena	May-2		
Melanerpes carolinus	Jan-18		
Melanophora roralis	Sep-19		
Melanoplus	Aug-8		
Melanoplus differentialis		Aug-18	
Melanoplus differentialis		Sep-6	
Melanostoma mellinum		May-6	
Melanostoma mellinum		May-7	
Melaphis rhois	Jul-11		
Melissa Blue	Jul-16		
Melissodes	Jul-28		
Melissodes bimaculatus		Jul-23	
Melissodes bimaculatus		Jul-29	
Melissodes trinodis	Aug-15		
Melittobia	Apr-6		
Meloe	Sep-23		
Melolonthini		May-27	
Melospiza georgiana	Apr-19		
Melospiza georgiana	Oct-25		
Melospiza melodia	Apr-4		
Melospiza melodia	Apr-5		
Melospiza melodia	May-19		
Menyanthes trifoliata		Jun-1	
Mertensia virginica	Apr-28		
Meshweavers	May-7		
Meshweavers	May-12		
Metallic Flea Beetles	Apr-5		
Metallic Sweat Bees	Apr-14		
Metallic Sweat Bees	Apr-15		
Metallic Sweat Bees	May-22		
Metallic Sweat Bees	Jun-4		
Metallic Sweat Bees	Jul-16		
Metallic Sweat Bees	Sep-1		
Metallic Sweat Bees	Sep-14		
Metcalfa pruinosa	Aug-24		
Micrathena mitrata	Oct-13		
Microbembex monodonta		Jun-21	
Microcrambus elegans		Jul-11	
Microdon	Jun-16		
Microlinyphia	Oct-4		
Micropterus salmoides		Aug-3	
Microtus	Nov-2		
Midland Clubtail	Jun-5		
Midland Clubtail	Jun-8		
Midland Clubtail	Jun-17		
Midland Clubtail	Jun-26		
Midland Clubtail	Jul-9		
Midwestern Salmonfly		Jan-2	
Midwestern Stone	Jan-2		
Milk-white Toothed Polypore		Jan-8	
Milkweed Assassin Bug		Jun-19	
Milkweed Leaf Beetle	May-26		
Milkweed Leaf Beetle	Jun-28		
Milkweed Leaf Beetle	Sep-15		
Milkweed Tussock Moth		Sep-3	
Mimosa strigillosa	Jun-19		
Mimulus ringens	Jul-29		
Mimus polyglottos	Jul-18		
Minnesota dwarf trout lily		May-4	
Minnows and Carps	Dec-4		
Minnows and Carps	Dec-4		
Minnows and Carps	Dec-4		
Miserable Mining Bee	Apr-23		
Mitella nuda	Jun-1		
Mites and Ticks	Dec-21		
Moehringia lateriflora		May-11	
Monarch	May-12		
Monarch	Jul-19		
Monarch	Aug-7		
Monarch	Aug-22		

Monarch Sep-2
Monarch Sep-16
Monarda punctata Aug-8
Money Spiders Mar-28
monument plant Jul-16
Moonseed Moth May-27
Morchella May-5
Mordellidae Jun-14
morels May-5
Mosaic Darners Feb-22
Mosaic Darners May-8
Mosaic Darners Jun-30
Mosquitoes Aug-10
mountain deathcamas Jun-15
mountain mint Feb-17
Mourning Cloak Mar-28
mud sedge Jun-1
musk thistle Jun-7
Muskrat Jan-9
Mustached Clubtail Dec-6
Mydas xanthopterus Jul-14
Myopa May-22
Myopa Jun-12
Myopa clausa May-11
Myopa clausa May-22
Myopinae Jun-11
Myotis lucifugus May-3
Myrica gale Jun-2
Myrmeleontidae Jun-18
Myrmeleontidae Jun-20
Myrmeleontidae Jun-22
Myrmica Oct-2
Myzinum Jun-20
Myzinum quinquecinctum Jul-22
Myzinum quinquecinctum Aug-8
Myzinum quinquecinctum Aug-25

N

Nabidae Apr-18
Nabis Feb-21
naked bishop's cap Jun-1
nannyberry Apr-1
nannyberry May-9
Naphrys pulex Apr-15
Naphrys pulex May-7
Narrow Lichen Case-bearer Apr-6
Narrow-banded Pond Fly Jun-1
Narrow-headed Sun Fly May-2
Narrow-headed Sun Fly May-2
Narrow-headed Sun Fly May-6
Narrow-headed Sun Fly May-9
Narrow-headed Sun Fly May-22
Narrow-headed Sun Fly Jun-2
Narrow-headed Sun Fly Sep-1
Narrow-headed Sun Fly Sep-7
Narrow-headed Sun Fly Sep-20
narrow-leaved cattail Nov-1
Narrow-leaved Hawk's-beard Jun-4
Narrow-winged Damselflies Jan-2
Narrow-winged Damselflies Mar-30
Narrow-winged Damselflies May-8
Narrow-winged Damselflies May-8
Narrow-winged Damselflies Oct-28
Narrow-winged Tree Cricket Sep-11
Nathalis iole Jun-21
Necrophila americana Sep-7
Nehalennia irene May-8
Nehalennia irene May-13
Nehalennia irene May-25
Neighborly Mining Bee May-3
Neighborly Mining Bee May-25
Nemotelus kansensis Jun-30
Neolasioptera convolvuli Jan-27
Neon Mar-29
Neophasia menapia Jul-16
Neotibicen canicularis Aug-16
Neotridactylus apicialis Sep-7
Neotridactylus apicialis Sep-7
Neovison vison Dec-23
Neoxabea bipunctata Aug-7
Neoxabea bipunctata Aug-12
Nerodia sipedon Jun-7
Net-spinning Caddisflies Jan-2
netleaf hackberry Jul-13
New England aster Sep-13
Nitidotachinus Mar-28
Noctua pronuba Aug-15
Noctuidae Aug-23
Nodding Beggarticks Sep-5
nodding trillium May-31
Nomad Bees Apr-15
Nomad Bees Apr-16
Nomad Bees Apr-17

Nomad Bees	Apr-17		Northern Water Snake	Jun-7	
Nomad Bees	Apr-22		Nut and Acorn Weevils	Jun-13	
Nomad Bees	Apr-22		Nut and Acorn Weevils	Jul-9	
Nomad Bees	May-3		Nymphalis antiopa	Mar-28	
Nomad Bees	May-4		Nysson lateralis	Jun-13	
Nomad Bees	May-17				
Nomad Bees	Jun-1		**O**		
Nomad Bees	Jun-12				
Nomad Bees	Sep-1		oaks	Jun-18	
Nomada	Apr-15		Oarisma garita	Jul-16	
Nomada	Apr-16		Oblique Stripetail	Jun-14	
Nomada	Apr-17		Oblique Stripetail	Jul-30	
Nomada	Apr-17		Ocyptamus fascipennis	Aug-31	
Nomada	Apr-22		Odocoileus virginianus	Jan-7	
Nomada	Apr-22		Odocoileus virginianus	Jan-17	
Nomada	May-3		Odocoileus virginianus	Jul-9	
Nomada	May-4		Oecanthus forbesi	Oct-9	
Nomada	May-17		Oecanthus niveus	Sep-11	
Nomada	Jun-1		Oedipodinae	Mar-4	
Nomada	Jun-12		Oenothera biennis	Jan-20	
Nomada	Sep-1		Oenothera biennis	Dec-3	
Nomada articulata	Jun-11		Oenothera cespitosa	Jul-15	
Nomada luteoloides	May-2		Oenothera gaura	Oct-4	
North American Racer	Jun-19		Oil Beetles	Sep-23	
Northern Amber Bumble Bee	Jun-6		Olethreutinae	Mar-19	
Northern Bluet	May-31		Olethreutinae	Apr-9	
Northern Cardinal	Mar-21		Olethreutine Leafroller Moths	Mar-19	
Northern Cardinal	Nov-13		Olethreutine Leafroller Moths	Apr-9	
Northern Cranesbill	Jun-2		Oligochaeta	Apr-4	
Northern Crescent	May-26		Oncopeltus fasciatus	May-22	
Northern Dog-day Cicada	Aug-16		Ondatra zibethicus	Jan-9	
Northern Flicker	Mar-26		Onoclea sensibilis	Mar-4	
Northern Leopard Frog	May-13		Onthophagus	May-15	
Northern Leopard Frog	Jul-25		Openfield Orbweaver	May-3	
Northern Mockingbird	Jul-18		Ophiogomphus	Dec-5	
Northern Pearly-eye	Jul-23		Ophiogomphus	Dec-6	
northern red oak	Feb-5		Ophiogomphus rupinsulensis	Jun-5	
northern red oak	Mar-4		Ophiogomphus rupinsulensis	Jun-26	
northern red oak	Oct-31		Ophion	Apr-9	
northern red oak	Dec-2		Ophion Wasps	Apr-9	
northern red oak	Dec-10		Ophioninae	Apr-9	
Northern Short-tailed Shrew	Oct-28		Opiliones	Oct-12	
Northern Shoveler	Mar-21		Opuntia	Jun-20	
Northern Shoveler	Oct-25		Orange Mint Moth	May-26	
Northern Spring Amphipod	Jan-2		Orange Skipperling	Jul-15	
Northern Spring Amphipod	Nov-30		Orange Sulphur	Sep-20	
Northern Water Plantain	Feb-5		Orange-legged Drone Fly	Jul-16	

Orange-legged Drone Fly	Aug-12	Painted Lady	Aug-22
Orange-legged Drone Fly	Sep-14	Painted Lady	Sep-6
Orange-spotted Drone Fly	May-2	Painted Lady	Sep-7
Orchard Orbweaver	Aug-19	Painted Turtle	Apr-4
Orchelimum nigripes	Aug-27	Painted Turtle	May-26
Orchelimum nigripes	Aug-30	Pale Beauty	Jun-6
Organ Pipe Mud Dauber	Aug-18	Pale Green Assassin Bug	Oct-21
Ornate Snipe Fly	Jun-14	Paleacrita merriccata	Apr-2
Orthocladiinae	Feb-27	Paleacrita vernata	Mar-7
Orthonama obstipata	Apr-23	Paleacrita vernata	Apr-1
Orthoneura nitida	Oct-18	Palpita	May-29
Orthoneura pulchella	Jun-12	Palthis angulalis	May-29
Orthoptera	Jul-14	Pandora Sphinx	Sep-2
Orthosia hibisci	Apr-1	panicgrasses	Nov-5
Orthosia hibisci	Apr-9	panicled aster	Aug-18
Orussus minutus	May-28	Panicum	Nov-5
Orussus minutus	Jun-8	Panorpa	Aug-22
Orussus terminalis	Jun-8	Pantala hymenaea	Jul-21
Osmia	Jun-1	paper birch	Mar-4
Osmia	Jun-10	paper birch	Nov-12
Osmorhiza claytonii	May-22	Papilio cresphontes	Aug-19
Osmorhiza longistylis	May-28	Papilio cresphontes	Aug-22
Osteopilus septentrionalis	Jun-22	Papilio glaucus	Aug-15
ostrich fern	Mar-10	Papilio glaucus	Aug-22
Ostrich-plume Moss	Mar-25	Papilio polyxenes	Jun-6
Ostrya	Nov-23	Papilio troilus	Jun-21
Othocallis siberica	Apr-4	Paraguayan Purslane	Jun-18
Otospermophilus variegatus	Jul-13	Paragus	May-22
Oxybelus	Jun-9	Paragus	Sep-8
Oxybelus	Jun-16	Paragus haemorrhous	Jun-14
Oxybelus	Jul-6	Paranthidium jugatorium	Jul-30
Oxybelus	Jul-9	Paranthidium jugatorium	Jul-31
Oxybelus	Jul-27	Parasteatoda	Apr-29
Oxybelus	Sep-2	Parasteatoda	Aug-24
Oxybelus	Sep-7	Parcoblatta virginica	Mar-28
Oxytelinae	Mar-6	Pardosa	Feb-21
		Parhelophilus	Jun-2

P

		Parhelophilus	Jun-27
		partridge pea	Aug-8
Pachydiplax longipennis	Jun-18	Passaloecus	Jun-23
Pachydiplax longipennis	Jun-24	Passaloecus	Dec-17
Pachydiplax longipennis	Jul-7	Passaloecus	Dec-21
Pachydiplax longipennis	Jul-19	Passaloecus	Dec-26
Pachypsylla celtidismamma	Oct-2	Passer domesticus	Jan-8
Pachypsylla celtidismamma	Nov-29	Passer domesticus	Jan-29
Packera paupercula	May-28	Passer domesticus	Feb-17
Painted Lady	Aug-7	Passer domesticus	Feb-20

Pavement Ants	Oct-12		Philodromidae	Feb-24
Pea Clams Feb-4			Philodromus	Mar-22
Pearl Crescent	Aug-27		Philodromus	Apr-18
Peck's Skipper	Jul-27		Philodromus	Jun-15
Peck's Skipper	Jul-30		Phlaeothripidae	Feb-13
Pedicularis canadensis		May-4	Phlox divaricata	Apr-25
Pelecinus polyturator	Aug-16		Phlox divaricata	May-4
Pelegrina proterva	May-12		Phoenix Jun-20	
Pelegrina proterva	May-22		Pholcidae Jun-22	
Pelegrina proterva	May-22		Pholcidae Nov-4	
Pellenes Apr-7			Pholcus phalangioides	Feb-8
Pennsylvania sedge	Apr-14		Phoradendron juniperinum	Jul-14
Penstemon rostriflorus		Jul-15	Phoridae Dec-22	
Peponapis pruinosa	Jul-22		Phormia regina	Feb-22
Pepper and Salt Skipper		Jun-2	Phormia regina	Apr-21
Percina maculata	Jan-2		Phragmites	Nov-22
Perdita halictoides	Jul-1		Phryma leptostachya	Jun-27
Pergid Sawflies	May-3		Phyciodes cocyta	May-26
Pergidae May-3			Phyciodes tharos	Aug-27
Periplaneta australasiae		Jun-22	Phyla lanceolata	Jul-29
Perithemis tenera	Jun-18		Phymata Jul-28	
Perithemis tenera	Jun-22		Phymata Aug-8	
Perithemis tenera	Jul-29		Phymata Sep-20	
Perithemis tenera	Aug-15		Phymata Oct-18	
Peromyscus leucopus	Nov-11		Physa acuta	Feb-7
Persicaria Aug-10			Physa acuta	Feb-7
Petrobiinae	Sep-21		Physidae Nov-30	
Phalaenophana pyramusalis		Aug-5	Physocephala tibialis	Aug-15
Phalaris arundinacea	Feb-5		Picea Jan-15	
Phaneta Jun-1			Picea mariana	Jun-1
Phanogomphus lividus		Jun-2	Picnic Beetle	Oct-16
Phantom Crane Fly	May-8		Picoides pubescens	Feb-26
Phantom Crane Fly	May-22		Picoides pubescens	Mar-18
Phantom Crane Fly	Sep-15		Pieris rapae	Jul-23
Phaonia sp. Feb-11			Pileated Woodpecker	Jun-9
Phellinus tremulae	Mar-5		Pine Siskin Oct-15	
Pherbellia parallela	Apr-21		Pine White Jul-16	
Pheucticus ludovicianus	May-18		Pineapple-weed	Jun-4
Pheucticus melanocephalus	Jul-15		Pinus ponderosa	Jul-17
Phidippus tyrrellii	Jul-16		Pinus strobus	Mar-16
Phigalia denticulata	Apr-9		Pipunculidae	Jun-14
Phigalia strigataria	Apr-1		pixie cup lichens	Mar-4
Philadelphia fleabane	May-25		pixie cup lichens	Mar-4
Philanthus gibbosus	Jun-29, Jun-30, Jul-2, Aug-25, Sep-2		pixie cup lichens	Mar-26
			pixie cup lichens	Dec-4
Philanthus sanbornii	Jul-6		Plagiomnium	May-4
Philanthus ventilabris	Jun-18		Plain Pocketbook	Dec-4

plains coreopsis	Aug-22	Podophyllum peltatum	May-4
Plains Snakecotton	Aug-8	Poecile atricapillus	Jan-1
Planorbidae	Jan-11	Poecile atricapillus	Sep-28
Planorbidae	Feb-4	Poecile atricapillus	Nov-30
Planorbidae	Nov-22	Pogonomyrmex badius	Jun-20
Planorbidae	Nov-24	Polished Lady Beetle	Jun-24
Plantae	Aug-4	Polistes	Feb-16
Plantae	Aug-8	Polistes bellicosus	Jun-18
Plantago lanceolata	Mar-21	Polistes dominula	Jun-30
Plantago lanceolata	Jul-2	Polistes fuscatus	Apr-1
Plantago rugelii	Nov-5	Polistes fuscatus	Apr-14
Plants	Aug-4	Polistes fuscatus	Apr-21
Plants	Aug-8	Polistes fuscatus	Jun-22
Plathemis lydia	Apr-24	Polistes fuscatus	Oct-3
Plathemis lydia	May-16	Polistes fuscatus	Nov-27
Plathemis lydia	May-20	Polites peckius	Jul-27
Plathemis lydia	May-23	Polites peckius	Jul-30
Plathemis lydia	Jun-6	Polites themistocles	Jun-15
Plathemis lydia	Jun-9	Pollenia	Feb-22
Plathemis lydia	Jun-10	Polydrusus formosus	May-24
Platycheirus quadratus	May-6	Polygonia comma	Apr-1
Platycheirus quadratus	May-6	Polygonia comma	Apr-21
Platycheirus quadratus	May-7	Polygonia comma	Jul-4
Platycheirus quadratus	May-12	Polygonia comma	Jul-21
Platycheirus quadratus	May-13	Polygonia comma	Sep-7
Platydracus	Jun-14	Polygonia interrogationis	Feb-18
Platydracus	Aug-27	Polypodiopsida	Jun-21
Platytipula	Sep-15	Polyporales	Jan-9
Plebejus melissa	Jul-16	Polytrichum juniperinum	Mar-4
Plecoptera s	Dec-4	Polytrichum juniperinum	Dec-4
Plecoptera s	Dec-5	Pompilidae	Jun-26
Pleuroceridae	Jan-11	Pompilidae	Jul-27
Pleurocerids	Jan-11	Pompilidae	Jul-27
Pleuroprucha insulsaria	Oct-7	Pompilidae	Sep-6
Pluchea baccharis	Jun-18	Pompilidae	Sep-7
Plume Moths	Jun-1	Pompilidae	Sep-22
plums, cherries, and allies	Jun-22	ponderosa pine	Jul-17
Plusiodonta compressipalpis	May-27	Pondweed	Oct-11
Plutella xylostella	Apr-28	Pontia protodice	Jun-18
Poanes hobomok	May-22	Populus alba	Nov-26
Poanes hobomok	May-24	Populus balsamifera	Jun-2
Poanes hobomok	Jun-6	Populus deltoides	Jan-8
Poanes taxiles	Jul-13	Populus deltoides	Feb-13
Poanes taxiles	Jul-16	Populus deltoides	Sep-19
Poanes taxiles	Jul-16	Populus grandidentata	Nov-23
Podalonia	Jul-16	Populus tremuloides	Mar-8
Podalonia	Jul-16	Portulaca amilis	Jun-18

Portulaca oleracea	Jun-18	Punctured Tiger Beetle	Sep-2
Potamogeton	Oct-11	Pure Green Augochlora	Jul-9
Potentilla	Jun-26	Pure Green Augochlora	Sep-14
Potter Wasps	Jun-18	Pure Green Augochlora	Oct-19
Potter Wasps	Jun-18	purple loosestrife	Jul-29
Powder Puff	Jun-19	purple pitcher plant	May-31
Powdered Geometer Moths	Jun-1	purple prairie clover	Jun-27
prairie alumroot	Jun-15	Purple Skimmer	Jun-18
prairie blazing star	Jul-27	Purple Skimmer	Jun-18
Prairie Buttercup	Apr-14	Purple Tiger Beetle	Sep-2
prairie clovers	Dec-1	pussy willow	Mar-3
prairie cordgrass	Nov-3	pussy willow	Mar-3
prairie smoke	Apr-7	pussy willow	Mar-4
Predaceous Diving Beetles	Mar-28	Pycnanthemum	Feb-17
Predaceous Diving Beetles	Mar-30	Pycnanthemum virginianum	Aug-11
Prenolepis imparis	Apr-18	Pyganodon grandis	Feb-15
prickly ash	Apr-4	Pygmy Grasshoppers	Jun-1
prickly ash	May-2	Pygmy Grasshoppers	Sep-7
Prickly pears	Jun-20	Pyralis farinalis	Aug-26
prickly tree-clubmoss	Dec-4	Pyrausta orphisalis	May-26
Primula pauciflora	Jul-15	Pyrginae	Jun-21
prince's plume	Jul-15	Pyrrharctia isabella	May-27
Priocnemis minorata	May-4		
Priocnemis minorata	May-12		

Q

Prionyx parkeri	Jun-18		
Proctacanthus	Jun-20	Queen Anne's lace	Oct-12
Procyon lotor	Feb-14	Queen Ant Kidnapper	Jul-1
Procyon lotor	Apr-15	Queen Ant Kidnapper	Jul-4
Progomphus alachuensis	Jun-20	Queen Ant Kidnapper	Jul-6
Promachus vertebratus	Oct-4	Quercus	Jun-18
Proteoteras aesculana	May-29	Quercus alba	Nov-30
Pruinose Squash Bee	Jul-22	Quercus bicolor	Jan-19
Prunus	Jun-22	Quercus gambelii	Jul-14
Prunus americana	Jan-24	Quercus gambelii	Jul-15
Prunus serotina	Mar-2	Quercus macrocarpa	Jan-24
Prunus virginiana	May-12	Quercus macrocarpa	Dec-30
Pseudacris maculata	Mar-31	Quercus rubra	Feb-5
Pseudomyrmex gracilis	Jun-22	Quercus rubra	Mar-4
Pseudopanurgus	Sep-14	Quercus rubra	Oct-31
Ptecticus trivittatus	Jul-20	Quercus rubra	Dec-2
Pteromalidae	Sep-24	Quercus rubra	Dec-10
Pteronarcys pictetii	Jan-2	Quercus turbinella	Jul-14
Pterophoridae	Jun-1	Quercus virginiana	Jun-18
Pterostichini	Mar-1	Question Mark	Feb-18
Ptilium crista-castrensis	Mar-25	Quick Gloss Snail	Sep-19
Ptinidae	Dec-24	Quiscalus mexicanus	Jul-18
Punctured Tiger Beetle	Jul-27		

R

Rabid Wolf Spider	Jun-22	Red-winged Blackbird	Feb-19
Rabidosa rabida	Jun-22	Red-winged Blackbird	Mar-24
Ramalina	Nov-18	Redbelly Snake	Oct-13
Ramalina	Nov-23	Reddish Speckled Dart	Apr-9
Ramalina	Nov-23	reed canary grass	Feb-5
Rambur's Forktail	Jun-18	Regulus calendula	Apr-19
Rambur's Forktail	Jun-18	Regulus calendula	Sep-6
Rambur's Forktail	Jun-20	Reptiles	Mar-27
Ramshorn snails	Jan-11	Reptilia	Mar-27
Ramshorn snails	Feb-4	Resin Bees	Jun-29
Ramshorn snails	Nov-22	Resin Bees	Jul-10
Ramshorn snails	Nov-24	Rhamnus cathartica	Feb-28
Ranunculus abortivus	Apr-23	Rhamphomyia	Apr-28
Ranunculus abortivus	Apr-25	Rhaphidophoridae	May-3
Ranunculus abortivus	May-4	Rhaphidophoridae	Sep-29
Ranunculus abortivus	May-5	Rhingia nasica	May-7
Ranunculus abortivus	May-19	Rhingia nasica	May-7
Ranunculus aquatilis	May-26	Rhingia nasica	May-15
Ranunculus hispidus	Apr-25	Rhinichthys obtusus	Dec-4
Ranunculus hispidus	May-4	Rhionaeschna multicolor	Jun-9
Ranunculus rhomboideus	Apr-14	Rhionaeschna multicolor	Jun-30
Ratibida pinnata	Jul-8	Rhionaeschna multicolor	Jul-11
rattlesnake master	Feb-17	Rhododendron	Jan-31
rattlesnake master	Jul-25	rhododendrons and azaleas	Jan-31
Red Admiral	Apr-14	Rhus glabra	Mar-25
Red Admiral	Sep-20	ribwort plantain	Mar-21
red baneberry	Jun-1	ribwort plantain	Jul-2
red columbine	May-11	Richardia brasiliensis	Jun-18
Red Flat Bark Beetle	Nov-18	Ring-necked Duck	Mar-19
Red Milkweed Beetle	Jun-17	riverbank grape	Dec-23
red osier dogwood	Jan-3	Robinia pseudoacacia	Aug-2
red osier dogwood	Jan-31	Rock Squirrel	Jul-13
red osier dogwood	Mar-4	Rodent and Lagomorph Bot Flies	Jul-13
red osier dogwood	Dec-31	Rose-breasted Grosbeak	May-18
Red Saddlebags	Jun-18	Rosy Camphorweed	Jun-18
Red Saddlebags	Jun-24	rough blazing star	Aug-8
Red Saddlebags	Sep-2	rough horsetail	Mar-26
Red-banded Leafroller Moth	Apr-26	rough horsetail	Oct-11
Red-bellied Woodpecker	Jan-18	round-headed bush clover	Jan-7
red-berried elder	Feb-11	Round-headed Katydids	Aug-22
Red-necked False Blister Beetle	Apr-14	round-leaved sundew	May-31
Red-spotted Purple	Aug-1	round-lobed hepatica	Nov-23
Red-tailed Hawk	Jan-7	Rubus	Jun-1
Red-veined Pennant	Jun-18	Rubus occidentalis	Jan-16
Red-veined Pennant	Jun-20	Rubus occidentalis	May-22
		Ruby Meadowhawk	Jul-27
		Ruby Meadowhawk	Aug-6

Ruby-crowned Kinglet	Apr-19
Ruby-crowned Kinglet	Sep-6
Rudbeckia hirta	Aug-8
Running Crab Spiders	Feb-24
russet alder leaf beetle	Jun-1
Rusty Snaketail	Jun-5
Rusty Snaketail	Jun-26
Rusty-patched Bumble Bee	Sep-12
Rusty-patched Bumble Bee	Sep-13
Rusty-patched Bumble Bee	Sep-14
Rusty-patched Bumble Bee	Sep-15
Rusty-patched Bumble Bee	Sep-16
Rusty-patched Bumble Bee	Sep-17

S

Sabatia grandiflora	Jun-18
Sachem	Jun-17
Sachem	Aug-22
Sachem	Sep-7
Saffron-winged Meadowhawk	Oct-4
Saffron-winged Meadowhawk	Oct-4
Saffron-winged Meadowhawk	Oct-4
Saffron-winged Meadowhawk	Oct-9
Saffron-winged Meadowhawk	Oct-9
Sagittaria	Jun-18
Sagittaria latifolia	Aug-27
Salix ×fragilis	May-25
Salix discolor	Mar-3
Salix discolor	Mar-3
Salix discolor	Mar-4
Salix interior	Mar-3
Salticidae	May-12
Salticidae	May-30
Sambucus racemosa	Feb-11
Sand Bittercress	Jun-1
Sandhill Crane	Feb-21
Sandhill Crane	Mar-24
Sandhill Crane	Jun-18
Sandhill Crane	Jun-19
Sandhill Crane	Jun-20
Sanguinaria canadensis	Apr-10
Sanguinaria canadensis	Apr-14
Sanguinaria canadensis	Apr-30
Sanguinaria canadensis	May-4
Sanicula odorata	May-25
Sarracenia purpurea	May-31
Satyrium edwardsii	Jun-27
Satyrium saepium	Jul-16
Satyrium saepium	Jul-17
Satyrium sylvinus	Jul-16
Satyrium titus	Jul-8
saw palmetto	Jun-18
Sawflies, Horntails, and Wood Wasps	May-22
Sawflies, Horntails, and Wood Wasps	May-26
Sayornis phoebe	Apr-4
Scaldweed	Aug-25
Scaptomyza	Apr-18
Scathophaga stercoraria	Apr-1
Scathophaga stercoraria	May-15
Sceliphron caementarium	Jun-16
Sceliphron caementarium	Jul-21
Sceloporus woodi	Jun-20
Schistocerca lineata	Aug-8
Schizachyrium scoparium	Feb-17
Schizachyrium scoparium	Nov-5
Schizomyia eupatoriflorae	Sep-2
Schizomyia racemicola	Sep-1
Schizophora	Feb-12
Schoenoplectus	Nov-22
Schoenoplectus acutus	Nov-22
Sciaridae	Apr-19
Sciomyzidae	Feb-21
Sciomyzidae	Nov-13
Sciurus carolinensis	Oct-15
Sciurus carolinensis	Dec-12
Scleroderma	Dec-4
Scolia bicincta	Sep-20
Scolia nobilitata	Jun-18
Scutellaria lateriflora	Jul-29
Seaside Grasshopper	Jun-22
Sedge Sprite	May-8
Sedge Sprite	May-13
Sedge Sprite	May-25
Semotilus atromaculatus	Sep-27
sensitive fern	Mar-4
Sepsis	Feb-21

Serenoa repens	Jun-18		Sleepy Duskywing	May-31
Sericomyia militaris	Jun-1		Sleepy Duskywing	Jun-1
sessile bellwort	May-31		Slender Spreadwing	Jun-16
Setaria	Jan-17		Slender Spreadwing	Jun-24
Setaria	Nov-5		Slender Spreadwing	Aug-6
Setophaga coronata		Apr-19	Slender Spreadwing	Sep-16
Setophaga coronata		Apr-26	Slow Mountain Grasshopper	Jul-17
Shadow Darner	Sep-1		Small Honey Ant	Apr-18
Shadow Darner	Sep-8		Small Phigalia Moth	Apr-1
Shadow Darner	Oct-9		small white leek	May-4
Shadow Darner	Oct-9		Small Wood-Nymph	Jul-16
shady horsetail	May-4		Small Wood-Nymph	Jul-16
Shamrock Orbweaver		Sep-3	small-flowered buttercup	Apr-23
Shanks, Tattlers, and Allies		Apr-25	small-flowered buttercup	Apr-25
sharp-lobed hepatica		Mar-28	small-flowered buttercup	May-4
sharp-lobed hepatica		Apr-5	small-flowered buttercup	May-5
sharp-lobed hepatica		Apr-14	small-flowered buttercup	May-19
sharp-lobed hepatica		May-4	Sminthurinus henshawi	Apr-15
shelf fungi	Jan-9		Smokey-winged Beetle Bandit	Jun-30
shepherd's-purse	May-9		Smokey-winged Beetle Bandit	Jun-30
Shield-handed Square-headed Wasps Sep-7			Smokey-winged Beetle Bandit	Jul-1
Shore Flies	Dec-19		Smokey-winged Beetle Bandit	Jul-2
Short-tailed Ichneumonid Wasps Apr-9			Smokey-winged Beetle Bandit	Jul-3
Short-winged Flower Beetles		Apr-14	Smokey-winged Beetle Bandit	Jul-5
showy goldenrod	Dec-1		Smokey-winged Beetle Bandit	Jul-5
showy tick-trefoil	Jul-9		Smokey-winged Beetle Bandit	Jul-6
Sialia sialis	Apr-4		Smokey-winged Beetle Bandit	Jul-7
Siberian elm	Mar-2		Smokey-winged Beetle Bandit	Jul-11
Siberian squill	Apr-4		Smokey-winged Beetle Bandit	Jul-19
side-flowering skullcap		Jul-29	Smokey-winged Beetle Bandit	Jul-23
Silene vulgaris	Jul-23		smooth sumac	Mar-25
Silphium laciniatum		Mar-25	Snaketails	Dec-5
Silphium laciniatum		Jul-9	Snaketails	Dec-6
Silphium perfoliatum		Jul-21	snakewort	May-4
silver maple	Oct-22		Snowberry Clearwing	Jun-1
Silver-bordered Fritillary		Jun-1		
Silver-bordered Fritillary		Jun-2		
Simuliidae	Feb-7			
Sisyrinchium	Dec-25			
Sitta carolinensis	Feb-14			
Six-spotted Fishing Spider		May-26		
Six-spotted Tiger Beetle		May-22		
Skimmers	Nov-15			
Skippers	Jul-16			
Slate Drakes	Dec-4			

Solanum nigrum	Sep-21	Sphecodes	May-22
Solidago	Jan-16	Sphecodes	Jun-11
Solidago	Jul-14	Sphecodes	Jul-9
Solidago	Sep-21	Sphecodes	Sep-20
Solidago canadensis	Aug-4	Sphex dorsalis	Jun-18
Solidago flexicaulis	Jan-17	Sphex ichneumoneus	Jul-3
Solidago flexicaulis	Sep-4	Sphex ichneumoneus	Aug-11
Solidago flexicaulis	Dec-20	Sphex ichneumoneus	Aug-25
Solidago rigida	Aug-11	Sphex ichneumoneus	Aug-27
Solidago speciosa	Dec-1	Sphex ichneumoneus	Sep-13
Somatochlora franklini	May-31	Sphex pensylvanicus	Aug-11
Somula decora	Jun-24	Sphex pensylvanicus	Aug-14
Song Sparrow	Apr-4	Sphinx kalmiae	Jul-4
Song Sparrow	Apr-5	Spicebush Swallowtail	Jun-21
Song Sparrow	May-19	Spider Wasps	Jun-26
Sonoran scrub oak	Jul-14	Spider Wasps	Jul-27
Sooty Dancer	Jul-13	Spider Wasps	Jul-27
Sooty Dancer	Jul-15	Spider Wasps	Sep-6
Sorghastrum nutans	Jan-5	Spider Wasps	Sep-7
Sorghastrum nutans	Feb-17	Spider Wasps	Sep-22
Southern Black Widow	Jun-18	Spiders	Mar-27
Southern Black Widow	Jun-19	Spiders	Mar-28
southern live oak	Jun-18	Spiders	May-30
Southern Spreadwing	May-22	Spiders	Jun-2
Southern Toad	Jun-22	Spiders	Nov-24
Spanish moss	Jun-18	Spiders	Dec-1
Sparganium	Aug-27	Spiketails	Dec-5
Sparganium eurycarpum	Feb-5	Spilomyia longicornis	Sep-14
Spatula clypeata	Mar-21	Spilomyia sayi	Sep-1
Spatula clypeata	Oct-25	Spinus pinus	Oct-15
Speckled Green Fruitworm Moth Apr-1		Spinus tristis	Aug-15
		Spinybacked Orbweaver	Jun-18
Speckled Green Fruitworm Moth Apr-9		Spittlebugs	May-28
		Spiza americana	Jun-15
Spectralia gracilipes	Jul-3	Spizella passerina	Apr-26
Spectralia gracilipes	Jul-6	Sporobolus michauxianus	Nov-3
Sphaeriidae	Feb-4	Spot-winged Glider	Jul-21
Sphaerophoria	Sep-11	Spotted Bird Grasshopper	Aug-8
Sphaerophoria	Sep-15	Spotted Cucumber Beetle	Oct-11
Sphaerophoria contigua	Jul-8	Spotted deadnettle	Apr-29
Sphaerophoria philanthus	May-25	spotted horse mint	Aug-8
Sphaerophoria philanthus	Sep-7	spotted knapweed	Aug-11
Sphecius speciosus	Jul-19	Spotted Lady Beetle	Mar-20
Sphecius speciosus	Jul-23	Spotted Lady Beetle	May-5
Sphecius speciosus	Jul-27	Spotted Lady Beetle	May-26
Sphecius speciosus	Jul-31	Spotted Spreadwing	Jul-16
Sphecodes	May-3	Spotted Spreadwing	Aug-18

Spotted Spreadwing	Aug-29	Strymon melinus	Jun-20
Spotted Wood Fly	Jun-24	Sturnus vulgaris	Mar-18
Spragueia onagrus	Jun-18	Stylogomphus albistylus	Jun-1
Spread-wing Skippers	Jun-21	Stylurus Dec-5	
Spring Azure	Jun-1	Stylurus notatus	Apr-25
Spring Azure	Jun-11	Stylurus notatus	Aug-25
Spring Beauty Mining Bee	May-4	Succineidae	Mar-29
Spring Cankerworm Moth	Mar-7	Succineidae	May-5
Spring Cankerworm Moth	Apr-1	sugar maple	Mar-3
Springtime Darner	Jun-9	Sumac Gall Aphid	Jul-11
Springwater Dancer	Jul-15	Sumac Leaf Blotch Miner Moth Oct-6	
spruces Jan-15			
Spurred Ceratina	May-22	Sumitrosis inaequalis	May-16
Spurthroat Grasshoppers	Aug-8	Sun Beetles	Apr-9
Squirrel Tree Frog	Jun-22	Sun Beetles	Nov-8
Stanleya pinnata	Jul-15	Sunburst Lichens	Jan-12
star-flowered lily-of-the-valley May-25		Sunflower Mining Bee	Aug-29
		Sunflower Mining Bee	Sep-1
starflower	Jun-1	swamp laurel	May-31
Steatoda triangulosa	Nov-6	swamp milkweed	Jul-29
Stellaria aquatica	May-26	swamp milkweed	Nov-2
Steller's rock-brake	May-4	swamp milkweed	Nov-22
Stenolophus comma	Apr-9	Swamp Sparrow	Apr-19
Stenoscelis brevis	Jun-8	Swamp Sparrow	Oct-25
Stichopogon trifasciatus	Jun-20	swamp white oak	Jan-19
stickseed Dec-18		sweetflag Jun-2	
Stictia carolina	Aug-8	Swift Feather-legged Fly	May-30
stiff-leaved goldenrod	Aug-11	Swift Feather-legged Fly	Jul-23
Stink Bug Hunter	Jul-5	Swift River Cruiser	Dec-27
Stink Bug Hunter	Jul-19	Sylvan Hairstreak	Jul-16
Stink Bug Hunter	Jul-27	Sylvicola Apr-1	
Stink Bug Hunter	Jul-28	Sylvilagus floridanus	May-18
Stink Bug Hunter	Aug-18	Symmorphus	Jul-4
Stink Bug Hunter	Sep-1	Symmorphus canadensis	Jul-10
Stoneflies	Dec-4	Sympetrum	Aug-4
Stoneflies	Dec-5	Sympetrum corruptum	Sep-8
Storeria dekayi	Sep-22	Sympetrum corruptum	Sep-8
Storeria occipitomaculata	Oct-13	Sympetrum costiferum	Oct-4
Stratiomys obesa	Jun-16	Sympetrum costiferum	Oct-4
Stream Bluet	Jun-15	Sympetrum costiferum	Oct-4
Stream Bluet	Jun-26	Sympetrum costiferum	Oct-9
Stream Cruiser	Dec-2	Sympetrum costiferum	Oct-9
Striped Fishing Spider	Sep-5	*Sympetrum obtrusum* Jun-24, Jul-7, Jul-19, Jul-27, Jul-28, Aug-4, Aug-6, Aug-11, Aug-22, Aug-22, Aug-25, Aug-27, Sep-7, Sep-8, Sep-9, Sep-18, Sep-19, Oct-3, Oct-4, Oct-9, Oct-11	
Striped Sweat Bees	Aug-7		
Striped Sweat Bees	Aug-18		
Striped Sweat Bees	Sep-17		
Striped Whitelip	Mar-28		

Sympetrum rubicundulum   Jul-27, Aug-6
Sympetrum semicinctum   Jul-19
Sympetrum vicinum   Jul-23, Jul-27,
  Aug-7, Aug-18, Aug-22, Aug-25, Sep-1,
  Sep-7, Sep-8, Sep-9, Sep-17, Sep-19,
  Sep-23, Sep-29, Oct-3, Oct-4, Oct-9,
  Oct-10, Oct-16, Oct-18
Symphyotrichum drummondii   Sep-13,
  Dec-8
Symphyotrichum lanceolatum   Aug-18
Symphyotrichum novae-angliae   Sep-13
Symphyotrichum oblongifolium   Sep-29
Symphyta May-22
Symphyta May-26
Synhalonia   Jun-1
Synhalonia   Jun-2
Syritta pipiens   Jul-20
Syritta pipiens   Sep-7
Syritta pipiens   Sep-9
Syritta pipiens   Sep-20
Syritta pipiens   Oct-18
Syrphidae   Sep-9
Syrphini   Jun-1
Syrphini   Jun-22
Syrphus   May-2
Syrphus   Sep-7
Syrphus   Sep-8
Syrphus   Sep-14
Syrphus   Sep-16
Syrphus   Sep-17
Syrphus   Sep-20
Syrphus   Oct-4
Syrphus   Oct-16
Syrphus   Oct-18
Syrphus   Oct-20

T

Tachycineta bicolor   Apr-4
Tachytes crassus   Jul-9
Taiga Bluet   Jun-1
tall bellflower   Jul-21
tall bellflower   Oct-26
tamarack Jan-20
tamarack May-31
Tamias striatus   Jun-23
Tamias umbrinus   Jul-13

Taraxacum officinale Apr-11
Taraxacum officinale Apr-21
Tardigrada Jan-21
Tardigrades   Jan-21
Tawny Sanddragon   Jun-20
Tawny-edged Skipper Jun-15
Taxiles Skipper   Jul-13
Taxiles Skipper   Jul-16
Taxiles Skipper   Jul-16
Taxodium distichum   Jun-19
Taxus canadensis   Mar-16
Teloschistes chrysophthalmus   Apr-19
Teloschistes chrysophthalmus   Nov-23
Teloschistes chrysophthalmus   Nov-23
Temnothorax   Apr-23
ten-angled pipewort   Jun-18
Tenthredinidae   May-9
Tenthredinidae   May-15
Tenthredinidae   Jun-9
Tetanocerini   Apr-22
Tetanocerini   May-24
Tetragnatha   May-5
Tetragnathidae   May-30
Tetramorium   Oct-12
Tetraopes tetrophthalmus   Jun-17
Tetrigidae Jun-1
Tetrigidae Sep-7
Tetrix   May-5
Tettigidea May-13
Teucrium canadense   Jul-19
Teucrium canadense   Dec-18
Thalictrum dioicum   May-4
Thaumatomyia   May-28
Theridion Jul-7
Theridion Oct-5
Thick-legged Hoverfly Jul-20
Thick-legged Hoverfly Sep-7
Thick-legged Hoverfly Sep-9
Thick-legged Hoverfly Sep-20
Thick-legged Hoverfly Oct-18
Thin-legged Wolf Spiders   Feb-21
thistles   Jul-16
thistles   Nov-24
thistles   Nov-25
Thlaspi arvense   Apr-21
Thomisidae   Jun-1
Thomisidae   Aug-8

thornless honey locust	Oct-14	Tree Swallow	Apr-4	
Thread-legged Bugs	Mar-27	trembling aspen	Mar-8	
Three-banded Robber Fly	Jun-20	Triangulate Comb-foot	Nov-6	
threeleaf goldthread	Jun-1	Trichodezia albovittata	Jul-23	
Thuja	Jan-28	Trichopoda pennipes	May-30	
thyme-leaved speedwell	May-15	Trichopoda pennipes	Jul-23	
Tilia americana	Feb-12	Trichoptera	Dec-4	
Tilia americana	Nov-24	Trichoptera	Dec-4	
Tillandsia usneoides	Jun-18	Tricolored Bumble Bee	May-31	
Timulla	Aug-25	Tricolored Bumble Bee	Jun-1	
Tipula	Apr-22	Trifolium	Nov-5	
Tipula	Apr-23	Trigonarthris minnesotana	Jun-11	
Tipula	Jul-17	Trillium cernuum	May-31	
Tipula	Nov-21	Trimerotropis maritima	Jun-22	
Tipulidae	Feb-11	Tringa	Apr-25	
Tipulomorpha	Feb-11	Trochosa	Mar-11	
Tipulomorpha	Oct-7	Trochosa terricola	Mar-19	
Toothed Phigalia Moth	Apr-9	Trochosa terricola	Apr-9	
Tortoise Scales	Feb-10	Troglodytes aedon	May-18	
Toxomerus geminatus	Apr-25	Trombidiidae	Feb-14	
Toxomerus geminatus	May-5	Tropical Mexican Clover	Jun-18	
Toxomerus geminatus	May-7	True Bugs	Mar-27	
Toxomerus geminatus	May-17	true mosses	Jan-17	
Toxomerus geminatus	Jul-18	true mosses	Mar-25	
Toxomerus geminatus	Sep-1	true sedges	Mar-2	
Toxomerus geminatus	Sep-6	true sedges	Dec-29	
Toxomerus geminatus	Oct-18	True Velvet Mites	Feb-14	
Toxomerus marginatus	May-6	Trypoxylon	Apr-6	
Toxomerus marginatus	May-6	Trypoxylon	Jul-24	
Toxomerus marginatus	Jun-11	Trypoxylon collinum	Feb-9	
Toxomerus marginatus	Jun-14	Trypoxylon politum	Aug-18	
Toxomerus politus	Sep-1	Tube-tailed Thrips	Feb-13	
Toxophora	Jun-18	Tuft-legged Orbweaver	Apr-9	
Tragopogon dubius	May-24	Tuft-legged Orbweaver	Apr-14	
Tramea carolina	Jun-18	Tufted Globetail	Jul-8	
Tramea lacerata	Sep-22	Tule Bluet	May-25	
Tramea onusta	Jun-18	Tumbling Flower Beetles	Jun-14	
Tramea onusta	Jun-24	Turdus migratorius	Jan-6	
Tramea onusta	Sep-2	Turdus migratorius	Mar-18	
Transverse Flower Fly	May-17	Turkey Vulture	Mar-24	
Transverse Flower Fly	May-25	tussock cottongrass	Jun-1	
Transverse Flower Fly	Jun-24	Twelve-spotted Skimmer	May-25	
Transverse Flower Fly	Aug-19	Twelve-spotted Skimmer	Jun-6	
Transverse Flower Fly	Sep-1	Twelve-spotted Skimmer	Jun-29	
Transverse Flower Fly	Oct-18	Twice-stabbed Lady Beetle	Apr-3	
Transverse Flower Fly	Oct-19	Two-lined Swamp Fly	May-8	
Transverse Flower Fly	Oct-20	Two-spotted Bumble Bee	Apr-16	

Two-spotted Bumble Bee Apr-28
Two-spotted Bumble Bee May-21
Two-spotted Bumble Bee Jul-20
Two-spotted Long-horned Bee Jul-23
Two-spotted Long-horned Bee Jul-29
Two-spotted Tree Cricket Aug-7
Two-spotted Tree Cricket Aug-12
Twomarked Treehopper Oct-9
Typha Jan-9
Typha angustifolia Nov-1
Typical Carps Aug-27
Typocerus Jun-30
Typocerus Jul-11

U

Uinta Chipmunk Jul-13
Ulmus Feb-1
Ulmus Oct-14
Ulmus americana Mar-13
Ulmus pumila Mar-2
Umbrella Paper Wasps Feb-16
Unequal Cellophane Bee Apr-14
Urocyon cinereoargenteus Mar-4
Uta stansburiana Jul-14
Uvularia grandiflora Apr-23
Uvularia grandiflora May-4
Uvularia sessilifolia May-31

V

Vaccinium May-31
Vanessa atalanta Apr-14
Vanessa atalanta Sep-20
Vanessa cardui Aug-7
Vanessa cardui Aug-22
Vanessa cardui Sep-6
Vanessa cardui Sep-7
Vanessa virginiensis May-3
Vanessa virginiensis May-3
Vanessa virginiensis Jun-15
Variable Dancer Jun-18
Variable Tiger Beetle Jul-16
Variegated Cuckoo Bees Jun-15
Variegated Meadowhawk Sep-8
Variegated Meadowhawk Sep-8
Verbascum thapsus Feb-4
Verbascum thapsus Feb-13
Verbascum thapsus Dec-3
Verbena Bud Moth Jun-10
Verbena hastata Jan-20
Veronica arvensis Apr-4
Veronica arvensis May-5
Veronica arvensis May-22
Veronica bullii May-3
Veronica catenata Jul-29
Veronica serpyllifolia May-15
Vespinae Oct-24
Vespula maculifrons Oct-3
Vespula maculifrons Oct-16
Vespula vidua Oct-1
Vespula vidua Oct-3
Viburnum lentago Apr-1
Viburnum lentago May-9
Viburnum opulus Nov-9
Viceroy Aug-22
Vinca minor Jan-31
Viola Jun-1
Viola pedata May-3
Viola pubescens Apr-25, May-2, May-4
Viola sororia Apr-21, May-4
violets Jun-1
Virginia bluebells Apr-28
Virginia mountain mint Aug-11
Virginia Opossum Nov-18
Virginia spring beauty Apr-14, May-4
Virginia wildrye Nov-16
Virginia wood cockroach Mar-28
Virile Crayfish Jan-2, Jan-16, Feb-14, May-31
Vitis riparia Dec-23
Vulgichneumon brevicinctor Nov-28

W

Wasp Mantidfly Jun-16, Jul-4
Water Chickweed May-26
Waterlily Leafcutter Moth May-27
Wavy Patterneye Oct-18
Waxcaps Oct-4
Webbhelix multilineata Mar-28
Western Blacknose Dace Dec-4

Western Branded Skipper   Jul-17
Western Honey Bee May-5
Western Red Damsel   Jul-8
Western Roundtail   May-6
Western Roundtail   May-7
Western Whiptail   Jul-13
white avens   Jul-7
White Bear Sedge   May-1
white fawnlily   Apr-25
white fir   Jul-14
white fir   Jul-15
White Micrathena   Oct-13
white oak Nov-30
white poplar   Nov-26
White Sucker   Dec-4
white turtlehead   Mar-4
White-bearded Robber Fly   Jun-15
White-belted Ringtail   Jul-14
White-breasted Nuthatch   Feb-14
White-faced Meadowhawk   Jun-24, Jul-7, Jul-19, Jul-27, Jul-28, Aug-4, Aug-6, Aug-11, Aug-22, Aug-22, Aug-25, Aug-27, Sep-7, Sep-8, Sep-9, Sep-18, Sep-19, Oct-3, Oct-4, Oct-9, Oct-11
White-footed Mouse   Nov-11
White-lined Sphinx Jun-16, Jul-8
White-spotted Cankerworm Moth   Apr-2
White-striped Black   Jul-23
White-tailed Deer   Jan-7, Jan-17, Jul-9
whitecedars   Jan-28
whorled milkweed   Aug-25, Oct-11, Dec-3
Widow Skimmer   Jun-8, Jun-26, Jul-9, Aug-4
Widow Yellowjacket   Oct-1
Widow Yellowjacket   Oct-3
wild asparagus   Nov-16
wild calla Jun-1
wild cucumber   Mar-9, Aug-2, Oct-27
wild geranium   May-4, May-25
wild sarsaparilla   Jun-1
Wilson's Snipe   Mar-23
winged euonymus   Mar-15
winged euonymus   Oct-29
Winter Firefly   Mar-28

Wolf Spiders   Mar-4, Mar-27, Mar-28, May-3, Nov-24
Wolf's Milk   Sep-21, Nov-18
wood anemone   May-4, May-31
Wood Duck   Oct-9
Wood Frog   Apr-1, Apr-4
Wood Ground-beetle   May-31
Wood, Mound, and Field Ants May-9
Woodland Ground Beetles   Mar-1
Woolly Aphids and Gall-making Aphids   Apr-12
wormseed wallflower   May-28

X

Xanthoria Jan-12
Xanthotype   Aug-25
Xysticus sp.   May-13, Aug-22

Y

Yellow Bullhead   May-13
Yellow Dung Fly   Apr-1, May-15
Yellow Garden Spider   Aug-4
yellow salsify   May-24
Yellow-collared Scape Moth   Sep-21
Yellow-marked Buprestids   Jul-16
Yellow-rumped Warbler   Apr-19, Apr-26
Yellow-shouldered Drone Fly   May-2, May-2, May-22

Z

Zadontomerus   Apr-16, Sep-16, Oct-18
Zanthoxylum americanum   Apr-4, May-2
Zea mays ss   Feb-20
Zebra Mussel   Nov-24
Zelus   Jun-24
Zelus longipes   Jun-19
Zelus luridus   Oct-21
Zodion   Jun-12
Zonitoides arboreus   Sep-19

"Our work here stays forever unfinished"
– *Iliad* (Book 2, lines 137-138)

www.ingramcontent.com/pod-product-compliance
Lightning Source LLC
Chambersburg PA
CBHW020718180526
45163CB00001B/20